Lecture Notes in Mathem

Edited by A. Dold, F. Takens and B. Teissier

Editorial Policy
for the publication of monographs

1. Lecture Notes aim to report new developments in all areas of mathematics – quickly, informally and at a high level. Monograph manuscripts should be reasonably self-contained and rounded off. Thus they may, and often will, present not only results of the author but also related work by other people. They may be based on specialized lecture courses. Furthermore, the manuscripts should provide sufficient motivation, examples and applications. This clearly distinguishes Lecture Notes from journal articles or technical reports which normally are very concise. Articles intended for a journal but too long to be accepted by most journals, usually do not have this "lecture notes" character. For similar reasons it is unusual for doctoral theses to be accepted for the Lecture Notes series.

2. Manuscripts should be submitted (preferably in duplicate) either to one of the series editors or to Springer-Verlag, Heidelberg. In general, manuscripts will be sent out to 2 external referees for evaluation. If a decision cannot yet be reached on the basis of the first 2 reports, further referees may be contacted: the author will be informed of this. A final decision to publish can be made only on the basis of the complete manuscript, however a refereeing process leading to a preliminary decision can be based on a pre-final or incomplete manuscript. The strict minimum amount of material that will be considered should include a detailed outline describing the planned contents of each chapter, a bibliography and several sample chapters.
Authors should be aware that incomplete or insufficiently close to final manuscripts almost always result in longer refereeing times and nevertheless unclear referees' recommendations, making further refereeing of a final draft necessary.
Authors should also be aware that parallel submission of their manuscript to another publisher while under consideration for LNM will in general lead to immediate rejection.

3. Manuscripts should in general be submitted in English.
Final manuscripts should contain at least 100 pages of mathematical text and should include
- a table of contents;
- an informative introduction, with adequate motivation and perhaps some historical remarks: it should be accessible to a reader not intimately familiar with the topic treated;
- a subject index: as a rule this is genuinely helpful for the reader.

Lecture Notes in Mathematics 1713

Editors:
A. Dold, Heidelberg
F. Takens, Groningen
B. Teissier, Paris

Subseries: Fondazione C. I. M. E., Firenze
Adviser: Roberto Conti

Springer
Berlin
Heidelberg
New York
Barcelona
Hong Kong
London
Milan
Paris
Singapore
Tokyo

F. Bethuel G. Huisken S. Müller K. Steffen

Calculus of Variations and Geometric Evolution Problems

Lectures given at the 2nd Session of the
Centro Internazionale Matematico Estivo
(C.I.M.E.) held in Cetraro, Italy,
June 15–22, 1996

Editor: S. Hildebrandt, M. Struwe

Fondazione
C.I.M.E.

Springer

Authors

Fabrice Bethuel
Université Paris-Sud
Laboratoire d'Analyse Numérique et EDP
URA CNRS 760, Bâtiment 425
91405 Orsay, France

Gerhard Huisken
Alexander Polden
Mathematisches Institut
Universität Tübingen
Auf der Morgenstelle
72076 Tübingen, Germany

Stefan Müller
Max-Planck Institute for Mathematics
in the Sciences
Inselstraße 22–26
04103 Leipzig, Germany

Klaus Steffen
Mathematisches Institut
Universität Düsseldorf
Universitätsstraße 1
40225 Düsseldorf, Germany

Editors

Stefan Hildebrandt
Mathematisches Institut der Universität
Beringstraße 6,
53115 Bonn, Germany

Michael Struwe
ETH-Zentrum, Rämistrasse 10
8092 Zürich, Switzerland

Cataloging-in-Publication Data applied for

Die Deutsche Bibliothek - CIP-Einheitsaufnahme

Calculus of variations and geometric evolution problems : held in
Cetraro, Italy, June 15 - 22, 1996 / Fondazione CIME. F. Bethuel ...
Ed.: S. Hildebrandt ; M. Struwe. - Berlin ; Heidelberg ; New York ;
Barcelona ; Hong Kong ; London ; Milan ; Paris ; Singapore ; Tokyo
: Springer, 1999
(Lectures given at the ... session of the Centro Internazionale
Matematico Estivo (CIME) ... ; 1996,2) (Lecture notes in mathematics
; Vol. 1713 : Subseries: Fondazione CIME)
ISBN 3-540-65977-3

Mathematics Subject Classification (1991): 35-06, 49-06, 58-06, 73-06

ISSN 0075-8434
ISBN 3-540-65977-3 Springer-Verlag Berlin Heidelberg New York

© Springer-Verlag Berlin Heidelberg 1999
Printed in Germany

Typesetting: Camera-ready TEX output by the authors
SPIN: 10650190   46/3143-543210 - Printed on acid-free paper

PREFACE

The international summer school on

Calculus of Variations and Geometric Evolution Problems

was held at Cetraro, Italy, June 15-23, 1996.

The lecturers, F. Bethuel, G. Huisken, S. Müller, K. Steffen had complete freedom in choosing the topics of their courses within the themes of the conference. The contributions to this volume reflect quite closely the lectures given at Cetraro which have provided an image of a fairly broad field in analysis where in recent years we have seen many important contributions. Among the topics treated in the courses were variational methods for Ginzburg-Landau equations, variational models for microstructure and phase transitions, a variational treatment of the Plateau problem for surfaces of prescribed mean curvature in Riemannian manifolds, - both from the classical point of view and in the setting of geometric measure theory. The second theme of the conference was presented in lectures on geometric evolution equations for hypersurfaces in a Riemannian manifold. G. Huisken has included his student A. Polden as co-author, because the notes presented in this volume present a hitherto unpublished part of Polden's thesis providing, for example, a shorttime existence proof for the gradient flow of the Willmore functional.

The organizers would like to express their gratitude to the speakers for their excellent lectures and to all participants for contributing to the success of the summer school.

S. Hildebrandt, M. Struwe

TABLE OF CONTENTS

TABLE OF CONTENTS

Variational methods for
Ginzburg-Landau equations

F. Bethuel

I. INTRODUCTION

Ginzburg-Landau functionals were first introduced by V. Ginzburg and L. Landau in 1950 [GL] in the context of superconductivity. They were aimed to model (on a macroscopic scale) the energy state of a superconducting sample, in presence of an exterior magnetic field. Similar energy functionals appeared thereafter in various contexts, and under different forms. In particle physics one may mention the Abelian Higgs model, the Poliakov-t'Hooft monopole, and more generally various models of chromodynamics.

A common feature of the above models is that they involve a nonconvex potential. A typical example for such a potential is the function $V(u) = \left(1 - |u|^2\right)^2$ ($u \in \mathbb{R}$ or $u \in \mathbb{R}^2$ for instance). The vacuum manifold is the set of point where V achieves its minimum. If $u \in \mathbb{R}$, in our example, then the vacuum manifold is $\{+1, -1\}$, whereas if $u \in \mathbb{R}^2$, then the vacuum manifold is S^1, the unit circle. The topology of the vacuum manifold turns out to be crucial in the study of the model : it will induce various topological defects, called in our context vortices.

To make things more precise, we will start with a very simple model situation, which was studied in particular in a joint book with H. Brezis and F. Hélein [BBH]. An important part of these notes will be devoted to the study of this model. Nevertheless, in the last sections, we will show how the technics introduced can be useful for attacking more realistic physical situations : as we will see, although progresses have been obtained, outstanding mathematical problems remain open in that direction.

II. A SIMPLE MODEL

Let Ω be a smooth bounded domain in $\mathbb{R}^2$ (throughout this paper, we will restrict ourselves to two-dimensional problems). We will consider complex-valued maps on Ω, that is maps from Ω to $\mathbb{R}^2$. The simplest Ginzburg-Landau functional for such maps takes the form

$$E_\epsilon(v) = \frac{1}{2} \int_\Omega |\nabla v|^2 + \frac{1}{4\epsilon^2} \int_\Omega \left(1 - |v|^2\right)^2.$$

Here ϵ is a positive parameter, homogeneous to a length. In the sequel, we will mainly be interested in the case ϵ is small : the asymptotic limit ϵ tends to zero will be central in our analysis. The nonconvex potential V is here

$$V(v) = \frac{1}{4\epsilon^2} \left(1 - |v|^2\right)^2,$$

and the vacuum manifold is the unit circle S^1. For critical maps v of the energy, the potential V forces (for small ϵ) $|v|$ to be close to 1 : hence there are almost S^1–valued. However, at some points v may have to vanish : this introduces defects of topological nature (which will be called vortices).

In order to have a well-posed mathematical problem, we have to prescribe boundary conditions. The simplest idea will be to impose Dirichlet boundary datas (although this might not correspond to any realistic physical situation...). For that purpose, let g be a smooth map from $\partial\Omega$ to the circle S^1. We prescribe v to be equal to g on $\partial\Omega$. It is then natural to introduce the Sobolev space

$$H_g^1(\Omega; \mathbb{R}^2) = \left\{v \in H^1(\Omega; \mathbb{R}^2),\ v = g \text{ on } \partial\Omega\right\}.$$

The functional E_ϵ is indeed well-defined, smooth on H_g^1, and satisfies moreover the Palais-Smale condition. Critical points of E_ϵ then verify the Ginzburg-Landau equation

(1)
$$\begin{cases} -\Delta v = \dfrac{1}{\epsilon^2} v \left(1 - |v|^2\right) & \text{in } \Omega \\ v = g & \text{on } \partial\Omega. \end{cases}$$

We will often refer to (1) as $(GL)_\epsilon$. Since the nonlinearity on the right-hand side is subcritical, solutions to (1) are smooth on $\bar{\Omega}$. Moreover one has

Proposition 1. *Any solution v to (1) verifies*

(2)
$$|v| \le 1 \text{ on } \Omega,$$

and

(3)
$$|\nabla v| \le \frac{C}{\epsilon}$$

where the constant C depends only on Ω and g.

<u>Proof</u> : Inequality (2) is a consequence of the maximum principle. Indeed we have

$$\frac{1}{2}\Delta|v|^2 = v \cdot \Delta v + |\nabla v|^2$$

$$= \frac{1}{\epsilon^2}|v|^2 \left(|v|^2 - 1\right) + |\nabla v|^2$$

$$\geq \frac{1}{\epsilon^2}|v|^2 \left(|v|^2 - 1\right).$$

Hence the function $w = |v|^2 - 1$ satisfies

$$-\Delta w + a(x)w \leq 0 \quad \text{on } \Omega$$

$$w = 0 \quad \text{on } \partial\Omega,$$

with $a(x) = \frac{2}{\epsilon^2}|v|^2 \geq 0$. By the maximum principle, we conclude that $w \leq 0$.

For (3), we note that, by (2)

$$|\Delta u| \leq \frac{1}{\epsilon^2}$$

and the conclusion follows by elliptic estimates (see [BBH2]).

In the next section, we will be interested in minimizing solutions. The existence of such solutions is easy to establish. Since E_ϵ is positive, one has

$$\kappa_\epsilon = \text{Inf } \left\{ E_\epsilon(v), \ v \in H_g^1 \right\} > 0,$$

and any minimizing sequence for κ_ϵ is bounded in H_g^1, hence converges weakly up to a subsequence to some map u_ϵ. By lower-semicontinuity of E_ϵ (for the weak topology), u_ϵ is a minimizer. In the sequel, we will always denote minimizers by u_ϵ.

<u>Remark</u> : Minimizers might not be unique (for sufficient small ϵ, at least). We will give later examples of nonuniqueness, when symmetries are present.

Next, we will study the asymptotic limit as ϵ tends to zero.

III. ASYMPTOTIC ANALYSIS OF MINIMIZERS

The winding number d of g (from $\partial\Omega$ to S^1) plays a crucial role in the asymptotic analysis, inducing, in the case $d \neq 0$, the appearance of vortices and the divergence of the minimal energy κ_ϵ, as $\epsilon \to 0$. This is deeply related to the following

Proposition 2. *Assume Ω is simply connected. Set*

$$H_g^1(\Omega; S^1) = \left\{ v \in H_g^1(\Omega; \mathbb{R}^2), \ |v| = 1 \right\}.$$

Then $H_g^1(\Omega; S^1)$ is nonempty if and only if $d = 0$.

With similar notations, the fact that $C_g^0(\Omega; S^1)$ is non-empty if and only if d is zero reduces to standard degree theory (and is of course well-known). For H^1 maps the proof is slightly more involved, and relies on the following.

Lemma 1. *Assume Ω is simply connected. Let v be a map in $H^1(\Omega; S^1)$, where*

$$H^1(\Omega; S^1) = \left\{ v \in H^1(\Omega; \mathbb{R}^2), \ |v| = 1 \right\}.$$

Then there exists a real-valued map $\varphi \in H^1(\Omega; \mathbb{R})$ such that

$$v = \exp i\varphi.$$

Moreover

(4)
$$|\nabla v| = |\nabla \varphi| \quad \text{a.e.}$$

<u>Proof of Lemma 1.</u> Since v is S^1-valued, if (x_1, x_2) are cartesian coordinates on Ω, then v_{x_1} is parallel to v_{x_2}. This writes

(5)
$$v_{x_1} \times v_{x_2} = 0.$$

[Here, we have embedded $\mathbb{R}^2$ into $\mathbb{R}^3$, and $\times$ denotes the cross-product in $\mathbb{R}^3$]. We may rewrite (4) in divergence form

(6)
$$\frac{\partial}{\partial x_1}(v \times v_{x_2}) + \frac{\partial}{\partial x_2}(-v \times v_{x_1}) = 0.$$

Since Ω is simply connected, by Poincaré's Lemma, there is some map ψ in $H^1(\Omega; \mathbb{R})$ such that

$$\begin{cases} v \times v_{x_1} = \psi_{x_1} \\ v \times v_{x_2} = \psi_{x_2}. \end{cases}$$

[Here $v \times v_{x_1}$ is orthogonal to $\mathbb{R}^2$ in $\mathbb{R}^3$, and considered as a scalar]. Next consider the S^1−valued map

$$w = \exp - i\psi \cdot v$$

where the multiplication stands for complex multiplication. Then, for $i = 1, 2$,

$$w_{x_i} = -i\psi_{x_i}(\exp - i\psi \cdot v) + (v \times v_{x_i} \exp - i\psi \cdot v)$$
$$= 0.$$

Hence $\nabla w = 0$, thus w is a constant. The conclusion follows.

Remark : The conclusion of Lemma 1 would be false if, instead of $H^1(\Omega; S^1)$, we had considered $W^{1,p}(\Omega; S^1)$, for $p < 2$. Take for instance $\Omega = D^2$ the unit disc and

$$v(x) = \frac{x}{|x|}.$$

Then v belongs to $W^{1,p}$ for any $p < 2$, but of course cannot be written as the exponential of a $W^{1,p}$ function. The proof fails because, although (5) still holds almost everywhere, (6) is no longer true (for proving (6) in Lemma 1, one may approximate v in $H^1(\Omega; \mathbb{R}^2)$ by smooth functions and pass to the limit : this uses H^1 bounds).

Proof of Proposition 2 : Assume $H_g^1(\Omega; S^1)$ is not empty and let v be in $H_g^1(\Omega; S^1)$. Then, by Lemma 1 there exists some function φ in $H_g^1(\Omega; S^1)$ such that

$$v = \exp i\varphi \quad \text{on } \Omega.$$

In particular $g = \exp i\varphi$ on $\partial\Omega$, which implies that the degree of g is zero. The proof is complete.

Next we are going to turn to the asymptotic analysis when $d = 0$.

III.1. The case d = 0.

In this case, there is no topological need for vortices, and indeed, they do not appear (for minimizers at least). First, we notice that κ_ϵ remains bounded independently of ϵ. To see this, let v_0 be any map in $H_g^1(\Omega; S^1)$ (this is possible by Proposition 2), and take v_0 as a comparison map. We have

$$(7) \qquad\qquad \kappa_\epsilon \le E_\epsilon(v_0) = \frac{1}{2}\int_\Omega |\nabla v_0|^2$$

(since $(|v_0| - 1)^2 = 0$). In particular κ_ϵ remains bounded independently of ϵ, and the set of minimizers u_ϵ remains bounded in H^1. Therefore for any sequence u_ϵ of minimizers ($\epsilon \to 0$), one may extract a subsequence u_{ϵ_n}, $\epsilon_n \to 0$, such that u_{ϵ_n} converges weakly in H^1 to some map u_*. On the other hand, by (7)

$$\int_\Omega \left(1 - |u_{\epsilon_n}|^2\right)^2 \le C\epsilon_n^2 \to 0,$$

which implies $|u_*| = 1$, and hence $u_* \in H_g^1(\Omega; S^1)$. By (7) and lower semi-continuity of the Dirichlet energy, we obtain, for any $v_0 \in H_g^1(\Omega; S^1)$

$$\int_\Omega |\nabla u_*|^2 \le \liminf_{\epsilon_n \to 0} \kappa_{\epsilon_n} \le \int_\Omega |\nabla v_0|^2,$$

from which we deduce that

$$u_{\varepsilon_n} \to u_* \quad \text{strongly in } H^1,$$

and u_* is a solution of the minimization problem

(8)
$$\text{Inf} \left\{ \int_\Omega |\nabla v|^2, \; v \in H_g^1(\Omega; S^1) \right\}.$$

We have

Proposition 3. *There is a unique solution to the minimization problem (8). This solution u_* writes*

$$u_* = \exp i\varphi_*$$

where $\varphi \in H^1(\Omega; \mathbb{R})$ is the solution to

$$\Delta \varphi_* = 0 \quad \text{in } \Omega$$

$$\exp i\varphi_* = g \quad \text{on } \partial\Omega.$$

<u>Proof of Proposition 3</u> : It is a direct consequence of Lemma 1. Indeed, for any $\psi \in H^1(\Omega; \mathbb{R})$ such that $\exp i\psi = g$ on $\partial\Omega$ we have

$$\int_\Omega |\nabla \varphi|^2 \leq \int_\Omega |\nabla \psi|^2,$$

and hence

$$\int_\Omega |\nabla u_*|^2 \leq \int_\Omega |\nabla(\exp i\psi)|^2$$

which yields the result.

Since the limiting map u_* is unique, it turns out that the full sequence u_ε converges strongly to that limit. Actually, the convergence rate can be estimated. In [BBH2] the following estimates are established :

$$\|u_\varepsilon - u_*\|_{L^\infty(\Omega)} \leq C\varepsilon^2, \quad \|\nabla(u_\varepsilon - u_*)\|_{L^\infty(\Omega)} \leq C\varepsilon,$$

and for any compact subset K of Ω, and every integer k,

$$\|u_\varepsilon - u_*\|_{C^k(K)} \leq C(k, K)\varepsilon^2.$$

We turn now to the more difficult case $d \neq 0$.

III.2. The case d ≠ 0

We may assume for instance that $d > 0$. This case is more involved, since

$$\kappa_\varepsilon \to +\infty, \quad \text{as } \varepsilon \to 0.$$

Indeed, assume by contradiction that κ_{ε_n} remains bounded for some subsequence ε_n tending to zero. Then, we may repeat the argument of the previous section to assert that $u_{\varepsilon_n} \to u_*$ in H^1 and that $u_* \in H_g^1(\Omega; S^1)$. This contradicts Proposition 2.

The problem becomes thus a problem of singular limit. Since u_ε is smooth, the fact that the degree is nonzero forces u_ε to vanish somewhere in Ω. The points where u_ε vanish will play an important role : the Dirichlet energy will concentrate in their neighborhood, accounting for the divergence of κ_ε, as ε tends to zero. In order to get some inside in the problem, we will start with an upper bound on κ_ε.

III.2.1. An upper bound for κ_ε. In this paragraph, we will show the following

Proposition 4. *There is a constant C depending on g, Ω, such that*

$$\kappa_\varepsilon \leq \pi d |\log \varepsilon| + C.$$

Proof. We will first prove the estimate in a special case.
A) The case $\Omega = D^2$ and $g(\exp i\theta) = \exp i\theta$. In this case we may construct a comparison map v_ε in the following way, using polar coordinates (r, θ) on D^2 :

$$v_\varepsilon = (r, \theta) = f_\varepsilon(r)\exp i\theta = f_\varepsilon(r)\frac{z}{|z|}.$$

Here f_ε is any function on $[0, 1]$ verifying the following conditions

$$f_\varepsilon(r) = 0 \quad \text{for } 0 < r < \frac{\varepsilon}{2},$$
$$f_\varepsilon(r) = 1 \quad \text{for } \varepsilon < r \leq 1,$$
$$\text{and} \quad |f_\varepsilon'(r)| \leq \frac{4}{\varepsilon} \quad \text{for any } 0 \leq r \leq 1.$$

Next, we compute the energy of v_ε. We have

$$\frac{1}{4\varepsilon^2}\int_{D^2}\left(1 - |v_\varepsilon|^2\right)^2 = \frac{1}{4\varepsilon^2}\int_{B(\varepsilon)}\left(1 - |v_\varepsilon|^2\right)^2 \leq \frac{\pi\varepsilon^2}{4\varepsilon^2} = \frac{\pi}{4},$$

and

$$\int_D |\nabla v_\varepsilon|^2 = I + II,$$

where

$$I = \int_{B(\epsilon)} |\nabla v_\epsilon|^2 \leq \frac{C}{\epsilon^2} \int_{B(\epsilon)} dx \leq \frac{C}{\epsilon^2} \pi \epsilon^2 = \pi C,$$

$$II = \int_{D \backslash B(\epsilon)} |\nabla v_\epsilon|^2 = \int_{D \backslash B(\epsilon)} |\nabla(\exp i\theta)|^2 = \int_{D \backslash B(\epsilon)} \frac{1}{r^2} = 2\pi \int_\epsilon^1 \frac{1}{r^2} r dr,$$

that is

$$II = 2\pi |\log \epsilon|.$$

The conclusion follows, in case A.

B) <u>The general case</u>. We will construct a similar comparison map. For that purpose, choose d points $a_1, ..., a_d$ in Ω. Consider the map

$$w_\epsilon = \prod_{i=1}^d f_\epsilon(|z - a_i|) \frac{z - a_i}{|z - a_i|}$$

which generalizes the construction of case 1 (here we have used complex notation). On $\partial \Omega$, we have, if ϵ is sufficiently small,

$$w_\epsilon(z) = \prod_{i=1}^d \frac{z - a_i}{|z - a_i|},$$

so that w_ϵ is S^1−valued on $\partial \Omega$, and has precisely degree d. Therefore, there exists some smooth map $\varphi : \partial \Omega \to \mathbb{R}$ such that

$$g = \prod_{i=1}^d \frac{z - a_i}{|z - a_i|} \exp i\varphi.$$

Next, we extend φ inside Ω, for instance by a harmonic extension and set

$$v_\epsilon = \prod_{i=1}^d f_\epsilon(|z - a_i|) \frac{z - a_i}{|z - a_i|} \exp i\varphi.$$

This map is smooth, equal to g on $\partial \Omega$ (provided ϵ is sufficiently small). Therefore v_ϵ can be used as a comparison map. Computing the energy of v_ϵ as in the case A leads to

$$E_\epsilon(v_\epsilon) \leq \pi d |\log \epsilon| + C.$$

III.2.2. Some results

It turns out that the comparison map v_ε constructed in the previous section is quite similar to the real minimizers u_ε (of course, for a suitable choice of the points a_i) and that the bound of Proposition 3 is sharp. More precisely in [BBH], we proved the following.

Theorem 1. *There exists a constant $C > 0$ depending only on g such that*

$$|\kappa_\varepsilon - \pi d|\log \varepsilon|| \leq C, \quad \text{for } 0 < \varepsilon < 1.$$

- *The map u_ε has exactly d zeroes, provided ε is sufficiently small.*
- *There exist exactly d points $a_1, ..., a_d$ in Ω such that, up to a subsequence ε_n tending to zero*

$$u_{\varepsilon_n} \to u_* \text{ on any compact subset of } \Omega \backslash \bigcup_{i=1}^d \{a_i\},$$

where u_* is the S^1-valued map, equal to g on $\partial\Omega$ given by

$$u_* = \prod_{i=1}^d \frac{z - a_i}{|z - a_i|} \exp i\varphi,$$

where φ is a harmonic function.
- *The configuration $(a_1, ..., a_d)$ is not arbitrary, but minimizes on $\Omega^d \backslash \Delta$ (where Δ denotes the diagonal, i.e. $\Delta = \{(x_1, ..., x_d) \in \Omega^d, \text{ such } \exists \, i \neq j \, x_i = x_j\}$) a renormalized energy, which has (roughly) the form*

$$W_g(a_1, ..., a_d) = -\pi \sum_{i \neq j} \log |a_i - a_j| + \text{boundary contributions.}$$

- *We have the expansion*

$$\kappa_\varepsilon = \pi d|\log \varepsilon| + W_g(a_1, ..., a_d) + d\gamma_0 + o(1)$$

where $o(1) \to 0$ as $\varepsilon \to 0$, and where γ_0 is some universal constant.

Remarks : 1. In [BBH], the theorem was proved under the additional assumption that Ω is starshaped. This assumption was removed by M. Struwe [Str] (see also a simple proof by Del Pino and Felmer [DF]).

2. The second assertion of the theorem (namely, the fact that u_ε has only d zeroes) relies heavily on a deep result by P. Baumann, N. Carlson and D. Philips [BCP]). This result can be stated as follows : let u_ε be a minimizing solution to $(GL)_\varepsilon$, on the disc

D^2, such that, on ∂D^2, u_ε is a monotone parametrization of S^1. Then u_ε has only <u>one</u> zero.

The renormalized energy will be defined later, and we will see how it is related to the problem. The proof of Theorem 1 is rather involved, and will not be given here. Instead, we will sketch in the next section the proof of the convergence of solutions to $(GL)_\varepsilon$ which are not necessarily minimizers. Let us emphasize at this stage that the difficult part in the proof of Theorem 1 is to prove that each singularity a_i has precisely degree $+1$: this is of course not true for non-minimizing solutions, as we will see later also.

IV. CONVERGENCE OF SOLUTIONS TO THE GINZBURG-LANDAU EQUATION

An asymptotic analysis, similar to Theorem 1, was carried out in [BBH] for solutions to the Ginzburg-Landau equation which are not necessarily minimizing. We have ([BBH], Section X)

Theorem 2. *Assume that Ω is starshaped. There exists a constant $C > 0$, depending only on g and Ω, such that for any solution v_ε to (GL_ε), we have*

$$(9) \qquad\qquad E_\varepsilon(v_\varepsilon) \leq C(|\log \varepsilon| + 1).$$

Let (v_ε) be a sequence of solutions to $(GL)_\varepsilon$ $(\varepsilon \to 0)$. Then there exist a subsequence $\varepsilon_n \to 0$, ℓ points $a_1, ..., a_\ell$ and ℓ integers $d_1, ..., d_\ell$, different from zero, such that

$$v_{\varepsilon_n} \to v_* = \prod_{i=1}^{\ell} \left(\frac{z - a_i}{|z - a_i|} \right)^{d_i} \exp i\varphi$$

(where φ is harmonic), on every compact subset of $\Omega \backslash \bigcup_{i=1}^{\ell} \{a_i\}$ and in $W^{1,p}$ for $p < 2$. Moreover, there exists a constant C, depending only on Ω and g such that

$$\sum_{i=1}^{\ell} |d_i| \leq C, \qquad \ell \leq C.$$

The configuration (a_i, d_i) is critical for a renormalized energy.

<u>Remarks</u> :
1. In Theorem 2 the assumption Ω starshaped is crucial in the proof. Actually the theorem is not true for domains which are not simply connected. For instance take

$\Omega = D^2 \backslash D^2 \left(\frac{1}{2}\right)$ and let $g = +1$ on ∂D^2, $g = -1$ on $\partial D^2 \left(\frac{1}{2}\right)$. Then we may construct a solution to $(GL)_\varepsilon$ of the form

$$v_\varepsilon(r, \theta) = f_\varepsilon(r),$$

where f_ε is a real valued function on $[\frac{1}{2}, 1]$ such that $f_\varepsilon\left(\frac{1}{2}\right) = -1$, $f_\varepsilon(1) = 1$. One may then prove that $E_\varepsilon(v_\varepsilon) \geq \frac{C}{\varepsilon}$, for some constants $C > 0$, contradicting the bound (9).

2. It is an open problem to determine if the assumption Ω starshaped can be replaced by the weaker assumption Ω is simply connected.

3. In Theorem 2, one may replace the assumption Ω starshaped assuming instead that the energy bound (9) holds for some constant C (see the Appendix of this section).

The aim of this section is to sketch the proof of Theorem 2, which is actually much simpler than the proof of Theorem 1. The main point is to establish estimate (9) as well as an a priori bound for solutions in $W^{1,p}$, for any $1 < p < 2$. To that aim the main ingredients in the proof are :

<u>Step 1</u> (localizing the vortices). There exist constants N and λ depending only on g, and ℓ points $x_1, ..., x_\ell$, with $\ell \leq N$, such that

$$|v_\varepsilon| \geq \frac{1}{2} \quad \text{on } \Omega_\varepsilon = \Omega \backslash \bigcup_{i=1}^{\ell} B(x_i, \lambda \varepsilon)$$

(In the limit $\varepsilon \to 0$, the points x_i will converge to the points a_i). The points x_i, as well as ℓ depend on the solution v_ε considered.

<u>Step 2</u> (rewriting the equation in form of a linear elliptic system). On Ω_ε we have

$$(10) \qquad \frac{\partial}{\partial x_1}(v \times v_{x_1}) + \frac{\partial}{\partial x_2}(v \times v_{x_2}) = 0$$

$$(11) \qquad \frac{\partial}{\partial x_1}\left(-\frac{1}{|v|^2}v \times v_{x_1}\right) + \frac{\partial}{\partial x_2}\left(-\frac{1}{|v|^2}v \times v_{x_1}\right) = 0.$$

<u>Step 3</u>. From Step 2, we deduce the bound (9) as well as $W^{1,p}$ estimates.

Next we turn to Step 1.

IV.1. Proof of Step 1. We begin with the following crucial estimate ([BBH], Section III).

Proposition 5. *There exists a constant $C \geq 0$, depending only on Ω and g such that, for any solution v_ϵ of* $(GL)_\epsilon$

$$\frac{1}{4\epsilon^2} \int_\Omega \left(1 - |v_\epsilon|^2\right)^2 \leq C.$$

The proof of Proposition 5 relies on Pohozaev's identity. We have

Lemma 2. *Let G be any subdomain of Ω. We have*

$$\frac{1}{2} \int_{\partial G} (x.\nu) \left|\frac{\partial v}{\partial \nu}\right|^2 + \frac{1}{2\epsilon^2} \int_G (1 - |v|^2)^2 = \int_{\partial G} \frac{1}{2}(x.\nu) \left|\frac{\partial g}{\partial \tau}\right|^2$$
$$- \int_{\partial G} x.\tau \frac{\partial u}{\partial \nu} \frac{\partial g}{\partial \tau} + \frac{1}{\epsilon^2} x.\nu \left(1 - |v|^2\right)^2,$$

where ν denotes the unit normal to ∂G, τ the unit tangent to ∂G, so that (ν, τ) is direct.

<u>Proof</u> : Multiply the equation by $x.\nabla v = x_1 \frac{\partial v}{\partial x_1} + x_2 \frac{\partial v}{\partial x_2}$ and integrate par parts.

<u>Proof of Proposition 5 completed</u> : Take $G = \Omega$. Since Ω is starshaped we may choose the origin so that $x.\nu > 0$. The conclusion then follows applying Lemma 2.

The next ingredient in the proof of Step 1 is the following ([BBH], Section III).

Proposition 6. *The exists positive constants λ_0 and μ_0 (depending only on Ω and g) such that if v is a solution of $(GL)_\epsilon$ satisfying*

$$\frac{1}{\epsilon^2} \int_{\Omega \cap B_{2\ell}} \left(1 - |v|^2\right)^2 \leq \mu_0,$$

where $B_{2\ell}$ is some disc of radius 2ℓ, with $\frac{\ell}{\epsilon} \geq \lambda_0$ and $\ell \leq 1$, then

$$|v(x)| \geq \frac{1}{2} \quad \forall x \in \Omega \cap B_\ell.$$

<u>Sketch of the proof</u> : we know that there is a constant C sucht that $|\nabla v| \leq \frac{C}{\epsilon}$. Suppose that there is a point x_0 in B_ℓ such that

$$|v(x_0)| < \frac{1}{2}.$$

Then, since $|\nabla v| \leq C\epsilon^{-1}$, $v(x) < \frac{3}{4}$ on the ball $B\left(x_0, \frac{\epsilon}{4C}\right)$ and therefore

$$\frac{1}{4\epsilon^2} \int_{B\left(x_0, \frac{\epsilon}{4C}\right)} \left(1 - |v|^2\right)^2 > \frac{1}{4\epsilon^2} \frac{1}{16} \pi \frac{\epsilon^2}{4C^2} = \frac{\pi}{256 C^2}.$$

The end of the proof is straightforward.

Proof of Step 1 completed : The proof of Step 1 is completed using a standard covering argument. We consider a family of disc $B(x_i, \lambda_0 \varepsilon)_{i \in I}$ such that $x_i \in \Omega$, $\forall i \in I$, and

$$\Omega \subset \bigcup_{i \in I} B(x_i, \lambda_0 \varepsilon)$$

$$B\left(x_i, \frac{\lambda_0 \varepsilon}{4}\right) \cap B\left(x_j, \frac{\lambda_0 \varepsilon}{4}\right) = \emptyset \quad \text{if } i \neq j.$$

Let J be the subset of the set of indices I such that, for $i \in J$

$$\frac{1}{\varepsilon^2} \int_{B(x_i, \lambda_0 \varepsilon)} \left(1 - |v|^2\right)^2 \geq \mu_0.$$

In view of Proposition 5, there is some constant N depending only on g and G such that

$$\text{card } J \leq N.$$

On the other hand, if $x \in \Omega \backslash \bigcup_{i \in J} B(x_i, \lambda_0 \varepsilon)$, then by Proposition 6, we have

$$|v(x)| \geq \frac{1}{2};$$

(indeed if $x \in \Omega \backslash \bigcup_{i \in J} B(x_i, \lambda_0 \varepsilon)$, then, there is some $j \in I \backslash J$ such that $x \in B(x_j, \lambda_0 \varepsilon)$ and $\frac{1}{\varepsilon^2} \int_{B(x_j, 2\lambda \varepsilon_0)} \left(1 - |v|^2\right)^2 \leq \mu_0$, and then we apply Proposition 6).
    This completes the proof of Step 1.

IV.2. Proof of Step 2. In order to prove (10), note that the right-hand side of (GL_ε) is colinear to v. Therefore

$$\Delta v \times v = 0$$

which yields (10).
    For (11), we notice that we may define on Ω_ε the map $\frac{v}{|v|}$ (since $|v| \geq \frac{1}{2}$ on Ω_ε). Since $\frac{v}{|v|}$ is S^1−valued, we have by the argument of Lemma 1, Section III

$$\left(\frac{v}{|v|}\right)_{x_1} \times \left(\frac{v}{|v|}\right)_{x_2} = 0$$

which yields (11), writing it in divergence form.

    Next, we will make use of equations (10) and (11). A natural idea would be to use Poincaré's Lemma on Ω_ε : unfortunately Ω_ε is not simply connected, and we cannot apply it directly. Nevertheless, we have the following version of Poincaré's Lemma.

Lemma 3. *Let G be any smooth open domain in $\mathbb{R}^2$, and D be a vector field on G such that*

$$\text{div } D = 0$$

and

$$\int_{\Gamma_i} D.\nu = 0,$$

for each connected component Γ_i of ∂G where ν is the exterior normal to Γ_i. Then there exists a function H on Ω such that

$$D = (H_{x_2}, -H_{x_1}).$$

The proof of Lemma 3 is standard (see for instance [BBH], p. 3). We will use Lemma 3 to obtain a Hodge de Rham decomposition of $v \times \nabla v$. For that purpose let Φ_ϵ be a function defined on Ω_ϵ, solution of the linear problem

$$\text{div}\left(\frac{1}{\rho^2}\nabla\Phi_\epsilon\right) = 0 \quad \text{in } \Omega_\epsilon$$

(12)
$$\Phi_\epsilon = 0 \qquad \text{on } \partial\Omega$$
$$\Phi_\epsilon = \text{const} C_i \qquad \text{on } \partial\omega_i$$
$$\int_{\partial\omega_i} \frac{1}{\rho^2}\frac{\partial\Phi}{\partial\nu} = 2\pi d_i \quad (\nu \text{ normal exterior to } \omega_i).$$

Here we have set $\rho = |v|$ and $\omega_i = B(x_i, \lambda_0\epsilon)$, where the points x_i have been obtained in Step 1. We assume moreover that the sets ω_i are disjoint (this is always possible, changing λ_0 to some other constant, see [BBH], Section IV.2), and that they do not intersect the boundary. We have also set

$$d_i = \deg\left(\frac{v}{|v|}, \partial\omega_i\right).$$

The existence of the solution Φ_ϵ is standard, and can be obtained by a minimization procedure. Next, combining (11) and (12), we have

$$\frac{\partial}{\partial x_1}\left(\frac{1}{\rho^2}(\Phi_{x_1} - v \times v_{x_2})\right) + \frac{\partial}{\partial x_2}\left(\frac{1}{\rho^2}(\Phi_{x_2} - v \times v_{x_1})\right) = 0.$$

Set $D = \left(\frac{1}{\rho^2}(\Phi_{x_1} - v \times v_{x_2}), \frac{1}{\rho^2}(\Phi_{x_1} - v \times v_{x_2})\right)$. We see that

$$\text{div } D = 0,$$

and we may verify that

$$\int_{\partial\omega_i} D.\nu = 0.$$

Hence we are in position to apply Lemma 3, that is, there is some function H on Ω_ϵ such that, on Ω_ϵ

$$\begin{cases} v \times v_{x_1} + \Phi_{x_2} = \rho^2 H_{x_1} \\ v \times v_{x_2} - \Phi_{x_1} = \rho^2 H_{x_2}. \end{cases}$$

By equation (11), we verify that

$$(13) \qquad\qquad \text{div}(\rho^2 \nabla H) = 0.$$

Finally, we verify that, on Ω_ϵ

$$|v \times \nabla v| \leq |\nabla \Phi| + |\nabla H|,$$

which yields

$$(14) \qquad\qquad |\nabla v| \leq 2(|\nabla \Phi| + |\nabla H| + |\nabla \rho|) \quad \text{on } \Omega_\epsilon.$$

Therefore, in order to obtain estimates for v, it suffices to obtain estimates for Φ, H and ρ. Since Φ and H verify linear elliptic systems this will be obtained using methods from that theory.

IV.3. Step 3. Estimates for Φ, H and ρ.

Let us first begin with H. We have

Proposition 7. *There is a constant C independent of ϵ such that*

$$\int_{\Omega_\epsilon} |\nabla H|^2 \leq C.$$

<u>Sketch of the proof</u> : By (13) we have multiplying the equation by H itself (setting $\omega_0 = \Omega$)

$$(15) \qquad\qquad \int_{\Omega_\epsilon} \rho^2 |\nabla H|^2 = \sum_{i=0} \int_{\partial \omega_i} \rho^2 H \frac{\partial H}{\partial \nu}.$$

By the maximum principle, we have (see Lemma X.4 of [BBH])

$$\text{Sup}_{\Omega_\epsilon} H - \text{Inf}_{\Omega_\epsilon} H \leq \sum_{i=0}^{\ell} \left(\text{Sup}_{\partial \omega_i} H - \text{Inf}_{\partial \omega_i} H \right),$$

which can be estimated using

$$\text{Sup}_{\partial \omega_i} H - \text{Inf}_{\partial \omega_i} H \leq \int_{\partial \omega_i} \left| \frac{\partial H}{\partial \tau} \right| \leq 4 \int_{\partial \omega_i} \left(\left| \frac{\partial v}{\partial \tau} \right| + \left| \frac{\partial \Phi}{\partial \nu} \right| \right).$$

Since $|\nabla v| \leq \frac{C}{\varepsilon}$, we deduce, since the length of $\partial\omega_i$ is less than $2\pi\lambda\varepsilon$,

$$\int_{\partial\omega_i} |\nabla v| \leq C.$$

On the other hand, linear estimates for Φ yield, for $i = 0, ...$

$$(16) \qquad \int_{\partial\omega_i} \left|\frac{\partial\Phi}{\partial\nu}\right| \leq \sum_{i=1}^{\ell} |d_i| \leq C$$

(see Lemma X.6 of [BBH]). Combining our previous results we obtain

$$\operatorname{Sup}_{\Omega_\varepsilon} H - \operatorname{Inf}_{\Omega_\varepsilon} H \leq C,$$

where C is some constant depending only on ρ and Ω. We may assume that $\operatorname{Inf}_{\Omega_\varepsilon} H = 0$, so that $|H|_{L^\infty} \leq C$. Going back to (15), we have

$$\int_{\partial\omega_i} \rho^2 \left|\frac{\partial H}{\partial\nu}\right| \leq \int_{\partial\omega_i} \left|\frac{\partial v}{\partial\nu}\right| + \left|\frac{\partial\Phi}{\partial\tau}\right| = \int_{\partial\omega_i} |\nabla v| \leq C.$$

Hence

$$\left|\sum_i \int_{\partial\omega_i} \rho^2 H \frac{\partial H}{\partial\nu}\right| \leq |H|_{L^\infty} \sum_i \int_{\partial\omega_i} \left|\frac{\partial H}{\partial\nu}\right| \leq C,$$

which completes the proof of Proposition 6.

Next we turn to Φ. We have

Proposition 8. *Let $1 \leq p < 2$. There is some constant C_p depending only on p and g such that*

$$(17) \qquad \int_\Omega |\nabla\Phi|^p \leq C_p.$$

<u>Sketch of the proof</u> : We extend Φ to Ω by setting

$$\Phi = C_i \quad \text{on } \omega_i$$

and we expend ρ to Ω, setting

$$\rho = 1 \quad \text{on } \omega_i.$$

For any map $\varphi \in C_0^1(\Omega)$, we have

$$\left|\int_\Omega \rho^2 \nabla\Phi \nabla\varphi\right| = \left|\sum_{i=1}^{\ell} \int_{\partial\omega_i} \varphi \frac{\partial\Phi}{\partial\nu}\right| \leq \|\varphi\|_{L^\infty} \sum_{i=1}^{\ell} \int_{\partial\omega_i} \left|\frac{\partial\Phi}{\partial\nu}\right|$$

$$\leq C\|\varphi\|_{L^\infty(\Omega)},$$

where we have used the estimate (16). This shows that $\operatorname{div}(\rho^2\nabla\Phi)$ is a measure on Ω, which is bounded independently of ε. Estimate (17) then directly follows from a theorem of Stampacchia [Stam] for elliptic equations in divergence form.

Proposition 9. *There is some constant C independent of ε such that*

$$\int_\Omega |\nabla\Phi|^2 \leq C(|\log\varepsilon| + 1).$$

<u>Sketch of the proof</u> : By the maximum principle, we have

$$\|\Phi\|_{L^\infty} = \max_i |C_i|.$$

Multiplying (12) by Φ, we obtain (setting $\Phi = C_i$ on ω_i)

$$(18) \qquad \frac{1}{4}\int_\Omega |\nabla\Phi|^2 \leq 2\pi d_i \sum_i |C_i|$$
$$\leq 2\pi C \max_i |C_i|,$$

where C is some absolute constant. Next we will use Trudinger's inequality, which asserts the existence of universal constants σ_1 and σ_2 such that

$$\int_\Omega \exp\left(\frac{|u|}{\sigma_1\|\nabla u\|_{L^2}}\right) \leq \sigma_2 |\Omega|, \quad \forall u \in H_0^1(\Omega).$$

In particular, for $i = 1, ..., \ell$,

$$\int_{\omega_i} \exp\left(\frac{\Phi}{\sigma_1\|\nabla\Phi\|_{L^2}}\right) \leq \sigma_2 |\Omega|,$$

which writes

$$|\omega_i| \exp\left(\frac{C_i}{\sigma_1\|\nabla\Phi\|_{L^2}}\right) \leq \sigma_2 |\Omega|.$$

This yields

$$|C_i|^2 \leq \sigma_1^2 \|\nabla\Phi\|_{L^2} \log\left(\frac{\sigma_2|\Omega|}{|\omega_i|}\right) \leq \sigma_1^2 \|\nabla\Phi\|_{L^2}^2 \log\left(\frac{\sigma_2|\Omega|}{\pi\lambda^2\varepsilon^2}\right).$$

Combining this with (18), we obtain the desired result.

We turn now to estimates for ρ. Note that the equation verified by ρ is

$$(19) \qquad -\Delta\rho + \frac{1}{\rho^3}|v \times \nabla v|^2 = \frac{1}{\varepsilon^2}\rho(1 - \rho^2).$$

Multiplying by $\rho - 1$ and integrating over Ω_ϵ, we obtain

$$\int_{\Omega_\epsilon} |\nabla \rho|^2 = \int_{\partial \Omega} \frac{\partial \rho}{\partial \nu}(\rho - 1) - \sum_i \int_{\partial \omega_i} \frac{\partial \rho}{\partial \nu}(\rho - 1)$$

$$+ \int_{\Omega_\epsilon} \frac{(1 - \rho)}{\rho^3} |v \times \nabla v|^2 - \frac{1}{\epsilon^2} \int_\Omega (1 - \rho)^2(1 + \rho)$$

$$\leq \int_{\partial \Omega_\epsilon} |\nabla v| + 8 \int_{\Omega_\epsilon} |v \times \nabla v|^2$$

$$\leq C + 16 \int_{\Omega_\epsilon} |\nabla \Phi|^2 + |\nabla H|^2.$$

Hence, we obtain

Proposition 10. *There is some constant C independent of ϵ, such that*

$$\int_\Omega |\nabla \rho|^2 \leq C(|\log \epsilon| + 1).$$

Next, we will prove

Proposition 11. *Given any $1 \leq p < 2$ there are constants α and C depending only on g, G and p such that*

$$\int_{\Omega_\epsilon} |\nabla \rho|^p \leq C \epsilon^\alpha.$$

Proof : Consider for $\beta \in (0, 1)$ the set

$$W = \{x \in \Omega_\epsilon, \ \rho > 1 - e^\beta\},$$

and define $\bar\rho$ by $\bar\rho = \text{Max}(\rho, 1 - \epsilon^\beta)$, so that $\rho = \bar\rho$ on W. Multiplying (19) by $(1 - \bar\rho)$ and integrating over Ω_ϵ yields

$$\int_W |\nabla \rho|^2 \leq 8\epsilon^\beta \left(\int_\Omega |v \times \nabla v|^2 + C \right),$$

and hence

$$\int_W |\nabla \rho|^2 \leq C \, \epsilon^\beta (|\log \epsilon| + 1) \to 0 \ \text{ as } \epsilon \to 0.$$

On the other hand, we have

$$\int_{\Omega \setminus W} |\nabla \rho|^p \leq \left(\int_\Omega |\nabla \rho|^2 \right)^{p/2} \text{meas}(\Omega \setminus W)^{1 - (p/2)}$$

$$\leq C(|\log \epsilon| + 1)^{p/2} \text{meas}(\Omega \setminus W)^{1 - (p/2)}.$$

Since $\frac{1}{\epsilon^2}\int_\Omega (1-|\rho|^2)^2 \le C$, we deduce that

$$\text{meas}(\Omega\backslash W) \le C\,\epsilon^{2-2\beta},$$

and this yields the desired conclusion.

Combining Proposition 7 to 11, together with (14), we deduce

Proposition 12. *There is a constant $C > 0$ depending only on g and Ω such that, for any solution v_ϵ to $(GL)_\epsilon$*

$$E_\epsilon(v_\epsilon) \le C(|\log \epsilon| + 1).$$

Moreover for any $1 \le p < 2$, there is a constant C_p such that

(20)
$$\int_\Omega |\nabla v_\epsilon|^p \le C_p.$$

Finally we may complete the proof of Theorem 2.

IV.4. Proof of Theorem 2 completed.

Since v_ϵ is bounded in $W^{1,p}$, there is a subsequence $\epsilon_n \to 0$ and a map v_* in $W^{1,p}$ such that

$$v_{\epsilon_n} \to v_* \text{ in } W^{1,p}(\Omega).$$

Since

$$\int \left(1 - |v_{\epsilon_n}|^2\right)^2 \le C\,\epsilon_n^2 \to 0,$$

we deduce that $|v_*| = 1$, i.e. v_* is S^1-valued. Passing to a subsequence if necessary, we may assume that the number ℓ of points x_i is independent of ϵ, and that, for $i = 1, ..., \ell$

$$x_i \to a_i \in \bar\Omega.$$

Let K be a compact subset of $\Omega\backslash\bigcup_{i=1}^\ell \{a_i\}$. We claim that there exists a constant C_K independent of ϵ such that

(21)
$$\int_K |\nabla v_{\epsilon_n}|^2 \le C_K.$$

To see that, we have to establish bounds for Φ, H and ρ on K. For H, we already have established the desired bound. For Φ, we use the elliptic equation

$$\text{div}\left(\frac{1}{\rho^2}\nabla\Phi_\epsilon\right) = 0.$$

Let $B(x_0, r)$ be a ball in $\Omega \backslash \bigcup_{i=1}^{\ell} \{a_i\}$. Since Φ_ϵ is bounded in $W^{1,p}$ there exists a radius $r_0 \in \left[r, \frac{r}{2}\right]$ such that

$$\int_{\partial B(x_0, r)} |\Phi_\epsilon|^p + |\nabla \Phi_\epsilon|^p \leq C$$

where C is independent of ϵ (but depends on r). Multiplying the equation by Φ and integrating by parts, using the previous bound, we see that

$$\int_K |\nabla \Phi_\epsilon|^2 \leq C,$$

and a similar estimate holds for ρ : this establishes (21). From (21) we deduce that $v_* \in H^1(K)$, for any compact subset K of $\Omega \backslash \bigcup_{i=1}^{\ell} \{a_i\}$. On the other hand, since v_{ϵ_n} is bounded in $W^{1,p}$ we may pass to the limit in (11), to assert that

$$(22) \qquad \frac{\partial}{\partial x_i} (v_* \times v_{*x_1}) + \frac{\partial}{\partial x_2} (v_* \times v_{*x_2}) = 0.$$

Let $B(x_0, r)$ be a ball in $\Omega \backslash \bigcup_{i=1}^{\ell} \{a_i\}$. Since $v_* \in H^1(B(x_0, r))$ there is some $\varphi \in H^1(B(x_0, r), \mathbb{R})$ such that

$$v_* = \exp i\varphi.$$

Relation (22) thus implies that

$$\Delta \varphi_* = 0, \quad \text{on } B(x_0, r),$$

i.e. $\varphi_* \in C^\infty(B(x_0, r))$, and thus $v_* \in C^\infty \left(\Omega \backslash \bigcup_{i=1}^{\ell} \{a_i\}\right)$. Finally going back to (22), a similar argument together with the $W^{1,1}$ bound allows to assert that there are integer $d_i \in \mathbb{Z}$ and a harmonic function φ on Ω such that

$$v_* = \prod_{i=1}^{\ell} \left(\frac{z - a_i}{|z - a_i|}\right)^{d_i} \exp i\varphi$$

(see [BBH, Section I]).

IV.5. Appendix : a variant of Theorem 2

We may replace the assumption Ω starshaped in Theorem 2 by the energy bound (9) on solutions. More precisely, we have

Theorem 2 bis. *Let v_ϵ be a sequence of solutions of (GL_ϵ) such that there exists some constant $K > 0$ such that*

$$(23) \qquad E_\epsilon(v_\epsilon) \leq K(|\log \epsilon| + 1).$$

Then there exists a subsequence $\varepsilon_n \to 0$, ℓ points $a_1, ..., a_\ell$ in Ω, and ℓ integers $d_1, ..., d_\ell$ such that

$$v_{\varepsilon_n} \to v_* = \prod_{i=1}^{\ell} \left(\frac{z - a_i}{|z - a_i|} \right)^{d_i} \exp i\varphi \quad \text{in } W^{1,p}(\Omega), \text{ for any } d_i \le p < 2,$$

and in $C^k(K)$ for any compact subset K of $\Omega \backslash \bigcup_{i=1}^{\ell} \{a_i\}$. Here φ is a harmonic function on Ω.

Sketch of the proof : The only point to establish is to prove that Step 1 of Theorem 2 is still valid for Theorem 2bis. Since we do not assume that Ω is starshaped, we cannot use Pohozaev's identity directly on Ω. Instead, we will use a local version of it : this technic was introduced independently in [BR] and in [Str].

Let $\alpha \in (0,1)$. We will consider balls of radius r in $[\varepsilon^\alpha, \varepsilon^{\alpha/2}]$. We have

Lemma 4. *There is some constant C depending only on g, K and α such that, for any point x_0 in Ω, there is some r_0 in $(\varepsilon^\alpha, \varepsilon^{\alpha/2})$ such that*

$$(24) \qquad \int_{\partial B(x_0, r_0) \cap \Omega} |\nabla v|^2 + \frac{1}{\varepsilon^2} \left(1 - |v|^2 \right)^2 \le \frac{C}{r_0}.$$

Proof : Assume (24) is false for any r_0 in $(\varepsilon^\alpha, \varepsilon^{\alpha/2})$ integrating on $(\varepsilon^\alpha, \varepsilon^{\alpha/2})$ we obtain

$$\int_{B(x_0, \varepsilon^\alpha) \cap \Omega} |\nabla v_\varepsilon|^2 + \frac{1}{2\varepsilon^2} \left(1 - |v|^2 \right)^2 \ge C \frac{\alpha}{2} |\log \varepsilon|$$

which contradicts (23), if C is choosen large enough.

Applying Pohazaev's identity to the ball $B(x_0, r_0)$, we are led to

Lemma 5. *There is a constant C_α depending only on α, g and Ω such that*

$$\frac{1}{4\varepsilon^2} \int_{B(x_0, \varepsilon^\alpha) \cap \Omega} \left(1 - |v_\varepsilon|^2 \right)^2 \le C_\alpha.$$

Using then the proof of Step 1 of Theorem 2, we obtain the local version :

Proposition 13. *There is a constant N_α depending only on α, g and Ω, such that there exist ℓ points $x_1, ..., x_\ell$ in $B(x_0, \varepsilon^\alpha)$ with $\ell \le N$, such that*

$$|v_\varepsilon(x)| \ge \frac{1}{2} \quad \text{on } \Omega \cap B(x_0, \varepsilon^\alpha) \backslash B(x_i, \lambda \varepsilon)$$

(λ being the constant of Step 1).

The crucial observation is now

Proposition 14. Let x_0 be a point in Ω, such that

(25)
$$|v_\epsilon(x_0)| \leq \frac{1}{2}.$$

Let $\beta \in [0,1]$. Then, there is a constant $\eta > 0$, β, g and Ω, such that

(26)
$$\int_{B(x_0,\epsilon^\beta) \cap \Omega} |\nabla v_\epsilon|^2 \geq \eta(|\log \epsilon| + 1).$$

Proof : The proof relies again on Pohazaev's identity. Assume that (26) is contradicted (for a given η). Then there exists some $r_0 \in (\epsilon^\beta, \epsilon^{\beta/2})$ such that

$$\int_{\partial B(x_0,r_0) \cap \Omega} |\nabla v_\epsilon|^2 + \frac{1}{4\epsilon^2} \left(1 - |v|^2\right)^2 \leq \frac{2\eta}{\beta r_0}.$$

By Pohazaev's identity, this implies that

$$\frac{1}{4\epsilon^2} \int_{B(x_0,\epsilon^\beta) \cap \Omega} \left(1 - |v|^2\right)^2 \leq \frac{C\eta}{\beta}$$

for some constant C. If we choose η so small that

$$0 < \frac{C\eta}{\beta} \leq \mu_0,$$

where μ_0 is the constant in Proposition 5, and apply then Proposition 6, we see that we obtain a contradiction with (25).

Proof of Theorem 2bis completed : Let $\beta \in (0,1)$ and consider a finite collection of points $(x_i)_{i \in I}$ such that

$$\Omega \subset \bigcup_{i \in I} B\left(x_i, \epsilon^\beta\right),$$

and

$$B\left(x_i, \epsilon^\beta\right) \cap B\left(x_j, \epsilon^\beta\right) = \emptyset \quad \text{if} \quad i \neq j.$$

By assumption (23), we have, for some universal constant C,

(27)
$$\sum_{i \in I} \int_{B(x_i,2\epsilon^\beta)} |\nabla v_\epsilon|^2 \leq CK(|\log \epsilon| + 1).$$

Let J be the subset of I, such that $i \in J$ if and only if

$$\int_{B(x_i, 2\epsilon^\beta)} |\nabla v_\epsilon|^2 \geq \eta(|\log \epsilon| + 1).$$

By (27) the cardinal of J is bounded independently of ϵ by a constant $\tilde{N}$ depending only on g, Ω and β. By Proposition 13, we see that

$$|v(x)| \geq \frac{1}{2} \quad \text{if} \quad x \in B\left(x_i, \epsilon^\beta\right), \quad i \notin J,$$

that is

$$|v(x)| \geq \frac{1}{2} \quad \text{on } \Omega \backslash \bigcup_{i \in J} \left(x_i, \epsilon^\beta\right).$$

Hence, in order to seek our "vortices", we only have to locate them on $\bigcup_{i \in J} \left(x_i, \epsilon^\beta\right)$. To complete the proof of Step 1 for Theorem 2bis it then suffices to invoke Proposition 2 with $\beta = \alpha$. The rest of the proof is then similar to the proof of Theorem 2.

V. RENORMALIZED ENERGIES

V.1. An auxiliary problem

In order to get some insight to this notion, we will start with some auxiliary problem for S^1−valued map (see [BBH], Section 1). Consider first a smooth bounded domain Ω in $\mathbb{R}^2$, a smooth map g from $\partial\Omega$ to S^1, of degree d. Fix d points in Ω, say $a_1, a_2, ..., a_d$ and let $\epsilon > 0$ be a small parameter. We set

$$\omega_i^\epsilon = B(a_i, \epsilon)$$

and

$$\Omega_\epsilon = \Omega \backslash \bigcup_{i=1}^{d} \omega_i.$$

Consider the class of S^1−valued functions v on Ω_ϵ defined by

$$V = \{v \in H^1(\Omega_\epsilon, S^1) \ \deg(v, \partial\omega_i) = 1 \text{ for } i = 1, .., d, \ v = g \text{ on } \partial\Omega\}.$$

We wish to determine minimizers for

(28) $$\gamma_\epsilon = \text{Inf}\left\{\int_{\Omega_\epsilon} |\nabla v|^2, \ v \in V\right\},$$

as well as estimates for γ_ϵ. We will show that this problem is related to a linear problem.

Theorem 3. *We have*

$$\gamma_\epsilon = \int_{\Omega_\epsilon} |\nabla \Phi|^2$$

where Φ is the solution to

(29)
$$\begin{cases} \Delta \Phi = 0 \ \text{ on } \partial \Omega \\ \int_{\partial \omega_i} \dfrac{\partial \Phi}{\partial \nu} = 2\pi \ \ \forall \, i \in \{1, ..., d\}, \ \Phi = \text{cte} = C_i \ \text{ on } \partial \omega_i \\ \dfrac{\partial \Phi}{\partial \nu} = g \times g_r \ \text{ on } \partial \Omega \\ \int_{\Omega_\epsilon} \Phi = 0. \end{cases}$$

(the constants C_i are <u>not</u> prescribed but free).

Moreover, the infimum (28) is achieved by a map $u \in V$ such that

$$u \times u_{x_1} = -\Phi_{x_2}, \quad u \times u_{x_2} = \Phi_{x_1}.$$

<u>Proof</u> : The solution to (29) can be found by minimizing the functional

$$G(\psi) = \frac{1}{2} \int_{\Omega_\epsilon} |\nabla \psi|^2 - 2\pi C_i - \int_{\partial \Omega} g \times g_r \psi$$

on the space

$$W = \left\{ \psi \in H^1(\Omega_\epsilon), \ \psi = \text{cte} = C_i \ \text{on } \partial \omega_i, \ \int_{\Omega_\epsilon} \psi = 0 \right\},$$

the constants C_i are free to vary. Let v be a map in V. Arguing as in Section IV, we have

$$\frac{\partial}{\partial x_1}(v \times v_{x_2}) - \frac{\partial}{\partial x_2}(v \times v_{x_1}) = 0$$

so that

$$\text{Div } D = 0 \quad \text{on } \Omega_\epsilon$$

$$\int_{\partial \omega_i} D.\nu = 0 \quad \text{for } i = 1, ..., d, \ \ D.\nu = 0 \text{ on } \partial \Omega$$

for $D = (v \times v_{x_2} - \Phi_{x_1}, \ v \times v_{x_1} + \Phi_{x_2})$.

By Poincaré's Lemma, there is some $H \in H^1(\Omega_\epsilon)$ such that $D = \nabla H$, i.e.

$$v \times v_{x_2} = H_{x_2} + \Phi_{x_1}, \quad v \times v_{x_1} = H_{x_1} - \Phi_{x_2}.$$

We deduce that

$$\int_{\Omega_\epsilon} |\nabla v|^2 = \int_{\Omega_\epsilon} |v \times \nabla v|^2$$

$$(30) \qquad = \int_{\Omega_\epsilon} |\nabla \Phi|^2 + \int_{\Omega_\epsilon} |\nabla H|^2 + 2 \int_{\Omega_\epsilon} H_{x_2} \Phi_{x_1} - H_{x_1} \Phi_{x_2}$$

$$= \int_{\Omega_\epsilon} |\nabla \Phi|^2 + |\nabla H|^2.$$

Indeed

$$\int_{\Omega_\epsilon} H_{x_2} \Phi_{x_1} - H_{x_2} \Phi_{x_1} = \int_\Omega \frac{\partial}{\partial x_1} (\Phi H_{x_2}) - \frac{\partial}{\partial x_2} (\Phi H_{x_1})$$

$$= \sum_{i=1}^d \int_{\partial \omega_i} \Phi H_\tau - \int_{\partial \Omega} \Phi H_\tau$$

Since $D \cdot \nu = 0$ on $\partial \Omega$, $H_\tau = 0$ on $\partial \Omega$. Integrating by parts we have

$$\int_{\partial \omega_i} \Phi H_\tau = \int_{\partial \omega_i} \Phi_\tau H = 0 \quad (\text{since } \Phi = \text{Cte on } \partial \omega_i).$$

This yields (30). We have already proved

$$\gamma_\epsilon \geq \int_{\Omega_\epsilon} |\nabla \Phi|^2.$$

Next consider $u = v \exp iH$ (we take $H = 0$ on $\partial \Omega$). A simple computation shows that $\int_{\Omega_\epsilon} |\nabla u|^2 = \int_{\Omega_\epsilon} |\nabla \Phi|^2$. This completes the proof.

In order to estimate γ_ϵ it suffices therefore to estimate the L^2- norm of Φ_ϵ. As $\epsilon \to 0$, one may verify that Φ_ϵ converges to Φ_* the solution of

$$\begin{cases} \Delta \Phi_* = 2\pi \sum_{i=1}^d \delta_{a_i} \text{ in } \Omega \\[2mm] \dfrac{\partial \Phi_*}{\partial \nu} = g \times g_\tau \text{ on } \partial \Omega \\[2mm] \displaystyle\int_\Omega \Phi_* = 0. \end{cases}$$

Actually, using linear estimate, one may prove (see [BBH], Section I)

$$\int_{\Omega_\epsilon} |\nabla (\Phi_\epsilon - \Phi_*)|^2 \to 0 \quad \text{as } \epsilon \to 0.$$

Hence

$$\gamma_\epsilon = \int_{\Omega_\epsilon} |\nabla \Phi_*|^2 + o(1), \quad \text{where } o(1) \to 0 \text{ as } \epsilon \to 0.$$

To derive an estimate for the right-hand side, we first notice that $\Delta R = 0$ for the function R given by

$$R = \Phi_* - \sum_{i=1}^{d} \log |x - a_i|.$$

In particular R is a smooth function on $\bar{\Omega}$. Going back to γ_e, and writing $\Phi_* = R + \sum_{i=1}^{d} \log |x - a_i|$, we see that

$$|\gamma_e - \pi d|\log \varepsilon|| \leq C$$

where C is some constant. Actually, we

Proposition 15. *We have*

$$\gamma_e |\log \varepsilon| + W_g(a_1, ..., a_d) + o(1), \text{ where } o(1) \to 0 \text{ as } \varepsilon \to 0,$$

where the function W_g is defined on $\Omega^d \backslash \Delta$ by the formula : for $(b_1, b_2, ..., b_d)$ in $\Omega^d \backslash \Delta$

$$W_g(b_1, ..., b_d) = -\pi \sum_{i \neq j} \log |b_i - b_j| + \sum R(b_i) + \int_{\partial \Omega} g \times g_\tau . \Phi_*.$$

The function W_g will be called the **renormalized energy** *(associated to the configuration $b_1, ..., b_d$).*

Remarks : Here we have assumed that the winding number around each singulairty is exactly $+1$, and that therefore there are exactly d singular points. More generally, one may handle a similar problem for ℓ points $a_1, ..., a_\ell$ in Ω, with winding numbers $d_1, d_2, ..., d_\ell$ such that $\sum_{i=1}^{\ell} d_i = d$. Then the asymptotic expansion of γ_e (for such a situation) would be

$$\gamma_e = \pi \left(\sum_{i=1}^{\ell} d_i^2 \right) |\log \varepsilon| + W_g((a_1, d_1), ..., (a_\ell, d_\ell)),$$

where the renormalized energy W_g is now given by

$$W_g((a_1, d_1), ..., (a_\ell, d_\ell)) = -\pi \sum_{i < j} d_i d_j \log |a_i - a_j|$$

$$+ \pi \sum_{i=1}^{\ell} d_i R(a_i) + \int_{\partial \Omega} g \times g_\tau . \Phi_*.$$

V.2. Back to Ginzburg-Landau

We are going to show how the previous analysis connects to Ginzburg-Landau functionals, and Theorem 1 : we will sketch the proof of the fact that the configuration $(a_1, ..., a_d)$ in Theorem 1 has to be minimizing for the renormalized energy.

As seen in Theorem 1, any sequence u_{ε_n} converges, up to a subsequence, to a map u_* of the form

$$u_* = \prod_{i=1}^{d} \frac{z - a_i}{|z - a_i|} \exp i\varphi$$

where φ is harmonic. Moreover, the convergence is very good on $\Omega \backslash \bigcup_{i=1}^{d} \{a_i\}$. For instance for any compact subset K of $\Omega \backslash \bigcup_{i=1}^{d} \{a_i\}$, we have

$$(31) \qquad \|u_\varepsilon - u_*\|_{C^1(K)} \leq C_K \varepsilon^2,$$

where C_K depends on K. Given $\rho > 0$, one may therefore write

$$(32) \qquad \begin{aligned} E(u_\varepsilon) &= \sum_{i=1}^{d} \int_{B(a_i, \rho)} \frac{1}{2} |\nabla u_{\varepsilon_n}|^2 + \frac{1}{4\varepsilon^2} (1 - |u_\varepsilon|)^2 \\ &= \int_{\Omega \backslash \bigcup_{i=1}^{d} B(a_i, \rho)} |\nabla u_*|^2 + o(1), \quad \text{where } o(1) \to 0 \text{ as } \varepsilon \to 0. \end{aligned}$$

Set

$$I(\varepsilon, \rho) = \inf \left\{ \int_{B(\rho)} |\nabla v|^2 + \frac{1}{4\varepsilon^2} (1 - |u|^2)^2, \ v \in H^1(B(\rho)), \ v = \frac{z}{|z|} \text{ on } \partial B(\rho) \right\}.$$

Note that

$$I(\varepsilon, \rho) = I\left(\frac{\varepsilon}{\rho}, 1\right).$$

Then, in view of (31), (32) writes

$$E(u_\varepsilon) = dI\left(\frac{\varepsilon}{\rho}, 1\right) + \int_{\Omega \backslash \bigcup_{i=1}^{d} (a_i, \rho)} |\nabla \Phi_*|^2 + R_1 + R_2,$$

where $R_1 \to 0$ as $\rho \to 0$ (and is independent of ε) and $R_2 \to 0$ for fixed ρ, as $\varepsilon \to 0$. From the computations of Section V.1, we have

$$\int_{\Omega \backslash \bigcup_{i=1}^{d} B(a_i, \rho)} |\nabla \Phi|^2 = \pi d |\log \rho| + W_g(a_1, ..., a_d) + o(1)$$

where $o(1) \to 0$ as $\rho \to 0$. This writes

$$(33) \qquad E_\varepsilon(u_\varepsilon) = dI\left(\frac{\varepsilon}{\rho}, 1\right) + \pi d |\log \rho| + W_g(a_1, ..., a_d) + R_1 + R_2.$$

The first two terms do not depend on the configuration $a_1, ..., a_d$. Therefore (since u_ϵ is minimizing), we claim that $(a_1, ..., a_d)$ has to be minimizing for the renormalized energy. Indeed assume not : i.e. that there is a configuration $(b_1, ..., b_d)$ with smaller energy. Then arguing as in Section III, Proposition 3, we construct the map

$$v_* = \prod_{i=1}^{d} \frac{z - b_i}{|z - b_i|} \exp i\varphi$$

where φ is harmonic, and so that $v_* = g$ on $\partial\Omega$. Next, for given $\rho > 0$, we construct v_ϵ to that v_ϵ is continuous and

$$v_\epsilon = v_* \quad \text{on } \Omega \setminus \bigcup_{i=1}^{d} (a_i, \rho)$$

$$v_\epsilon \text{ minimizes } \int_{B(a_i, \rho)} \frac{1}{2}|\nabla v|^2 + \frac{1}{4\epsilon^2} \left(1 - |v|^2\right)^2 \text{ on } B(a_i, \rho).$$

Then a computation similar to (33) yields

(34) $$E_\epsilon(v_\epsilon) = dI\left(\frac{\epsilon}{\rho}, 1\right) + \pi d \log \rho + W_g(b_1, ..., b_d) + o(1)$$

where $o(1) \to 0$ as $\rho \to 0$. Of course, for sufficiently small ϵ, a suitable choice of ρ shows that $E_\epsilon(v_\epsilon) < E_\epsilon(u_\epsilon)$, which contradicts the fact that u_ϵ is minimizers.

Let us finally complete this section saying a word about non-minimizing solutions. In Theorem 2 we asserted that the configuration $(a_1, ..., a_\ell)$ has to be critical for the renormalized energy. In order to prove this, we have, as in the previous argument, to move a little the vortices. To do so, there is a standard technic (very useful for instance in conformal problems). Consider a vector field X on Ω, with compact support, and let $\varphi(t, x)$: $\mathbb{R} \times \Omega \to \Omega$ be the 1-parameter family of diffeomorphism of Ω, obtained by integrating the flow X. Next, for v solution to (GL_ϵ), consider the family on maps $v_t(\cdot) = v(\varphi(t, \cdot))$. Basically the vortices of v_t are the vortices of v, moved along the direction of the flow X. Moreover since v is solution to GL_ϵ then

(35) $$\frac{d}{dt} E_\epsilon(v_t(\cdot)) = 0.$$

Relation (35) gives rise to a very interesting equation (see [BBH], Chapter VIII) which can be exploited to conclude that the configuration $a_1, ..., a_\ell$ has to be critical for the renormalized energy.

In the next section we will pursue a rather different issue, namely we will try to construct non-minimizing solutions. In view of Theorem 2 a natural question is also to know if one is able to prescribe the multiplicities of the vortices.

VI. CONSTRUCTION OF NON-MINIMIZING SOLUTIONS

VI.1. An example

Take $\Omega = D^2$ and the boundary value g of the form $g(\theta) = \exp id\theta$ (for $d \in \mathbb{N}^*$). In view of the symmetry, one can find solutions of the Ginzburg-Landau equation of the form (in polar coordinates)

$$v_\epsilon(r, \theta) = f_{d,\epsilon}(r) \exp id\theta,$$

where the function $f_{d,\epsilon}$ is solution on [0,1] of the ODE

$$(36) \qquad \begin{aligned} & r^2 f'' + r f' - d^2 f + \frac{1}{\epsilon^2} f(1 - f^2) = 0 \quad \text{on } [0, 1] \\ & f(0) = 0, \quad f(1) = 1. \end{aligned}$$

One may quite easily check that a solution f_d to (36) exists (by minimization for instance), and that f_d is non-decreasing. Moreover, computing $E_\epsilon(v_\epsilon)$ one finds that this solution has an energy of order $\pi d^2 |\log \epsilon|$. Hence if d is larger than 2, and ϵ is sufficiently small, v is a **non-minimizing** solution. Moreover

$$v_\epsilon \to v_* = \left(\frac{z}{|z|} \right)^d,$$

that is we have constructed solutions of multiplicity d. In the case $d = 1$, it is conjectured that the radially symmetric solution is minimizing : important progresses to establish that conjecture have been made by Mironescu [M1, M2].

There is a natural action of the group S^1, which leaves both the space and the functional invariant. Define $T : S^1 \to H_g^1(\Omega; \mathbb{R}^2)$ by

$$T_\alpha v(z) = \exp(-id\alpha).v(\exp i\alpha.z), \quad \forall \alpha \in [0, 2\pi]$$

for any function v in H_g^1, and any point z in H_g^1. Consider next the functions which are left invariant by T, that is the subset H_0 of H_g^1 defined by

$$H_0 = \{v \in H_g^1, \ T_\alpha v = v, \ \forall \alpha \in [0, 2\pi]\}.$$

It turns out that H_0 is precisely the set of radially symmetric functions, that is

$$H_0 = \{v \in H_g^1, \ \text{such that } v(r, \theta) = f(r) \exp id\theta\},$$

and that

(37)
$$\operatorname*{Inf}_{v \in H_0} H_0$$

is achieved by the solution v_ε given by (36). In presence of a $S^1 - group$ action, the index theory of Faddell and Rabinowitz can be applied, and allows to find more solutions [FR]. Actually, a general principal (see [AB]) asserts the we have at least as many orbits of solutions to (GL_ε) as the Morse Index of the minimizing solution for (37), and that these solutions have an energy strictly lower than (37), and therefore are not radially symmetric (Note that, if a solution is not in H_0, we have actually a full orbit of solutions, by the action of T_α). The computation of the Morse Index of v_ε can be estimated as follows (see [AB]).

Proposition 16. *There is an absolute constant $\mu_0 > 0$ such that the Morse Index of v_ε is larger than $\mu_0 |d|^2$ provided $d \geq 2$, and ε is sufficiently small.*

Sketch of the proof ([AB]) : In the neighborhood of v, we may write

$$E_\varepsilon(v_\varepsilon + \omega) = E_\varepsilon(v_\varepsilon) + Q_\varepsilon(w) + O(\|w\|^3), \quad \forall \, w \in H_0^1(\Omega).$$

Here Q_ε is the quadratic form given by

$$Q_\varepsilon(w) = \frac{1}{2} \int |\nabla w|^2 - \frac{1}{2\varepsilon^2} \int_{D^2} \left(1 - |v_\varepsilon|^2\right)^2 w^2 + \frac{1}{\varepsilon^2} \int_{D^2} (\hat{u}_d.w)^2,$$

that is

$$Q_\varepsilon(w) = \langle L_\varepsilon w, w \rangle,$$

where L_ε is the linear operator given by

$$L_\varepsilon(w) = -\Delta w - \frac{1}{\varepsilon^2}\left(1 - |f_d|^2\right) w + \frac{2}{\varepsilon^2}(w.v_\varepsilon)v_\varepsilon.$$

The Morse Index of v_ε is given by the number of negative eigenvalues of L_ε : this number is finite by standard Riesz-Fredholm theory. Moreover, if V is a subspace of $H_0^1(\Omega)$ such that

$$Q(x) < 0, \quad \forall \, x \in V$$

then

(38)
$$\dim V \leq \dim H_-,$$

where H_- is the space spanned by the eigenvectors with negative eigenvalues. We are going therefore to construct a space V_ε with the previous property.

First, consider the unique $\rho_\epsilon \in [0,1]$ such that

$$f_d(\rho_\epsilon) = \frac{1}{4}.$$

We have for any $w \in H_0^1(D(\rho_\epsilon)) \subset H^1(D^2)$

(39)
$$Q(w) \leq \tilde{Q}(w) \equiv \frac{1}{2} \int_{D(\rho_\epsilon)} |\nabla v|^2 - \frac{13}{16\epsilon^2} |\bar{v}|^2.$$

In particular, if V_ϵ is the finite dimensional subspace of $H_0^1(D(\rho_\epsilon))$ spanned by the eigenfunction of $-\Delta$ on $D(\rho_\epsilon)$ with eigenvalues less than $\frac{13}{16\epsilon^2}$, then

$$\tilde{Q}(w) < 0, \quad \forall\, w \in V_\epsilon.$$

Hence by (38) and (39)

(40)
$$\dim H \geq \dim V_\epsilon.^-$$

By standard spectral theory, for some constant $C > 0$,

$$\dim V_\epsilon \geq C\left(\frac{\rho_\epsilon^2}{\epsilon^2} - 1\right).$$

Using equation (36), one proves that

$$\rho_\epsilon \geq Cd\epsilon,$$

which yields

$$\dim H^- \geq C(d^2 - 1)$$

and completes the proof of Proposition 16.

Combining Proposition 16 with the Index theory of Faddell and Rabinowitz, we are led to

Theorem 4. *Let $d \geq 2$. There is an absolute constant $\mu_0 \geq 0$, such that, for sufficiently small ϵ, (GL_ϵ) on D^2 with $g = \exp id\theta$ has at least $\mu_0|d|^2$ of solutions.*

<u>Remark</u> : Recently, in a joint work with B. Helffer, we have been able to show that the estimate in Proposition 15 is in some sense optimal. More precisely, we proved (see [BH]) that there is some constant $\mu_1 > 0$ such that the Morse Index of v_ϵ is less than $\mu_1|d|^2$.

VI.2. Variational methods

Here we come back to the general case, without symmetries. In view of the previous analysis one might expect to find more solutions as d increases. This hope is also consistent with the following.

Proposition 17. *Assume Ω is starshaped, and that $g = 1$ on $\partial\Omega$; then the only solution to GL_ϵ is $v = 1$.*

Proof : Use Pohazaev's identity to assert that

$$\int_\Omega \left(1 - |v|^2\right) = 0$$

so that $|v| = 1$ on Ω, and GL_ϵ writes

$$\Delta v = 0.$$

The conclusion follows.

In the sequel of this section, we will therefore assume that $d \geq 2$, and will use Morse theory to construct solution. At this point, only a few results have been obtained. We will sketch the proof of the following theorem (see Almeida-B [AB2]).

Theorem 5. *Assume $d \geq 2$. If ϵ is sufficiently small, then (GL_ϵ) has at least three distinct solutions, among which one at least is non-minimizing.*

Remark. Other non-minimizing solutions have been produced by F.H. Lin [Li1]. For special boundary conditions g, he was able to produce solutions with vortices of opposite sign, which are local-minimizers, using heat flow methods. In contrast the solution produced in Theorem 5 has probably a non-zero Morse index.

The proof of Theorem 5 is based on Morse theory. We consider the level sets

$$E_\epsilon^a = \{v \in H_g^1(\Omega; \mathbb{R}^2),\ E_\epsilon(v) \leq a\}.$$

If E^a and E^b have different topologies, for some a and b in $\mathbb{R}$, then standard arguments of Morse theory assert that there is a critical value in (a, b), hence a solution to (GL_ϵ) (recall that the functional E_ϵ satisfies the Palais-Smale condition). Since $E^\infty = H_g^1$ is a contractible space (it is an affine space), we will apply the previous argument for $b = +\infty$ and show that for some $a > \kappa_\epsilon$, E^a has a non trivial topology. More precisely, we will prove :

Proposition 18. *There exists a constant $\chi_0 > 0$ such that, for $a = \kappa_e + \chi_0$, and ε sufficiently small there exists a loop in E^a which is not contractible, i.e. a continuous map $\gamma : S^1 \to E^a$ which cannot be extended to D^2 in a continuous way.*

Proposition 18 is of course the main ingredient of the proof of Theorem 5. As in many other variational problems in PDE's (see for instance J.M. Coron [C], Bahri and Coron [BC], C. Taubes [T], ...) the topology of level sets can partially be reduced to a finite dimensional problem. In our case, we already saw (at least for minimizers) that the energy functional (which is defined on an infinite dimensional space) is deeply related to the renormalized energy which is defined on a finite dimensional space : for minimizers on $\Sigma = \Omega^d \backslash \Delta$, where Δ is the diagonal. The bottom idea in the proof of Proposition 17 is that the topological properties of level sets E^a as stated above, that is for a close (but yet not too close !) to the infimum of the energy, are related to the topological properties of Σ. In particular, we will use the fact that $\pi_1(\Sigma) \neq 0$. However two new difficulties appear in the procedure above, which are mainly of analytical nature.
1) First, we have to define the notion of vortices for maps in E^a. Indeed, this notion was only defined at this stage for critical points, and the equation (in particular Pohozaev's identity) played a very important role in the analysis. Moreover some continuity in H^1 for the singularities has to be derived.
2) Second, we have to relate the energy of a map u to the renormalized energy of its vortices, as for instance in Theorem 1.

VI.2.1. Vortices for maps in E^a

In order to define vortices for maps u in E^a, we will proceed indirectly. Let $0 < \gamma < 1$ be given, and set $h = \varepsilon^\gamma$. Consider, for a given u in E^a, the minimization problem

(41)
$$\inf_{v \in H^1_g} F_h(v)$$

where F_h is given by

$$F_h(v) = E_\varepsilon(v) + \frac{1}{2h^2} \int_\Omega |u - v|^2.$$

Clearly F_h is achieved by some map u_h (we do not claim uniqueness) which verifies the "perturbed" Ginzburg-Landau equation

$$\frac{u_h - u}{h^2} - \Delta u_h = \frac{1}{\varepsilon^2} u_h \left(1 - |u_h|^2\right).$$

In view of our choice $h = \varepsilon^\gamma$, the perturbation is small (in some appropriate sense). Adapting the method of local estimates (cf. Section IV.5, proof of Theorem 2 bis), we may prove

Proposition 19. *Let K be an arbitrary constant, and assume that a verifies the bound*

(42)
$$a \leq K(|\log \varepsilon| + 1).$$

Let u be in E^a, and u_h be a minimizer for (41). There exist constants $N \in \mathbb{N}^*$, $\lambda > 0$, $\varepsilon_0 > 0$, $C_1 > 0$, depending only on g, K, and γ such that if $\varepsilon < \varepsilon_0$, then there exist ℓ points $a_1, ..., a_\ell$ in Ω, such that

$$\ell \leq N$$

$$|u^h(x)| \geq \frac{1}{2} \quad \text{on } \Omega \backslash \bigcup_{i=1}^{\ell} B(a_i, \lambda\varepsilon)$$

$$B(a_i, 2\lambda\varepsilon) \cap B(a_j, 2\lambda\varepsilon) = \emptyset \quad \text{if } i \neq j$$

and

$$\sum_{i=1}^{\ell} |d_i| \leq C_1, \quad d_i \neq 0 \quad \text{where } d_i = \deg\left(\frac{u_h}{|u_h|}, \partial B(a_i, \lambda\varepsilon)\right).$$

Note first that assumption (42) is much weaker than the assumption of Proposition 18, leading to the hope that more solutions (of higher energy and Morse index) can be found. For arbitrary maps u in E^a, one might have many (that is a number diverging with ε), regions where u vanishes : for instance, for a given map, one may insert a very large number of dipoles, i.e. a pair of vortices of opposite charge separated by a distance, say of order ε^γ, for some $0 < \gamma < 1$. At the end, this leads to a very blurred image, a map with many "details" on a smale scale (of order ε^α). Nevertheless, these details are basically unrelevant for Morse Theory. The idea behind Proposition 18 is that, if we omit the details h occuring on a length scale less than $h = \varepsilon^\gamma$, then things look more or less as in Theorems 1, 2, or 2bis. In other words, our approximation u^h (which is of parabolic type) smooths out details of small scale (of order $< h$).

However, there is a price to pay : the vortices themselves are only defined up to a small error, which corresponds to the scale of resolution we (arbitrarily) introduced. Therefore the map which assigns to an element in E^a its vortices (or more precisely the vortices of u^h) can certainly not be continuous. Nevertheless, it is η-almost continuous in the following sense.

Definition 1. *Let F and G be two metric spaces. Let $\eta \geq 0$ and f be a function from F to G. We say that f is η-almost continous at a point u_0 in F if, given any $\delta > 0$, there exists $\theta > 0$ such that if $d(u_0, v) \leq \theta$ implies $d(f(u_0), f(v)) \leq \eta + \delta$.*

The classical notion of continuity corresponds of course to the case $\eta = 0$. One can then prove that the map $\Phi : E^a \to W$ where W is the configuration space of charged vortices defined in [AB2]) is η-continuous for some $\eta = \varepsilon^\gamma$, where $0 < \gamma < 1$. It turns out that this property is sufficient to make use of the tools of algebraic topology we have in mind.

VI.2.2. Sketch of the proof of Proposition 18

First we construct the loop $\gamma_\varepsilon : S^1 \to E^a$ (with here $a = \kappa_\varepsilon + \chi_0$). The idea is of course to consider a loop γ in Σ which is not contractible, and then to construct a loop in E^a which has precisely the vortices given by γ. For that purpose assume for sake of simplicity that $0 \in \Omega$ and $B(2\delta_0) \subset \Omega$ for some $\delta_0 > 0$. Let $b_3, ..., b_d$ be $d - 2$ distinct points in $\Omega \backslash \overline{B(2(\delta_0))}$, and set

$$\gamma(\exp i\theta) = (0, \delta_0 \exp i\theta, b_3, ..., b_d) \text{ for } \theta \in [0, 2\pi].$$

Clearly γ is a continuous map from S^1 to Σ. We set

$$b_1(\theta) = 0, \ b_2(\theta) = \delta_0 \exp \theta, \ b_i(\theta) = b_i \text{ for } 3 \leq i \leq d - 2.$$

Next we construct γ_ε. For $\varepsilon > 0$ let f_ε be a smooth map from $(0,1)$ to $\mathbb{R}$ such that

$$f_\varepsilon(r) = 0 \quad \text{for } 0 < r < \frac{\varepsilon}{2}$$
$$f_\varepsilon(r) = 1 \quad \text{for } r > \varepsilon$$
$$|f_\varepsilon'(v)| \leq \frac{8}{\varepsilon}.$$

Let γ_ε be defined by

$$\gamma_\varepsilon(\exp i\theta) = \prod_{i=1}^{d} f_\varepsilon(|z - b_i(\theta)|) \frac{z - b_i(\theta)}{|z - b_i(\theta)|} \exp i\varphi_\theta,$$

where $\varphi(\theta)$ is a harmonic function, such that (for small ε)

$$\gamma_\varepsilon(\exp i\theta) = g \text{ on } \partial\Omega.$$

Clearly γ_ε is a continuous map from S^1 to H_g^1. Using the computations of Section V, we conclude that $\forall \, \theta \in [0, 2\pi]$.

$$E_0(\gamma_\varepsilon(\exp i\theta)) = -\pi d|\log \varepsilon| + W_g(b_1(\theta), ..., b_d(\theta)) + C_2 + o(1),$$

where C_2 is some absolute constant and $o(1)$ tends to zero as $\varepsilon \to 0$. In particular, for sufficiently small ε, there is a constant χ_0 such that, $\forall \, \theta \in [0, 2\pi]$

$$E_\varepsilon(\gamma_\varepsilon(\exp i\theta)) \leq \kappa_\varepsilon + \chi_0 \equiv a.$$

Hence γ_ϵ is a continuous loop in E^a.

The last (and most difficult) step in the proof of Proposition 18 is to prove that γ_ϵ is not contractible in E^a. For that purpose, we use Proposition 19. Here however a satisfies a better bound than (42), since it is close to the minimum value κ_ϵ, and as for minimizers, we may prove that all vortices have degree $+1$ (and hence, that their number is exactly d). More precisely, among the collection $(a_1, ..., a_\ell)$ provided by Proposition 19, we may find d vortices, say $a_1, ..., a_d$ and a radius ρ satisfying $\rho \leq \epsilon^\mu$ (where μ is some constant $0 < \mu < 1$) such that

$$|u^h(x)| \geq \frac{1}{2} \ \text{ on } \Omega \backslash \bigcup_{i=1}^{d} B(a_i, \rho)$$

$$\deg\left(\frac{u^h}{|u^h|}, \partial B(a_i, \rho)\right) = +1.$$

Moreover, one has

$$\text{dist}(a_i, a_j) \geq \mu_2, \ \forall i \neq j,$$

where μ_2 is some constant. Finally arguing as previously we may prove that the map $\Phi : E^a \to \Sigma$,

$$u \to \Phi(u) = (a_1, ..., a_d)$$

is η-almost continuous, form some $\eta \to 0$ as $\epsilon \to 0$. In particular

$$\Phi(\gamma_\epsilon(\exp i\theta)) = (b_1(\theta), ..., b_d(\theta)),$$

and $\Phi \circ \gamma_\epsilon$ is not contractible in Σ. This completes the proof.

VI.2.4. Sketch of the proof of Theorem 5

Since $E^\infty = H_g^1$ any loop in E^∞ is contractible (i.e. $\pi_1(E^\infty) = 0$). On the other hand, by Proposition 17, $\pi_1(E^a) \neq 0$, for $a = \kappa_\epsilon + \chi_0$. Hence there is a critical value

$$C > \kappa_\epsilon + \chi_0$$

hence a non minimizing solution to GL_ϵ. A third solution can then be constructed using a mountain-pass type argument (see [AB2]). Note that c can be defined using a Min-Max definition

(43) $$c = \text{Inf}\left\{ \underset{x \in D^2}{\text{Max}} \, E_\epsilon(f(x)), \ f \in T \right\}$$

where

$$T = \{ f \in C^0(\bar{D}^2, H_g^1(\Omega; \mathbb{R}^2), \ f = \gamma_\epsilon \text{ on } \partial D^2 \}.$$

In particular

$$c \le \pi(d+2)|\log \varepsilon| + C$$

for some constant C.

VI.3. Some open questions

The analysis above raises many open questions and suggests new directions.

Question 1. Is it possible to extend Theorem 2 to the case of simply connected domain, i.e. if Ω is simply connected, is there a constant $C > 0$, such that all solutions of GL_ε verify $E_\varepsilon(v_\varepsilon) \le C(|\log \varepsilon| + 1)$?

Question 2. What are the multiplicities of the vortices of the non-minimizing solution constructed in Theorem 5 ? (In [AB2] it is conjectured that the solution has either a vortex of multiplicity 2, and $d-2$ vortices of degree 1, or $d+1$ vortices of degree $+1$, and one vortex of degree -1).

Question 3. What are the multiplicities of the vortices in Theorem 4 ? In particular are there vortices of negative charge ?

Question 4. Is it possible to extend the methods of Theorem 4 (or Theorem 5) to the Abelian Higgs model on $\mathbb{R}^2$ (see [JT]) ?

Question 5. Let $(d_1, d_2, ..., d_k)$ be k positive number such that $d = d_1 + d_2 + ... + d_k$. Is it possible to construct a solution to GL_ε which has vortices of multiplicity $d_1, ..., d_k$?

Question 6. Consider the standard sphere S^3, equipped with a riemannian metric h. Consider the G.L functional defined one (S^3, h). Then the set of minimizers is the set of constant functions of norm 1, hence is diffeomorphic to S^1. Using arguments as in Theorem 5, there is a critical value $c > 0$, defined as (43) by a Min-Max. We conjecture that $c > k(|\log \varepsilon| + 1)$ for some constant $k > 0$. Moreover, if u is a solution corresponding to c, we conjecture that the set where u vanishes converges as $\varepsilon \to 0$ to two geodesics on (S^3, h) which are linked.

We hope this quite of ideas to lead to new geometrical properties.

VII. THE HEAT FLOW

Consider Ω a smooth bounded domain in $\mathbb{R}^2$ and g a smooth S^1-valued map from

$\partial\Omega$ to S^1. The heat flow equation for the Ginzburg-Landau equation writes

$$
(44) \qquad
\begin{cases}
\dfrac{\partial u}{\partial t} - \Delta u = \dfrac{1}{\varepsilon^2} u \left(1 - |u|^2\right) & \text{on } [0, +\infty[\times \Omega \\[2mm]
u(t, x) = g & \forall\, t \geq 0,\ \forall\, x \in \partial\Omega \\[2mm]
u(0, x) = u_0(x) & \forall\, x \in \Omega,
\end{cases}
$$

where $u : [0, +\infty[\times \Omega \to \mathbb{R}^2$, and the initial data u_0 is smooth and in H_g^1. By standard arguments a solution exists for all time, and is unique. Moreover, we have the equality

$$
(45) \qquad = \int_0^t \int_\Omega \left| \frac{\partial u}{\partial t} \right|^2 + E_\varepsilon(u(t)) = E_\varepsilon(u_0),
$$

hence the energy decreases along the flow (44). When we are able to define vortices for u_0, an important question is to derive the motion law for these vortices. In the case the energy of u_0 is close to κ_ε (i.e. is less than $\kappa_\varepsilon + C$, for some constant C independent of ε), this question was settled by F.H. Lin. He proved that, if the time t is scaled by $|\log \varepsilon|$, then the vortices move (in the limit $\varepsilon \to 0$) according to the opposite of the gradient of the renormalized energy (see also Jerrard and Soner for related results [GS]).

VIII. THE SCHRODINGER EQUATION

Here we assume that the domain is $\mathbb{R}^2$. The Schrödinger equation related to the Ginzburg-Landau functional

$$
\begin{cases}
i u_t = \Delta u + u \left(1 - |u|^2\right) & \text{on } [0, +\infty[\times \Omega \\[2mm]
u(x, 0) = u_0(x).
\end{cases}
$$

It appears in various models in physics, for instance superfluidity, nonlinear optics, or fluid dynamics. It is often termed Gross-Pitaevskii equation. Many problems remain open, as existence, motion low for vortices...

In a joint paper with J.C. Saut [BS], we have studied the existence problem for travelling wave solutions of the form. These solutions have the form

$$
u(x_1, x_2, t) = v(x_1 - ct, x_2)
$$

where (x_1, x_2) are cartesian coordinates on $\mathbb{R}^2$, v is a function on $\mathbb{R}^2$, and $c > 0$ is the speed of the wave. The equation for v reads

$$
ic \frac{\partial v}{\partial x_1} = \Delta v + v \left(1 - |v|^2\right).
$$

We establish the existence of a solution for small speeds. These solutions have been studied on a more formal level in a serie of papers (see for instance for references, Jones, Putterman [JPR], or Pismen and Nepomnyashchy [PN]).

The existence proof is based on the Mountain-Pass theorem for the functional

$$F(u) = \frac{1}{2} \int_{\mathbb{R}^2} |\nabla u|^2 + \frac{1}{4} \int_{\mathbb{R}^2} \left(1 - |u|^2\right)^2 - c \int_{\mathbb{R}^2} \left(i \frac{\partial u}{\partial x_1}, u\right).$$

The small parameter c plays here the role of the small parameter ε in our previous analysis.

IX. SUPERCONDUCTIVITY

As mentionned in the introduction, Ginzburg-Landau functionals have been first introduced to model superconductivity. The functional is however slightly more involved than the simple model we have considered so far. In order to account for electromagnetic effects one has to introduce a vector potential A, which can be considered as a 1-form

$$A = A_1 dx_1 + A_2 dx_2$$

where the functions A_1 and A_2 defined on Ω are real-valued. The Ginzburg-Landau functionals involve u and A and write

(45)
$$F_\varepsilon(u, A) = \frac{1}{2} \int_\Omega |\nabla_A u|^2 + |dA - H_0|^2 + \frac{1}{4\varepsilon^2} \left(1 - |u|^2\right)^2$$

where H_0 is a given function on Ω (the exterior applied field), u is complex-valued and

$$\nabla_A u = \left(\frac{\partial u}{\partial x_1} - iA_1 u, \ \frac{\partial u}{\partial x_2} - iA_2 u\right),$$

$$h = dA = \frac{\partial A_1}{\partial x_2} - \frac{\partial A_2}{\partial x_1},$$

and $\varepsilon > 0$ is a parameter (which depends on the material).

Some words on physics are in order. At low temperature, some material exhibits very special properties : they lose electric resistivity, and repel magnetic fluxes. This phenomenon is termed superconductivity. It turns out (according to the theory developped by Bardeen, Schaeffer and Cooper) that the electric current is not mediated by isolated electrons (as in usual conductors), but by pairs of electrons (with opposite sign), which behave like bosons. On a macroscopic level, these pairs of electrons (called Cooper pairs) are modelled by a complex-valued wave function u. The norm of u squared, $|u|^2$ represents the density of superconducting pairs of electrons : after some renormalizations, one may assert that

if $|u(x)| \simeq 1$, the sample is superconducting at the point $x \in \Omega$,

if $|u(x)| \simeq 0$, the sample is in the normal state (i.e. <u>not</u> superconducting).

Hence a sample may have regions where it is superconducting, and others where it superconductivity is lost.

For (45) we have restricted to the situation the sample is two-dimensional, and all magnetic fields are perpendicular to the sample. H_0 represents the exterior applied magnetic field. Stable configurations are supposed to be local minimizers for $F_\epsilon(u, A)$, on all possible configurations in $H^1(\Omega, \mathbb{R}^2) \times (H^1(\Omega, \mathbb{R}^2)$, ϵ is a parameter depending on the material. The function $h = dA = \frac{\partial A_2}{\partial x_1} - \frac{\partial A_1}{\partial x_2}$ represents the induced magnetic flux, and the electric current is given by

$$J = (iu, \nabla_A u) = ((iu, u_{x_1} - iA_1 u), ((iu, u_{x_2} - iA_2 u)).$$

An important feature of the functional F_ϵ is that it is gauge-invariant. More precisely, for every function $\varphi \in H^2(\Omega)$, we have

$$F_\epsilon(u, A) = F_\epsilon(v, B),$$

where

$$v = \exp i\varphi \cdot u$$
$$B = A + d\varphi.$$

All physically relevant quantities like $|u|$, J, h are gauge-invariant. In order to remove the invariance one may impose a condition on A, like the Coulomb gauge

(46)
$$\begin{cases} \text{div } A = 0 & \text{on } \Omega \\ A.\nu = 0 & \text{on } \partial\Omega. \end{cases}$$

Then, (45) and (46) define an elliptic problem.

When H_0 is small, the minimizing solution to (45) verifies (in the Coulomb gauge) $u(x) \simeq 1$ and $h = dA$ satisfies (approximatively) the London equation

$$\begin{cases} -\Delta h + h = 0 & \text{in } \Omega \\ h = H_0 & \text{on } \partial\Omega. \end{cases}$$

Hence all the material is superconducting.

When H_0 is large, and ϵ is small, vortices appear : they trap regions where $u(x) \simeq 0$, i.e. where the material is in the normal state.

An interesting problem is to determine the critical value H_c of H_0 for which vortices appear. A computation by the physicist Abrikosov shows that

$$H_c \simeq \pi |\log \varepsilon|.$$

However this estimate has not been completely rigorously proved on a mathematical level (see [BR2] for a discussion). Another interesting question is to describe the location (and the number) of vortices, when $H_0 > H_c$, and to prove the (observed) fact that they have all winding number $+1$ (as in Theorem 1).

The asymptotic analysis of [BBH] (for ε tending to zero) has been extended to F_ε, in the case $H_0 = 0$, and a Dirichlet type of boundary condition is imposed, that is

$$|u| = 1 \text{ on } \partial\Omega$$
$$\deg(u, \partial\Omega) = d \text{ is prescribed}$$

and

$$\tau . \nabla_A u = g \text{ on } \partial\Omega$$

where τ is the unit tangent vector to $\partial\Omega$, and $g : \partial\Omega \to \mathbb{R}$ is a smooth real valued function.

In an other direction, an important physical experiment has attracted much work from mathematicians. Consider a superconducting sample that has the shape of annulus, or a ring. The experiment is the following : put the sample at ambiant temperature in a magnetic field H_0. This magnetic field induces (by the standard rules of electromagnetism) a current, that circles around the annulus. Next cool down the sample, and later remove the magnetic field : a current persists. In view of the previous discussion, this phenomenon is related to the existence of local minimizers of F_ε, which are not constants (for $H_0 = 0$). This was investigated in work by Jimbo, Morita, and Zhaï [JMZ], Rubinstein and Sternberg [RS], and Almeida [A]. The fact that the topology of Ω (and more precisely, not trivial π_1) enters into the discussion is related to the following

Proposition 20. *Let Ω be a smooth bounded domain in $\mathbb{R}^n$ (for $n \geq 3$). If $\pi_1(\Omega) \neq \{0\}$, then $H^1(\Omega; S^1)$ has many connected components.*

The proof is essentially similar to the proof of Proposition 2. For instance if Ω is the annulus $D^2 \backslash D\left(\frac{1}{2}\right)$ then the different connected components are labelled by the degree on ∂D^2. A consequence is that the Dirichlet energy has infinitely many local minimizers. For small ε, some of these minimizers yield local minimizers for F_ε, and

L. Almeida proved that the levels set F_ε^a of F_ε have many components for small ε (see [A]).

REFERENCES

[A] L. Almeida, Thesis.

[AB1] L. Almeida and F. Bethuel, Multiplicity results for the Ginzburg-Landau equation in presence of symmetries, to appear in *Houston J. of Math.*

[AB2] L. Almeida and F. Bethuel, Topological methods for the Ginzburg-Landau equation, preprint.

[BBH] F. Bethuel, H. Brezis and F. Hélein, Ginzburg-Landau vortices, Birkhaüser, (1994).

[BBH2] F. Bethuel, H. Brezis and F. Hélein, Asymptotics for the minimization of a Ginzburg-Landau functional, *Calc. Var. and PDE, 1,* (1993) 123-148.

[BCP] P. Bauman, N. Carlson and D. Philipps, On the zeroes of solutions to Ginzburg-Landau type systems, to appear.

[BHe] F. Bethuel and B. Helffer, preprint.

[BR] F. Bethuel and T. Rivière, A minimization problem related to superconductivity, *Annales IHP, Analyse Non Linéaire,* (1995), 243-303.

[BR2] F. Bethuel and T. Rivière, Vorticité dans les modèles de Ginzburg-Landau pour la supraconductivité, Séminaire Ecole Polytechnique 1993-1994, exposé n° XV.

[BS] F. Bethuel and J.C. Saut, Travelling waves for the Gross-Putaevskii equation, preprint.

[DF] M. Del Pino and P. Felmer, preprint.

[GL] V. Ginzburg and L. Landau, On the theory of superconductivity, *Zh Eksper. Teoret. Fiz, 20* (1950) 1064-1082.

[JMZ] S. Jimbo, Y. Morita and J. Zhaï, Ginzburg-Landau equation and stable steady state solutions in a non-trivial domain, preprint.

[JS] LR.L. Jerrard and H.M. Soner, Asymptotic heat-flow dynamics for Ginzburg-Landau vortices, preprint, (1995).

[Li1] F.H. Lin, Solutions of Ginzburg-Landau equations and critical points of the renormalized energy, *Annales IHP, Analyse Non Linéaire, 12* (1995) 599-622.

[Li2] F.H. Lin, Some dynamical properties of Ginzburg-Landau vortices, to appear in *CPAM.*

[MCd] D. Mac Duff, Configuration spaces of positive and negative particles, *Topology, 14* (1974) 91-107.

[Mi1] P. Mironescu, On the stability of radial solutions of the Ginzburg-Landau equation, *J. Funct. Anal.*, *130* (1995) 334-344.

[Mi2] P. Mironescu, Les minimiseurs locaux pour l'équation de Ginzburg-Landau sont à symétrie radiale, *C. R. Acad. Sci. Paris*, *6*, (323), 593-598.

[PN] L. Pismen and A. Nepomnyashechy, Stability of vortex rings in a model of super-flow, *Physica D*, (1993) 163-171.

[RS] J. Rubinstein and P. Sternberg, Homotopy classification of minimizers of the Ginzburg-Landau energy and the existence of permanent currents, to appear.

[Sta] G. Stampacchia, Equations elliptiques du second ordre à coefficients discontinus, Presses Université de Montréal (1966).

[Str] M. Struwe, On the asymptotic behavior of the Ginzburg-Landau model in 2 dimensions, *J. Diff. Int. Equ.*, *7* (1994) 1613-1324 ; Erratum *8*, (1995) 224.

Geometric evolution equations for hypersurfaces

GERHARD HUISKEN AND ALEXANDER POLDEN

1 Introduction

Let $F_0 : \mathcal{M}^n \to (N^{n+1}, \bar{g})$ be a smooth immersion of a hypersurface $\mathcal{M}_0^n = F_0(\mathcal{M}^n)$ in a smooth Riemannian manifold $(N^{n+1}, \bar{g})$. We study one–parameter families $F : \mathcal{M}^n \times [0, T] \to (N^{n+1}, \bar{g})$ of hypersurfaces $\mathcal{M}_t^n = F(\cdot, t)(\mathcal{M}^n)$ satisfying an initial value problem

$$\frac{\partial F}{\partial t}(p, t) = -f\nu(p, t), \qquad p \in \mathcal{M}^n, \ t \in [0, T], \qquad (1.1)$$

$$F(p, 0) = F_0, \qquad p \in \mathcal{M}^n, \qquad (1.2)$$

where $\nu(p, t)$ is a choice of unit normal at $F(p, t)$ and $f(p, t)$ is some smooth homogeneous symmetric function of the principal curvatures of the hypersurface at $F(p, t)$.

We will consider examples where $f = f(\lambda_1, \cdots, \lambda_n)$ is monotone with respect to the principal curvatures $\lambda_1, \cdots, \lambda_n$ such that (1.1) is a nonlinear parabolic system of second order. Although there are some similarities to the harmonic map heatflow, this deformation law is more nonlinear in nature since the leading second order operator depends on the geometry of the solution at each time rather than the initial geometry. There is a very direct interplay between geometric properties of the underlying manifold $(N^{n+1}, \bar{g})$ and the geometry of the evolving hypersurface which leads to applications both in differential geometry and mathematical physics.

Here we investigate some of the general properties of (1.1) and then concentrate on the mean curvature flow $f = -H = -(\lambda_1 + \cdots + \lambda_n)$, the inverse mean curvature flow $f = H^{-1}$ and fully nonlinear flows such as the the Gauss curvature flow $f = -K = -(\lambda_1 \cdots \lambda_n)$ or the harmonic mean curvature flow, $f = -(\lambda_1^{-1} + \cdots + \lambda_n^{-1})^{-1}$. We discuss some new developments in the mathematical understanding of these evolution equations and include applications such as the use of the inverse mean curvature flow for the study of asymptotically flat manifolds in General Relativity.

In section 2 we introduce notation for the geometry of hypersurfaces in Riemannian manifolds and derive the crucial commutator relations for the second derivatives of the second fundamental form.

In section 3 we study the general evolution equation (1.1) and obtain evolution equations for metric, normal, second fundamental form and related geometric quantities. We discuss the parabolic nature of the evolution equations, a shorttime existence result and introduce the main examples.

We study the mean curvature flow in section 4. In this case the evolution law is quasilinear and the knowledge of the flow is more advanced than for all other cases. We give some examples of known results concerning regularity, longtime existence and asymptotic behaviour. In particular we discuss the formation of singularities and give an update of recent new results (joint with C.Sinestrari) concerning the classification of singularities in the mean convex case. The section concludes with an isoperimetric estimate for the one-dimensional case, ie the curve shortening flow.

Section 5 deals with fully nonlinear flows such as the Gauss curvature flow and the harmonic mean curvature flow. Without proof we review in particular results of Ben Andrews concerning an elegant proof of the 1/4-pinching theorem, the affine mean curvature flow, and a conjecture of Firey on the asymptotics of the Gauss curvature flow.

The inverse mean curvature flow is discussed in section 6. We explain the basic properties of this flow in its classical form relating it to the Willmore energy and Hawking mass of a twodimensional surface. In view of these properties the inverse mean curvature flow is particularly interesting in asymptotically flat 3-manifolds which appear as models for isolated gravitating systems in General Relativity. It is briefly explained how in recent joint work with T.Ilmanen an extended notion of the inverse mean curvature flow was used to prove a Riemannian version of the so called Penrose inequality for the total energy of an isolated gravitating system represented by an asymptotically flat 3-manifold.

While the first part of this article just described stems from lectures given by the first author at the CIME meeting at Cetraro 1996, the last section of the article is a previously unpublished part of the doctoral dissertation of Alexander Polden. It provides a selfcontained proof of shorttime existence for a variety of geometric evolution equations including hypersurface evolutions as above, conformal deformations of metrics and higher order flows such as the L^2-gradient flow for the Willmore functional.

The author wishes to thank the organisers of the Cetraro meeting for the opportunity to participate in this stimulating conference triggering joint work with Tom Ilmanen on inverse mean curvature flow, as well as for their patience in waiting for this manuscript.

2 Hypersurfaces in Riemannian manifolds

Let $(N^{n+1}, \bar{g})$ be a smooth complete Riemannian manifold without boundary. We denote by a bar all quantities on N, for example by $\bar{g} = \{\bar{g}_{\alpha\beta}\}$, $0 \leq \alpha, \beta \leq n$, the metric, by $\bar{y} = \{\bar{y}^\alpha\}$ coordinates, by $\bar{\Gamma} = \{\bar{\Gamma}^\gamma_{\alpha\beta}\}$ the Levi-Civita connection, by $\bar{\nabla}$ the covariant derivative and by $\bar{\mathrm{Riem}} = \{\bar{\mathrm{Riem}}_{\alpha\beta\gamma\delta}\}$ the Riemann curvature tensor. Components are sometimes taken with respect to the tangent vectorfields $(\partial/\partial y^\alpha)$, $0 \leq \alpha \leq n$ associated with a local coordinate chart $y = \{y^\alpha\}$ and sometimes with respect to a moving orthonormal frame $\{e_\alpha\}$, $0 \leq \alpha \leq n$, where $\bar{g}(e_\alpha, e_\beta) = \delta_{\alpha\beta}$. We write $\bar{g}^{-1} = \{\bar{g}^{\alpha\beta}\}$ for the inverse of the metric and use the Einstein summation convention for the sum of repeated indices. The Ricci curvature $\bar{\mathrm{Ric}} = \{\bar{R}_{\alpha\beta}\}$ and scalar curvature $\bar{R}$ of $(N^{n+1}, \bar{g})$ are then given by

$$\bar{R}_{\alpha\beta} = \bar{g}^{\gamma\delta} \bar{R}_{\alpha\gamma\beta\delta},$$

and the sectional curvatures (in an orthonormal frame) are given by $\bar{\sigma}_{\alpha\beta} = \bar{R}_{\alpha\beta\alpha\beta}$.

Now let $F : \mathcal{M}^n \to N^{n+1}$ be a smooth hypersurface immersion. For simplicity we restrict attention to closed surfaces, ie compact without boundary. The induced metric on $\mathcal{M}^n$ will be denoted by g, in local coordinates we have

$$\begin{aligned} g_{ij}(p) &= \langle \frac{\partial F}{\partial x^i}(p), \frac{\partial F}{\partial x^j}(p) \rangle_N \\ &= \bar{g}_{\alpha\beta}(F(p)) \frac{\partial F^\alpha}{\partial x^i}(p) \frac{\partial F^\beta}{\partial x^j}(p), \qquad p \in \mathcal{M}^n. \end{aligned}$$

Furthermore, $\{\Gamma^i_{jk}\}$, ∇ and $\mathrm{Riem} = \{R_{ijkl}\}$ with latin indices i, j, k, l ranging from 1 to n describe the intrinsic geometry of the induced metric g on the hypersurface.

If ν is a local choice of unit normal for $F(\mathcal{M}^n)$, we often work in an adapted othonormal frame $\nu, e_1, \cdots, e_n$ in a neighbourhood of $F(\mathcal{M}^n)$ such that $e_1(p), \cdots, e_n(p) \in T_p\mathcal{M}^n \subset T_pN^{n+1}$ and $g(p)(e_i(p), e_j(p)) = \delta_{ij}$ for $p \in \mathcal{M}^n$, $1 \leq i, j \leq n$.
The second fundamental form $A = \{h_{ij}\}$ as a bilinear form

$$A(p) : T_p\mathcal{M}^n \times T_p\mathcal{M}^n \to \mathbb{R}$$

and the Weingarten map $W = \{h^i_j\} = \{g^{ik}h_{kj}\}$ as an operator

$$W : T_p\mathcal{M}^n \to T_p\mathcal{M}^n$$

are then given by

$$h_{ij} = \langle \bar{\nabla}_{e_i}\nu, e_j \rangle = -\langle \nu, \bar{\nabla}_{e_i}e_j \rangle.$$

In local coordinates $\{x^i\}$, $1 \leq i \leq n$, near $p \in \mathcal{M}^n$ and $\{y^\alpha\}$, $0 \leq \alpha \leq n$, near $F(p) \in N$ these relations are equivalent to the Weingarten equations

$$\begin{aligned} \frac{\partial^2 F^\alpha}{\partial x^i \partial x^j} - \Gamma^k_{ij} \frac{\partial F^\alpha}{\partial x^k} + \bar{\Gamma}^\alpha_{\beta\delta} \frac{\partial F^\beta}{\partial x^i} \frac{\partial F^\delta}{\partial x^j} &= -h_{ij}\nu^\alpha, \\ \frac{\partial \nu^\alpha}{\partial x^i} + \bar{\Gamma}^\alpha_{\beta\delta} \frac{\partial F^\beta}{\partial x^i} \nu^\delta &= h_{ij}g^{jl} \frac{\partial F^\alpha}{\partial x^l}. \end{aligned}$$

Recall that $A(p)$ is symmetric, ie W is selfadjoint, and the eigenvalues $\lambda_1(p), \cdots, \lambda_n(p)$ are called the principal curvatures of $F(\mathcal{M}^n)$ at $F(p)$. Also note that at a given point $p \in \mathcal{M}^n$ by choosing normal coordinates and then possibly rotating them we can always arrange that at this point

$$g_{ij} = \delta_{ij}, \qquad \bar{\nabla}^T_{e_i}e_j = 0, \qquad h_{ij} = h^i_j = \mathrm{diag}(\lambda_1, \cdots, \lambda_n).$$

The classical scalar invariants of the second fundamental form are then symmetric homogeneous polynomials in the principal curvatures:
The mean curvature is given by

$$H := tr(W) = h_i^i = g^{ij}h_{ij} = \lambda_1 + \cdots + \lambda_n,$$

the Gauss–Kronecker curvature by

$$K := det(W) = det\{h_j^i\} = \frac{det\{h_{ij}\}}{det\{g_{ij}\}} = \lambda_1 \cdots \lambda_n,$$

the total curvature by

$$|A|^2 := tr(W^t W) = h_j^i h_i^j = h^{ij}h_{ij} = g^{ik}g^{jl}h_{ij}h_{kl} = \lambda_1^2 + \cdots + \lambda_n^2,$$

and the scalar curvature (in Euclidean space $\mathbb{R}^{n+1}$) by

$$R = H^2 - |A|^2 = 2(\lambda_1\lambda_2 + \lambda_1\lambda_3 + \cdots + \lambda_{n-1}\lambda_n).$$

More general, the mixed mean curvatures S_m, $1 \leq m \leq n$, are given by the elementary symmetric functions of the λ_i,

$$S_m := \sum_{i_1 < \cdots < i_m} \lambda_{i_1}\lambda_{i_2} \cdots \lambda_{i_m},$$

such that $S_1 = H$, $S_2 = (1/2)R$, $S_n = G$. Other interesting invariants include the harmonic mean curvature

$$\tilde{H} := (\lambda_1^{-1} + \cdots + \lambda_n^{-1})^{-1} = S_n/S_{n-1}$$

as well as other symmetric functions of the principal radii λ_i^{-1}. All the invariants mentioned or powers thereof are candidates for the speed f in our evolution problem (1.1).

For the purposes of analysis it is crucial to know the rules of computation involving the covariant derivatives, the second fundamental form of the hypersurface and the curvature of the ambient space. We assume the reader to have some background in differential geometry, but restate the formulas used in this article for convenience (in an adapted orthonormal frame).
The commutator of second derivatives of a vectorfield X on $\mathcal{M}^n$ is given by

$$\nabla_i\nabla_j X^k - \nabla_j\nabla_i X^k = R_{ijlm}g^{kl}X^m,$$

and for a one-form ω on $\mathcal{M}^n$ by

$$\nabla_i\nabla_j\omega_k - \nabla_j\nabla_i\omega_k = R_{ijkl}g^{lm}\omega_m.$$

More generally, the commutator of second derivatives for an arbitrary tensor involves one curvature term as above for each of the indices of the tensor. The corresponding laws of course also hold for the metric $\bar{g}$.

The curvature of the hypersurface and ambient manifold are related by the equations of Gauss

$$R_{ijkl} = \bar{R}_{ijkl} + h_{ik}h_{jl} - h_{il}hjk, \qquad 1 \le i,j,k,l \le n,$$
$$R_{ik} = \bar{R}_{ik} - \bar{R}_{oiok} + Hh_{ik} - h_{il}h_k^l, \qquad 1 \le i,k \le n,$$
$$R = \bar{R} - 2\bar{R}_{oo} + H^2 - |A|^2,$$

and the equations of Codazzi-Mainardi

$$\nabla_i h_{jk} - \nabla_k h_{ij} = \bar{R}_{ojki},$$
$$\nabla_i h_{ik} - \nabla_k H = \bar{R}_{ok}.$$

The following commutator identities for the second derivatives of the second fundamental form were first found by Simons [48] and provide the crucial link between analytical methods and geometric properties of $\mathcal{M}^n$ and N^{n+1}. See also [47] for a derivation of the following facts from the structure equations.

Theorem 2.1 *The second derivatives of A satisfy the identities*

$$\begin{aligned}\nabla_k\nabla_l h_{ij} = &\ \nabla_i\nabla_j h_{kl} + h_{kl}h_{im}h_{mj} - h_{km}h_{il}h_{mj} + h_{kj}h_{im}h_{ml}\\ &- h_{km}h_{ij}h_{ml} + \bar{R}_{kilm}h_{mj} + \bar{R}_{kijm}h_{ml}\\ &+ \bar{R}_{mjil}h_{km} + \bar{R}_{oioj}h_{kl} - \bar{R}_{okol}h_{ij} + \bar{R}_{mljk}h_{im}\\ &+ \bar{\nabla}_k\bar{R}_{ojil} + \bar{\nabla}_i\bar{R}_{oljk}.\end{aligned}$$

The trace of these identities plays an important role in mimimal surface theory and is of particular importance for mean curvature flow and inverse mean curvature flow:

Corollary 2.2 *The Laplacian $\Delta = \sum_i \nabla_i\nabla_i$ of the second fundamental form satisfies*

$$\begin{aligned}\Delta h_{ij} = &\ \nabla_i\nabla_j H + Hh_{im}h_{mj} - h_{ij}|A|^2 + H\bar{R}_{oioj}\\ &- \bar{R}_{oo}h_{ij} + \bar{R}_{kikm}h_{mj} + \bar{R}_{kjkm}h_{im}\\ &+ \bar{R}_{kijm}h_{km} + \bar{R}_{mjik}h_{km} + \bar{\nabla}_k\bar{R}_{ojik} + \bar{\nabla}_i\bar{R}_{okjk},\end{aligned}$$

$$\begin{aligned}\frac{1}{2}\Delta|A|^2 = &\ h_{ij}\nabla_i\nabla_j H + |\nabla A|^2 + Htr(A^3) - |A|^4\\ &+ Hh_{ij}\bar{R}_{oioj} - \bar{R}_{oo}|A|^2 + 2\bar{R}_{kikm}h_{mj}h_{ij} - 2\bar{R}_{kimj}h_{km}h_{ij}\\ &+ h_{ij}(\bar{\nabla}_k\bar{R}_{ojik} + \bar{\nabla}\bar{R}_{okjk}).\end{aligned}$$

Proof. By the Codazzi equations we first get

$$\nabla_k\nabla_l h_{ij} = \nabla_k(\nabla_i h_{lj} + \bar{R}_{ojil}).$$

Then compute from the definition of h_{ij}

$$\begin{aligned}\nabla_k(\bar{R}_{ojil}) = &\ \bar{\nabla}_k\bar{R}_{ojil} + h_{km}\bar{R}_{mjil}\\ &- h_{ik}\bar{R}_{ojol} - h_{lk}\bar{R}_{ojio}\end{aligned}$$

and commute ∇_i and ∇_k to derive

$$
\begin{aligned}
\nabla_k \nabla_l h_{ij} &= \nabla_i \nabla_k h_{lj} + R_{kilm} h_{mj} + R_{kijm} h_{ml} \\
&\quad + \bar{\nabla}_k \bar{R}_{ojil} + h_{km} \bar{R}_{mjil} - h_{ik} \bar{R}_{ojol} - h_{lk} \bar{R}_{ojio}.
\end{aligned}
$$

Then use the Codazzi equations again to get

$$
\begin{aligned}
\nabla_i \nabla_k h_{lj} &= \nabla_i (\nabla_j h_{kl} + \bar{R}_{oljk}) \\
&= \nabla_i \nabla_j h_{kl} + \bar{\nabla}_i \bar{R}_{oljk} \\
&\quad + h_{im} \bar{R}_{mljk} - h_{ij} \bar{R}_{olok} - h_{ik} \bar{R}_{oljo}.
\end{aligned}
$$

Employing the Gauss equations we finally conclude

$$
\begin{aligned}
\nabla_k \nabla_l h_{ij} &= \nabla_i \nabla_j h_{kl} + \bar{R}_{kilm} h_{mj} + \bar{R}_{kijm} h_{ml} \\
&\quad + \bar{R}_{mjil} h_{km} - \bar{R}_{ojol} h_{ik} - \bar{R}_{ojio} h_{kl} \\
&\quad + \bar{R}_{mljk} h_{im} - \bar{R}_{olok} h_{ij} - \bar{R}_{oljo} h_{ik} \\
&\quad + \bar{\nabla}_k \bar{R}_{ojil} + \bar{\nabla}_i \bar{R}_{oljk} \\
&\quad + h_{kl} h_{im} h_{mj} - h_{km} h_{il} h_{mj} + h_{kj} h_{im} h_{ml} - h_{km} h_{ij} h_{ml}
\end{aligned}
$$

and the conclusion follows from the symmetries of $\bar{R}_{\alpha\beta\gamma\delta}$.

3 The evolution equations

Let $F_0 : \mathcal{M}^n \to \mathbb{R}^{n+1}$ be a smooth closed hypersurfaceas as in the introduction in a smooth Riemannian manifold $(N^{n+1}, \bar{g})$, $n \geq 2$. Assume for simplicity that N, M are orientable and choose a unit normal field ν on $\mathcal{M}$. If $\mathcal{M}^n \subset \mathbb{R}^{n+1}$, we choose the exterior unit normal such that the mean curvature of a sphere is positive. We then consider the initial value problem (1.1), where f is a smooth, homogeneous function of the principal curvatures λ_i.

Shorttime existence for (1.1) can in general only be expected when the system is parabolic. to investigate the linearisation of (1.1), notice that due to the symmetry of f in an equivalent setting we may consider f as a function $\tilde{f}$ of the Weingarten map W or as a function $\hat{f}$ of the second fundamental form A:

$$
\tilde{f}(W) = \tilde{f}(\{h_j^i\}) = \hat{f}(A) = \hat{f}(\{h_{ij}\}) = f(\lambda_1 \cdots \lambda_n).
$$

In view of the Weingarten equations the linearisation of (1.1) is then an equation of the form

$$
\frac{\partial}{\partial t} G = -\frac{\partial \hat{f}}{\partial h_{ij}} g^{ik} g^{jl} \langle \frac{\partial^2 G}{\partial x^k \partial x^l}, \nu \rangle \nu + \text{lower order}.
$$

Thus the "symbol"

$$
\sigma_\beta^\alpha(\xi) = -\frac{\partial \hat{f}}{\partial h_{ij}} \xi^i \xi^j \nu^\alpha \nu^\beta
$$

of the RHS is always degenerate in tangential directions, reflecting the invariance of the original equation under tangential diffeomorphisms. It is strictly positive definite in normal direction if

$$-\frac{\partial \hat{f}}{\partial h_{ij}}(p)\,\xi^i \xi^j > 0 \qquad \forall\, 0 \neq \xi \in \mathbb{R}^n, \quad p \in \mathcal{M}^n,$$

or equivalently

$$-\frac{\partial f}{\partial \lambda_i}(p) > 0 \qquad \forall\, 1 \leq i \leq n, \quad p \in \mathcal{M}^n. \tag{3.1}$$

The problem with the degeneracy in tangential direction can be overcome in various ways: In [30] Hamilton solves degenerate parabolic equations satisfying an integrability condition. In our case the normal projection $\Pi^N : T_p N^{n+1} \to T_p \mathcal{M}^n$ yields an integrability condition since

$$\frac{\partial}{\partial t} F = f \nu \in \text{Kernel}\, \Pi^N,$$

such that Hamilton's result applies when (3.1) holds. The other approach consists in the choice of some vectorfield transversal to the initial surface. This breaks the gauge invariance of the equation and changes (1.1) to a scalar uniformly parabolic equation provided (3.1) holds. This approach was originally used by DeTurck [14] for the Ricci flow and has been used for the evolution of hypersurfaces in [33], [16]. In the last chapter of the present paper the second author gives a selfcontained proof of shorttime existence for a large class of geometric evolution equations including equations of higher order. For our purposes we note:

Theorem 3.1 *If $F_0 : \mathcal{M}^n \to (N^{n+1}, \bar{g})$ is a smooth, closed hypersurface such that*

$$-\frac{\partial f}{\partial \lambda_i}(p) > 0, \qquad 1 \leq i \leq n, \tag{3.2}$$

holds everywhere on $F_0(\mathcal{M}^n)$, then (1.1) has a smooth solution at least on some short time interval $[0, T)$, $T > 0$.

Examples. i) In the case of mean curvature flow $f = -H$ we have $-(\partial f / \partial \lambda_i) = 1$ and the flow admits a shorttime solution for any smooth initial data.

ii) For Gauss curvature flow $f = -G$ we get $-(\partial f / \partial \lambda_i) = \lambda_i^{-1} G$ and we have shorttime existence if the initial data are convex. More generally, the elementary symmetric functions $-f = S_m$, $1 \leq m \leq n$, satisfy $-(\partial f / \partial \lambda_i) > 0$ on the convex cone $\Gamma_m = \{\lambda \in \mathbb{R}^n | S_l(\lambda) > 0, 1 \leq l \leq m\}$, yielding shorttime existence for corresponding initial data.

iii) The quotients $Q_{k,l} = S_k / S_l$ for $1 \leq l < k \leq n$ again satisfy (3.1) on Γ_k. In particular, this yields a shorttime existence result for the harmonic mean curvature flow on convex initial data, since $-f = \tilde{H} = S_n / S_{n-1}$.

iv) The inverse mean curvature flow with $f = H^{-1}$ satisfies $-(\partial f / \partial \lambda_i) = H^{-2}$, yielding shorttime existence of a classical smooth solution for any initial data of positive mean curvature.

Working in the class of surfaces where shorttime existence is guaranteed, the interesting task is now to understand the longterm change in the shape of solutions, and to characterise their asymptotic behaviour both for large times and near singularities. For this purpose evolution equations have to be established for all relevant geometric quantities, in particular for the second fundamental form.

Theorem 3.2 *On any solution $M_t^n = F(\cdot, t)(\mathcal{M}^n)$ of (1.1) the following equations hold:*

(i) $\frac{\partial}{\partial t} g_{ij} = 2f h_{ij}$,

(ii) $\frac{\partial}{\partial t}(d\mu) = fH(d\mu)$,

(iii) $\frac{\partial}{\partial t} \nu = -\nabla f$,

(iv) $\frac{\partial}{\partial t} h_{ij} = -\nabla_i \nabla_j f + f(h_{ik} h_j^k - \bar{R}_{oioj})$,

(v) $\frac{\partial}{\partial t} H = -\Delta f - f(|A|^2 + \bar{R}\mathrm{ic}(\nu, \nu))$.

Here $d\mu$ is the induced measure on the hypersurface and Δ is the Laplace–Beltrami operator with respect to the time-dependent induced metric on the hypersurface.

Notice that $-\Delta f - f(|A|^2 + \bar{R}\mathrm{ic}(\nu, \nu)) = Jf$ is the Jacobi operator acting on f, as is wellknown from the second variation formula for the area.

Proof. The computations are best done in local coordinates $\{x^i\}$ near $p \in \mathcal{M}^n$ and $\{y^\alpha\}$ near $F(p)$ in N. Arranging coordinates at a fixed point p such that $g_{ij}(p) = \delta_{ij}$, $(\partial/\partial x^i) g_{jk}(p) = 0$, $\bar{g}_{\alpha\beta}(F(p)) = \delta_{\alpha\beta}$, $(\partial/\partial y^\alpha) \bar{g}_{\beta\delta}(F(p)) = 0$ all identities are straightforward consequences of the definitions and the Gauss–Weingarten relations. The computations have been carried out in detail for $f = -H$ in [34], [35] and [4]. A short derivation by the second author is also contained in the last section of this article.

We will now use the commutator identities in Theorem 2.1 to convert the evolution equations for the curvature into parabolic systems on the hypersurface. For this pupose we introduce for each speed function f the nonlinear operator L_f by setting

$$L_f u = L_f^{ij} \nabla_i \nabla_j u := -\frac{\partial \hat{f}}{\partial h_{ij}} \nabla_i \nabla_j u,$$

where $\hat{f}$ as before is the symmetric function f considered as a function of the h_{ij}. Note that for mean curvature flow $L_H = \Delta$ is the Laplace–Beltrami operator, for inverse mean curvature flow with $f = H^{-1}$ we have $L_f = (1/H^2)\Delta$ and in general L_f is an elliptic operator exactly when f is elliptic, ie satisfies (3.1). The following form of the evolution equations exhibits their parabolic nature.

Corollary 3.3 *On any solution $M_t^n = F(\cdot, t)(\mathcal{M}^n)$ of (1.1) the second fundamental form h_{ij} and the speed f satisfy*

$$\frac{\partial}{\partial t} h_{ij} = L_f^{kl} \nabla_k \nabla_l h_{ij} - \frac{\partial^2 f}{\partial h_{kl} \partial h_{pq}} \nabla_i h_{kl} \nabla_j h_{pq}$$

$$+\frac{\partial f}{\partial h_{kl}}\{h_{kl}h_{im}h_{mj}-h_{km}h_{il}h_{mj}+h_{kj}h_{im}h_{ml}-h_{km}h_{ij}h_{ml}$$

$$+\bar{R}_{kilm}h_{mj}+\bar{R}_{kijm}h_{ml}+\bar{R}_{mijl}h_{km}+\bar{R}_{oioj}h_{kl}-\bar{R}_{okol}h_{ij}+\bar{R}_{mljk}h_{im}$$

$$+\bar{\nabla}_k\bar{R}_{ojil}+\bar{\nabla}_i\bar{R}_{oljk}\}+f(h_{ik}h_j^k-\bar{R}_{oioj}),$$

$$\frac{\partial}{\partial t}f = L_j^{ij}\nabla_i\nabla_j f - f\frac{\partial f}{\partial h_{ij}}(h_{ik}h_j^k+\bar{R}_{oioj}).$$

Proof. From $\nabla_i f = (\partial f/\partial h_{kl})\nabla_i h_{kl}$ we see that

$$\nabla_i\nabla_j f = \frac{\partial f}{\partial h_{kl}}\nabla_i\nabla_j h_{kl} + \frac{\partial^2 f}{\partial h_{kl}\partial h_{pq}}\nabla_j h_{kl}\nabla_i h_{pq}.$$

This yields the first identity in view of Theorem 2.1 and Theorem 3.2(iv). Similarly we get

$$\frac{\partial}{\partial t}f = \frac{\partial\hat{f}}{\partial h_j^i}\frac{\partial}{\partial t}h_j^i = \frac{\partial\hat{f}}{\partial h_{ij}}g_{li}\frac{\partial}{\partial t}(g^{ik}h_{kj})$$

$$= -2fh_i^k h_{kj}\frac{\partial\hat{f}}{\partial h_{ij}} + \frac{\partial\hat{f}}{\partial h_{ij}}\frac{\partial}{\partial t}h_{ij}$$

$$= L_j^{ij}\nabla_i\nabla_j f + f\frac{\partial f}{\partial h_{ij}}(h_{ik}h_j^k-\bar{R}_{oioj}) - 2f\frac{\partial f}{\partial h_{ij}}h_i^k h_{kj}$$

$$= L_j^{ij}\nabla_i\nabla_j f - f\frac{\partial f}{\partial h_{ij}}(h_{ik}h_j^k+\bar{R}_{oioj}),$$

as required.

The curvature terms in this nonlinear reaction–diffusion system provide the key for understanding the interaction between geometric properties of the hypersurface and the ambient manifold. They are the tool to study these geometric phenomena with analytical means. For some choices of f we will now describe recent developments.

4 Mean curvature flow

In the case of mean curvature flow $f = -H$ it is well known [34] that for closed initial surfaces the solution of (1.1)–(1.2) exists on a maximal time interval $[0, T[, 0 < T \leq \infty$. In some cases the behaviour of the flow has been completely understood: For closed convex surfaces in $\mathbb{R}^{n+1}$, $n \geq 2$, it was shown in [34] that the solution contracts smoothly to a point, becoming more and more spherical at the end of the evolution. In [35] this was extended to general Riemannian manifolds under the assumption that the initial hypersurface is sufficiently convex: Each principal curvature λ_i of the initial surface has to be bounded below by a constant depending on the curvature and the derivative of the curvature in the ambient manifold. While the constants are optimal in locally symmetric spaces, the dependence on the derivatives of curvature in the general case is not desirable form a geometric point of view. Some of the fully nonlinear flows discussed in the next section have a better behaviour from this point of view.

Regularity and longtime existence was also obtained for surfaces that can be written as graphs, compare the joint work of the first author with Ecker in [16] and [17].

In the one–dimensional case Grayson proved that any embedded closed curve on a 2–surface of bounded geometry will either smoothly contract to a point in finite time or converge to a geodesic in infinite time, compare [25], [26] and earlier work of Gage and Hamilton in [22].

In higher dimensions it is well known that singularities will in usually occur before the area of the evolving surface tends to zero. If $T < \infty$, as is always the case in Euclidean space, the curvature of the surfaces becomes unbounded for $t \to T$. One would like to understand the singular behaviour for $t \to T$ in detail, having in mind a possible controlled extension of the flow beyond such a singularity. See [37] for a review of earlier results concerning local and global properties of mean curvature flow and its singularities. We will not discuss singularities in weak formulations of the flow, a good reference in this direction is [54] and [43].

Since the shape of possible singularities is a purely local question, we may restrict attention to the case where the target manifold is Euclidean space. Nevertheless, in the light of an abundance even of homothetically shrinking examples with symmetries, the possible limiting behaviour near singularities seems in general beyond classification at this stage. In recent joint work of C. Sinestrari and the author [41], [42] the additional assumption of nonnegative mean curvature is used to restrict the range of possible phenomena, while still retaining an interestingly large class of surfaces. We derive new a priori estimates from below for all elementary symmetric functions of the principal curvatures, exploiting the one-sided bound on the mean curvature. The estimates turn out to be strong enough to conclude that any rescaled limit of a singularity is (weakly) convex.

Recall that

$$S_k(\lambda) = \sum_{1 \le i_1 < i_2 < \ldots < i_k \le n} \lambda_{i_1} \lambda_{i_2} \cdots \lambda_{i_k}$$

are the elementary symmetric functions of the principal curvatures with $S_1 = H$. Then [42] establishes the estimates

Theorem 4.1 *(H.-Sinestrari) Suppose $F_0 : \mathcal{M} \to \mathbb{R}^{n+1}$ is a smooth closed hypersurface immersion with nonnegative mean curvature. For each k, $2 \le k \le n$, and any $\eta > 0$ there is a constant $C_{\eta,k}$ depending only on n, k, η and the initial data, such that everywhere on $\mathcal{M} \times [0, T[$ the estimate*

$$S_k \ge -\eta H^k - C_{\eta,k} \tag{4.1}$$

holds uniformly in space and time.

The proof proceeds by induction on the degree k of S_k and relies heavily on the algebraic properties of the elementary symmetric functions, the structure of the curvature evolution in this particular flow and the Sobolev inequality for hypersurfaces. In each step of the iteration an a priori estimate is proved for a quotient

$$Q_k = \frac{S_k}{S_{k-1}}$$

of consecutive elementary symmetric polynmials, making use of the concavity properties of this function. Using techniques in [35] and [49] the result can be extended to starshaped surfaces in $\mathbb{R}^{n+1}$ and to hypersurfaces in Riemannian manifolds.

Similarly as in the theory of minimal surfaces the structure of singularities is then studied by blowup methods, in this case by parabolic rescaling in space and time, compare [28], [36], [41]. Since η is arbitrary in the above estimate and the mean curvature $S_1 = H$ tends to infinity near a singularity, the scaling invariance is broken in inequality (4.1) and implies that near a singularity each S_k becomes nonnegative after appropriate rescaling:

Corollary 4.2 *(H.-Sinestrari) Let M_t be a mean convex solution of mean curvature flow on the maximal time interval $[0, T[$ as in Theorem 1.1. Then any smooth rescaling of the singularity for $t \to T$ is convex.*

The structure of the rescaled limit depends on the blowup rate of the singularity: If the quantity $\sup(T - t)|A|^2$ is uniformly bounded (type I singularity), the rescaling will yield a selfsimilar, homothetically shrinking solution of the flow which is completely classified in the case of positive mean curvature, see [36] and [37]. If the quantity $\sup(T - t)|A|^2$ is unbounded (type II singularity), the rescaling of the singularity can be done in such a way that an "eternal solution" (ie defined for all time) of mean curvature flow results where the maximum of the curvature is attained on the surface. In the convex case such solutions were shown by Hamilton to move isometrically by translations, [29]. Hence, combining the classification of type I singularities in [37], the result of Hamilton and the convexity result in Corollary 4.2, one derives a description of all possible singularities (type I and type II) in the mean convex case, compare [42].

Open problems which have to be adressed for the future goal of continuing the flow by surgery concern the classification of convex translating solutions, Harnack estimates for the mean curvature, more precise estimates on the rate of convergence as well as higher order asymptotics near singularities. Some guidance on the possible higher order behaviour near singularities can be taken from the degenerate examples constructed in [7]. A Harnack estimate for the mean curvature has so far only been obtained in the convex case [29], which is too restrictive for many applications. The work of Hamilton on the Ricci flow [28] has a close relation to the mean curvature flow and indicates a strategy for the extension of the flow past singularities once stronger estimates are available [31].

We conclude this section with the one-dimensional case, where an embedded curve is evolving in the plane or on some smooth surface by the curve shortening flow. The remarkable articles of Grayson on this flow [25],[26] show by a number of global arguments that for embedded curves no finite time singularity can occur unless the whole curve contracts to a single point.

The structure of all possible singularities in this case is now well understood: There are no embedded type I singularities except the shrinking circle, which is the desired outcome, and the only possible rescaling of a type II singularity is a so called grim reaper curve given by $y = \log \cos x$. To prove Graysons result it is therefore sufficient to give an argument excluding this last curve as a possible limiting shape. Such an argument is provided both by Hamilton in [32], where an isoperimetric estimate for the area in subdivisions of the enclosed region is shown, and by the first author in [38], where a lower bound for the ratio between extrinsic and intrinsic distances on the evolving curve is proved.

To describe the last result, let $F: S^1 \times [0,T] \to \mathbb{R}^2$ be a closed embedded curve moving by the curve shortening flow. If $L = L(t)$ is the total length of the curve, the intrinsic distance function l along the curve is smoothly defined only for $0 \le l < L/2$, with conjugate points where $l = L/2$. We therefore define a smooth function $\psi: S^1 \times S^1 \times [0,T] \to \mathbb{R}$ by setting

$$\psi := \frac{L}{\pi} \sin(\frac{l\pi}{L}).$$

With this choice of ψ, and with d being the extrinsic distance between two points on the curve, the isoperimetric ratio d/ψ approaches 1 on the diagonal of $S^1 \times S^1$ for any smooth embedding of S^1 in $\mathbb{R}^2$ and the ratio d/ψ is identically one on any round circle.

Theorem 4.3 *Let $F: S^1 \times [0,T] \to \mathbb{R}^2$ be a smooth embedded solution of the curve shortening flow (1.1). Then the minimum of d/ψ on S^1 is nondecreasing; it is strictly increasing unless $d/\psi \equiv 1$ and $F(S^1)$ is a round circle.*

Clearly the estimate prevents a grim reaper type singularity. The proof uses the maximum principle on the cross product of the curve with itself. It is an open problem whether similar lower order estimates can be used for the study or exclusion of certain singularities in higherdimensional flows.

5 Fully nonlinear flows

The Gauss curvature flow, where the speed $f = -K = -(\lambda_1 \cdots \lambda_n)$ is the product of the principle curvatures, was first introduced by Firey [20] as a model for the changing shape of a tumbling stone being worn from all directions with uniform intensity. The flow is parabolic only in the class of convex surfaces and much more nonlinear in its analytic behaviour than the mean curvature flow. Tso [52] proved existence, uniqueness and convergence of closed convex hypersurfaces to a point for this flow without however determining the limiting shape of the contracting surface. The conjecture of Firey (1974) that the limiting shape is that of a sphere regardless of the initial data, was only recently confirmed by Andrews [2]:

Theorem 5.1 *(Andrews) Let $\mathcal{M}_0^2$ be a smooth closed strictly convex initial surface in $\mathbb{R}^3$. Then there is a unique smooth solution of (1.1) with $f = -K$ on the time interval $[0,T[$, where $T = V(\mathcal{M}_0^2)/4\pi$ is determined by the enclosed volume of the initial surface, and the surfaces converge to a round sphere after appropriate rescaling.*

The corresponding result for mean curvature flow was obtained earlier by the author in [34] and for a large class of speed functions f including the harmonic mean curvature flow by Andrews in [1]. If the Gauss curvature K is replaced by some power K^α, a whole new range of interesting phenomena appears. If the homogeneity is 1, ie $\alpha = 1/n$, Chow proved contraction to a point and roundness of the limiting shape, [12]. In [5] Andrews shows that in the interval $1/(n+2) < \alpha \leq 1/n$ there is at least some smooth limiting shape at the end of the contraction, while for small values of α a degeneration of the surface near the end of the contraction is expected.

In the special case $\alpha = 1/(n+2)$, the evolution equation (1.1) becomes affine invariant. In line with the results just mentioned Andrews [3] proves by an extension of Calabi's estimate on the cubic ground form that convex initial data contract smoothly to a point in finite time, with ellipsoids as the natural unique limiting shape. As a consequence he derives an elegant proof of the affine isoperimetric inequality. Compare also the work of Sapiro and Tannenbaum [46] on the affine evolution of curves, which has applications in image processing.

For convex hypersurfaces in general Riemannian manifolds speedfunctions f such as the harmonic mean curvature or other quotients of elementary symmetric functions seem to have the best algebraic behaviour. In mean curvature flow the derivatives of the ambient curvature in the evolution equations of Corollary 2.2 are analytically hard to control, compare the dependance of the main result in [35] on these terms. For harmonic mean curvature flow and flows of similar structure Andrews derives an optimal convergence result for hypersurfaces having sufficiently positive principal curvatures in relation to the ambient curvature, [4]. In particular, he shows that such flows contract convex hypersurfaces in manifolds of positive sectional curvature to a point and gives a new argument for the classical 1/4-pinching theorem.

All speedfunctions considered so far were pointing in the same direction as the mean curvature vector, corresponding to contractions in the case of convex surfaces. In the last section we consider an expanding version of the flow.

6 The inverse mean curvature flow

The inverse mean curvature flow $f = H^{-1}$ is well posed for surfaces of positive mean curvature and characterised by its property that the area element is growing exponentially at each point: From Theorem 1.1(i) we have $\partial/\partial t(d\mu) = d\mu$. In particular, the total area of a smooth closed evolving surface is completely determined by its initial area:

$$|M_t^n| = |M_0^n| \exp(t).$$

The standard example for this behaviour is the exponentially expanding sphere of radius $R(t) = R(0)\exp(t/n)$. Further interesting properties of the flow follow from the evolution equation for the mean curvature H, which we derive from the evolution equation for the speed $f = H^{-1}$.

$$\frac{\partial H}{\partial t} = \frac{\Delta H}{H^2} - \frac{2|\nabla H|^2}{H^3} - \frac{|A|^2}{H} - \frac{\bar{Ric}(\nu, \nu)}{H}.$$

Due to the negative sign of the $|A|^2$-term we get from this equation by a simple application of the parabolic maximum principle the remarkable property that the mean curvature H is uniformly bounded in terms of its initial data and the Ricci curvature of the ambient manifold. This is in strong contrast to the mean curvature flow, where the blowup of the mean curvature causes the singularities studied in section 2. For the inverse mean curvature flow the critical behaviour occurs where $H \to 0$ and the speed becomes infinite. In Euclidean space it is clear that the maximum of the mean curvature is decreasing and the same is true for any L^p-norm.

In case $n = 2$ this property of the flow can be extended to closed surfaces in arbitrary three-manifolds of nonnegative scalar curvature: For any two-surface $\Sigma^2 \subset (N^3, \bar{g})$ the so called Hawking quasi-local mass of Σ^2 is defined as the geometric quantity

$$m_H(\Sigma^2) := \frac{|\Sigma^2|^{1/2}}{(16\pi)^{3/2}} \left(16\pi - \int_{\Sigma^2} H^2 \, d\mu \right),$$

and a computation based on the evolution equation for the mean curvature, the area element of the surface and the Gauss-Bonnet formula shows that for a solution M_t^2 of the inverse mean curvature flow

$$\frac{d}{dt} \int_{M_t^2} H^2 \, d\mu = 4\pi\chi(M_t^2) + \int_{M_t^2} -2\frac{|\nabla H|^2}{H^2} - \frac{1}{2}H^2 - \frac{1}{2}(\lambda_1 - \lambda_2)^2 - \bar{R} \, d\mu.$$

Hence, if the surface M_t^2 is connected and the scalar curvature $\bar{R}$ of the three-manifold is nonnegative, we have

$$\frac{d}{dt} \int_{M_t^2} H^2 \, d\mu \leq \frac{1}{2} \left(16\pi - \int_{M_t^2} H^2 \, d\mu \right)$$

and the Hawking quasi-local mass is nondecreasing along the inverse mean curvature flow:

$$\frac{d}{dt} m_H(M_t^2) \geq 0.$$

A major reason for the interest in the inverse mean curvature flow comes from the interpretation of this purely geometric fact in General Relativity: The spatial part of the exterior of an isolated gravitating system (like a star, black hole or galaxy) is modelled by the end of an asymptotically flat Riemannian 3–manifold with nonnegative scalar curvature as above. Here an end of a Riemannian 3-manifold $(N^3, \bar{g})$ is called *asymptotically flat* if it is realized by an open set that is diffeomorphic to the complement of a compact set K in $\mathbb{R}^3$, and the metric tensor $\bar{g}$ of M satisfies

$$|\bar{g}_{ij} - \delta_{ij}| \leq \frac{C}{|x|}, \qquad |\bar{g}_{ij,k}| \leq \frac{C}{|x|^2}, \qquad Ric \geq -\frac{C\bar{g}}{|x|^2},$$

as $|x| \to \infty$. The derivatives are taken with respect to the Euclidean metric $\delta = \{\delta_{ij}\}$ on $\mathbb{R}^3 \setminus K$. On such asymptotically flat ends a concept of total mass or energy is defined by a flux integral through the sphere at infinity,

$$m := \lim_{r \to \infty} \frac{1}{16\pi} \int_{\partial B_r^\delta(0)} (\bar{g}_{ii,j} - \bar{g}_{ij,i}) n^j \, d\mu_\delta,$$

which is a geometric invariant, despite being expressed in coordinates. It is finite precisely when the scalar curvature $\bar{R}$ of $\bar{g}$ satisfies

$$\int_{N^3} |\bar{R}| < \infty,$$

and from a physical point of view it is meant to measure both matter content and gravitational energy of the isolated system. Compare the joint papers [39][40] of the author and T. Ilmanen for references to these facts. The Hawking quasi-local mass defined above is used as a geometric concept for the energy of a three-dimensional region contained inside a two-dimensional surface, motivated by the fact that for large approximately round spheres S_R^2 it is true that $m_H(S_R^2) \to m$. Furthermore, since in the physically simplest case the outer boundary of a black hole can be represented by a minimal two-surface inside the given three-manifold, the inverse mean curvature flow can provide a relation between the size of the black hole and the total energy m: If there is a smooth connected solution of the inverse mean curvature flow starting from a minimal surface $M_0^2 \subset N^3$, (the apparent horizon of the black hole) and expanding smoothly to large round spheres where $m_H(M_t^2) \to m$, then by the monotonicity result above we have the inequality

$$\frac{1}{4\sqrt{\pi}}|M_0^2|^{1/2} = m_H(M_0^2) \leq m.$$

This relation between the size of the outermost black hole and the total energy of an isolated gravitating system is the Riemannian Penrose inequality, which sharpens the positive mass theorem. The argument just described was first put forward by Geroch, [24]. Also note the many other contributions to this approach which are refered to in [39]. The crucial question concerns of course the existence of such a solution to the flow by inverse mean curvature. For starshaped surfaces of positive mean curvature in $\mathbb{R}^{n+1}$ Gerhardt [23] and Urbas [53] show that the necessary estimates for complete regularity of the flow can be established and they prove longterm existence as well as asymptotic roundness in this class.

Without an assumption like starshapedness it is quite clear that singularities have to occur in certain situations. For example, the solution evolving from a thin symmetric torus can not exist forever, due to the upper bound on H some blowup in the speed H^{-1} must occur for such initial data. Similar examples can be constructed in the class of two-spheres making it clear that there cannot be a smooth solution for the flow in the general situations that are of natural interest in physics.

To overcome these difficulties, [39] introduces a weak concept of solution for the flow which still retains the crucial monotonicity of the Hawking mass. The weak concept is a level-set formulation of (1.1), where the evolving surfaces are given as level-sets of a scalar function u via

$$M_t^2 = \partial\{x|u(x) < t\},$$

and (1.1) is replaced by the degenerate elliptic equation

$$\text{div}_N \left(\frac{\nabla u}{|\nabla u|}\right) = |\nabla u|,$$

where the left hand side describes the mean curvature of the level-sets and the right hand side yields the inverse speed. This formulation in divergence form admits locally Lipschitz continuous solutions and is inspired by the work of Evans-Spruck [19] and Chen-Giga-Goto [11] on the mean curvature flow. Using elliptic regularisation and a minimization principle we show existence of a locally Lipschitz-continuous solution with level-sets of nonnegative mean curvature of class $C^{1,\alpha}$, still satisfying monotonicity of the Hawking quasi-local mass, compare [39]. The solution allows the phenomenon of fattening, which corresponds to jumps of the surfaces and is desirable for our main application. We thus succeed in adapting Geroch's original argument and derive the following sharp lower bound for the mass:

Theorem 6.1 *(H.-Ilmanen) Let N^3 be a complete, connected 3-manifold. Suppose that*

(i) *N^3 has nonnegative scalar curvature,*

(ii) *N^3 is asymptotically flat in the sense above with ADM mass m,*

(iii) *The boundary of N^3 is compact and consists of minimal surfaces, and N^3 contains no other compact minimal surfaces.*

Then $m \geq 0$, and

$$16\pi m^2 \geq |\Sigma^2|,$$

where $|\Sigma^2|$ is the area of any connected component of ∂N^3. Equality holds if and only if N^3 is one-half of the spatial Schwarzschild manifold.

The *spatial Schwarzschild manifold* is the manifold $\mathbb{R}^3 \setminus \{0\}$ equipped with the metric $\bar{g} := (1 + m/2|x|)^4 \delta$, representing the spatial exterior region of a single static black hole of mass m.

7 Short-Time Existence Theory

Classically, the existence theory for nonlinear parabolic equations is treated in two stages: first, the use of linearisation techniques to prove that a solution may be found for a short interval of time; and second, derivation of the all-important a priori estimates which enable us to extend the short-time solution to a maximal time interval. In this chapter, we carry out the first half of the process.

7.1 Evolution Equations for Manifolds and Hypersurfaces

This section introduces the primary concern of this work: evolution equations for geometric structures. Typically, we consider motions of manifolds and submanifolds driven by forces which stem from their curvature.

Specifically, we address two problems:

Conformal Deformation of a Manifold: Let (M^n, g) be a smooth Riemannian manifold, and consider the deformation process

$$\frac{\partial}{\partial t} g = \lambda(x, t) \cdot g, \tag{7.1}$$

for some function λ. This defines a continuous, conformal change in the metric — conformal because the metric changes only by a scaling factor; angles are not affected. The best-known examples of this are the Ricci flow on a compact 2-surface (described and solved completely in [27]) and the Yamabe flow on a manifold of dimension at least three [33]; in both these cases, the defining equation is $\frac{\partial}{\partial t} g = -R \cdot g$, where R is the scalar curvature of g.

Normal Deformation of a Hypersurface: Let $F : M^n \hookrightarrow (N^{n+1}, g)$ be a smooth immersion of a hypersurface in a Riemannian manifold. M^n is assumed orientable, so that there is a smoothly varying, globally defined unit normal vector. In this case, we consider deformation of F according to the equation

$$\frac{\partial}{\partial t} F = \lambda(x, t) \cdot n. \tag{7.2}$$

The best-known example of this is the mean-curvature flow of hypersurfaces, where the speed is (up to a sign) the mean curvature of F. This was first introduced by Mullins in [45] (an unjustly little-known work); it was later found independently by Brakke, who expressed the equation in the language of geometric measure theory in [9].

These are the standard examples of such problems, and they share a common structure. In both cases, the deformation process can be shown to be equivalent to a quasilinear scalar partial differential equation on M^n. When the impetus comes from the curvature, as in these examples, the scalar equations are strictly parabolic and of second order. Such are known to possess solutions under very general conditions, at least for some short period of time.

The total curvature problems we wish to study in this work also lead to quasilinear parabolic scalar equations, but of fourth or higher order. It will surprise nobody that such equations still admit short-time solutions. Nevertheless, when the setting is a manifold rather than a euclidean domain, this does not belong to the standard theory and requires proof.

The question of existence will be taken up in later sections; the question of how solutions actually behave will be taken up in later chapters. For the remainder of this section, we assume that we already have a solution to (7.1) or (7.2), and derive a handful of basic properties.

(7.1) and (7.2) imply evolution equations for the curvature and other geometric attributes of g and F. Consider first the conformal deformation.

Lemma 7.1 *Let M^n be a smooth manifold; let g_t be a one parameter family of metrics on M^n varying according to (7.1). Then, g_t can be written as $\exp 2u(x, t) \cdot g_0$, where the function u evolves by the equation $\frac{\partial}{\partial t} u = \frac{1}{2} \lambda$.*

Proof. It is obvious that g_t may be so represented. It follows that

$$\frac{\partial}{\partial t} g = \frac{\partial}{\partial t} \left(\exp 2u \cdot g_0 \right) = 2 \frac{\partial u}{\partial t} \cdot \exp 2u \cdot g_0 = 2 \frac{\partial u}{\partial t} \cdot g$$

and this provides the equation for u.

Any other metric $g^\natural$ in the same conformal class could take the place of g_0 in this lemma. The equation for the conformal factor is unaffected.

Notation: In what follows, we drop the subscript t from the time-dependent metric. The calculations will relate g to a fixed background metric; for convenience, we take this to be g_0. The covariant derivative and laplacian operators of g and g_0 will be represented as ∇, Δ and ∇^0, Δ^0; the curvatures will be represented in the same way. The zero may appear as a subscript or a superscript, whichever happens to be more convenient for typesetting; the meaning remains clear. This accords with the usage in subsequent chapters.

Lemma 7.2 *The Christoffel symbols and curvature of g may be expressed in terms of those of g_0 and the conformal factor u:*

$$\Gamma_{ij}^k = (\Gamma_0)_{ij}^k + (\delta_i^k \nabla_j^0 u + \delta_j^k \nabla_i^0 u - g_{ij}^0 g_0^{kl} \nabla_l^0 u)$$

$$R_{ij} = R_{ij}^0 - (n-2)\left(\nabla_i^0 \nabla_j^0 u - \nabla_i^0 u \nabla_j^0 u\right) - \left(\Delta^0 u + (n-2)\left|\nabla^0 u\right|^2\right) \cdot g_{ij}^0$$

$$R = e^{-2u}\left(R^0 - 2(n-1)\Delta^0 u - (n-1)(n-2)\left|\nabla^0 u\right|^2\right).$$

Similarly, the laplacian operator corresponding to g can be related to that of g_0: for any smooth function $\phi: M^n \to \mathbf{R}$,

$$\Delta\phi = e^{-2u}\Delta^0\phi + (n-2)g^0(\nabla^0 u, \nabla^0 \phi)$$

Proof. The first three equations may be found in the discussion of the Yamabe problem in [8]; the fourth follows easily. Let ϕ be a fixed smooth function on M^n; then, in local co-ordinates,

$$\Delta\phi = g^{ij}\left(\frac{\partial^2\phi}{\partial x^i \partial x^j} - \Gamma_{ij}^k \frac{\partial\phi}{\partial x^k}\right)$$

$$= e^{-2u}\left(g_0^{ij}\left(\frac{\partial^2\phi}{\partial x^i \partial x^j} - (\Gamma_0)_{ij}^k \frac{\partial\phi}{\partial x^k}\right) - g_0^{ij}(\Gamma - \Gamma_0)_{ij}^k \frac{\partial\phi}{\partial x^k}\right)$$

$$= e^{-2u}\Delta^0\phi - g_0^{ij}\left(\delta_i^k \frac{\partial u}{\partial x^j} + \delta_j^k \frac{\partial u}{\partial x^i} - g_{ij}^0 g_0^{kl} \frac{\partial u}{\partial x^l}\right)\frac{\partial\phi}{\partial x^k}$$

$$= e^{-2u}\Delta^0\phi + (n-2)g_0^{kl}\frac{\partial u}{\partial x^l}\frac{\partial\phi}{\partial x^k}$$

and this establishes the final equation. Combining the previous two lemmas gives the variation of the curvature under (7.1):

Lemma 7.3 *The change (7.1) produces in the curvature of g is given by:*

$$\frac{\partial}{\partial t}R_{ij} = -\frac{1}{2}\left(\Delta\lambda \cdot g_{ij} + (n-2)\nabla_i\nabla_j\lambda\right) \qquad \text{and} \qquad \frac{\partial}{\partial t}R = -(n-1)\Delta\lambda - R\lambda.$$

Proof. Differentiating the equation above for the Ricci curvature and substituting $\lambda = 2\frac{\partial}{\partial t}u$,

$$\frac{\partial}{\partial t}R_{ij} = \frac{1}{2}\left(-\left(\Delta^0\lambda \cdot g_{ij}^0 + (n-2)\nabla_i^0\nabla_j^0\lambda\right) + 2(n-2)\left(\nabla_i^0\lambda\nabla_j^0 u - g^0\left(\nabla^0\lambda, \nabla^0 u\right) \cdot g_{ij}^0\right)\right).$$

However, for any C^2 function f,

$$\Delta f \cdot g_{ij} + (n-2)\nabla_i\nabla_j f = \Delta^0 f \cdot g_{ij}^0 + (n-2)\nabla_i^0\nabla_j^0 f + L_f$$

in which the error term is given by

$$L_f = \left((\Gamma^0 - \Gamma)_{lm}^k (g^0)^{lm} g_{ij}^0 + (n-2)(\Gamma^0 - \Gamma)_{ij}^k\right) \cdot \nabla_k^0 f;$$

and using the transformation rule for the Christoffel symbols of Lemma 7.2, this simplifies to

$$L_f = 2(n-2)\left(g^0\left(\nabla^0\lambda, \nabla^0 u\right) \cdot g_{ij}^0 - \nabla_i^0\lambda\nabla_j^0 u\right).$$

But this matches precisely the final term in the evolution of R_{ij}, and so, cancelling,

$$\frac{\partial}{\partial t}R_{ij} = -\frac{1}{2}\left(\Delta\lambda \cdot g_{ij} + (n-2)\nabla_i\nabla_j\lambda\right),$$

which proves the first claim of the lemma. The evolution of the scalar curvature is simpler:

$$
\begin{aligned}
\frac{\partial}{\partial t}R &= \frac{\partial}{\partial t}\left(e^{-2u}\left(R^0 - 2(n-1)\Delta^0 u - (n-1)(n-2)|\nabla^0 u|^2\right)\right) \\
&= -2\frac{\partial u}{\partial t}\cdot R + e^{-2ut}\left(-2(n-1)\Delta^0\frac{\partial u}{\partial t} - 2(n-1)(n-2)\,g^0\left(\nabla^0 u, \nabla^0\frac{\partial u}{\partial t}\right)\right) \\
&= -(n-1)\Delta\lambda - R\lambda.
\end{aligned}
$$

and this proves the second claim. This is as much as we need say for now about the conformal problem.

Now let F_t be a one-parameter family of immersions $M^n \hookrightarrow N^{n+1}$ which vary in accordance with (7.2). Let g_t denote the induced metric $F_t^*\bar{g}$. As above, we shall drop the t subscripts wherever this would not lead to confusion.

From (7.2), we compute evolution equations for the geometric features of F. This is simplified immensely by the use of well-chosen co-ordinates.

These calculations are purely local in nature; so we focus on some point (x^*, t^*) in space-time. Let y^* be the image of x^* under F_{t^*}. We may assume that the co-ordinates on N^{n+1} are normal at y^*, and that those on M^n are normal at x^* relative to the metric induced at this one instant of time. In particular, the Christoffel symbols $\bar{\Gamma}_{\beta\gamma}^\alpha(y^*)$ and $\Gamma_{ij}^k(x^*, t^*)$ all vanish, and the Gauß and Weingarten equations reduce to

$$\frac{\partial^2 F^\alpha}{\partial x^i\partial x^j}(x^*, t^*) = -h_{ij}(x^*, t^*)\cdot n^\alpha(x^*, t^*), \qquad \frac{\partial n^\alpha}{\partial x^i}(x^*, t^*) = h_i^j(x^*, t^*)\cdot\frac{\partial F^\alpha}{\partial x^i}(x^*, t^*).$$

Lemma 7.4 *Under (7.2), the induced metric on M^n evolves according to*

$$\frac{\partial}{\partial t}g_{ij} = 2\lambda h_{ij}.$$

It follows directly that the inverse of the metric and the measure evolve by

$$\frac{\partial}{\partial t}g^{ij} = -2\lambda h^{ij} \qquad \text{and} \qquad \frac{\partial}{\partial t}d\mu = \lambda H\,d\mu.$$

Proof. Let (x^*, t^*) be a given point of spacetime, and assume the co-ordinate systems on M^n and N^{n+1} are normal at (x^*, t^*) and $F_{t^*}(x^*)$, as above. We compute the evolution of g_{ij} at the point (x^*, t^*).

The induced metric is by nature given by

$$g_{ij} = \bar{g}\left(\frac{\partial F}{\partial x^i}, \frac{\partial F}{\partial x^j}\right),$$

and hence, noting that $\bar{g}$ is symmetric and has no covariant derivative of its own,

$$\frac{\partial}{\partial t} g_{ij} = 2\bar{g}\left(\frac{\partial}{\partial t}\left\{\frac{\partial F}{\partial x^i}\right\}, \frac{\partial F}{\partial x^j}\right) = 2\bar{g}\left(\frac{\partial}{\partial x^i}\{\lambda n\}, \frac{\partial F}{\partial x^j}\right),$$

Now expand the product derivative. The derivative term in λ clearly vanishes because of orthogonality, and all that remains is

$$\frac{\partial}{\partial t} g_{ij} = 2\lambda\bar{g}\left(\frac{\partial n}{\partial x^i}, \frac{\partial F}{\partial x^j}\right).$$

and it follows directly from the reduced Weingarten equation at (x^*, t^*) that the final factor is simply h_{ij}; with that, the first claim of the lemma is proved. The rest is easy. To compute the evolution of the inverse of the metric, we differentiate the equation $g^{ik} \cdot g_{kl} = \delta^i_j$:

$$0 = \frac{\partial(g^{ik} \cdot g_{kl})}{\partial t} = \frac{\partial g^{ik}}{\partial t} g_{kl} + g^{ik}\frac{\partial g_{kl}}{\partial t} = \frac{\partial g^{ik}}{\partial t} g_{kl} + 2\lambda h^i_l.$$

Tracing with g^{lj} gives $\frac{\partial}{\partial t} g^{ij} = -2\lambda h^{ij}$, which establishes the the second claim. The final part follows from the rule for differentiating a determinant:

$$\frac{\partial}{\partial t} d\mu = \frac{\partial}{\partial t}(\sqrt{\det g}\, dx) = \frac{1}{2}\sqrt{\det g} \cdot g^{ij}\frac{\partial}{\partial t} g_{ij}\, dx = \lambda H\, d\mu,$$

and this completes the proof.

Next we derive the variation of the normal vector:

Lemma 7.5 *The change in the normal is given by*

$$\frac{\partial n}{\partial t} = -F_*(\mathrm{grad}^{M^n}\lambda).$$

Proof. n is a unit normal vector; thus, $\bar{g}(n, n) = 1$ everywhere. Differentiating this equation, we see that the derivative (*any* derivative) of n must be normal to n itself, and hence tangential to $F(M^n)$. It may therefore be represented in the form

$$\frac{\partial n}{\partial t} = g^{ij}\bar{g}\left(\frac{\partial n}{\partial t}, \frac{\partial F}{\partial x^i}\right) \cdot \frac{\partial F}{\partial x^j}. \tag{7.3}$$

Now differentiating the equation $\bar{g}(n, \frac{\partial}{\partial x^i} F) = 0$, we have

$$0 = \bar{g}\left(\frac{\partial n}{\partial t}, \frac{\partial F}{\partial x^i}\right) + \bar{g}\left(n, \frac{\partial}{\partial t}\frac{\partial F}{\partial x^i}\right) = \bar{g}\left(\frac{\partial n}{\partial t}, \frac{\partial F}{\partial x^i}\right) + \bar{g}\left(n, \frac{\partial(\lambda n)}{\partial x^i}\right) = \bar{g}\left(\frac{\partial n}{\partial t}, \frac{\partial F}{\partial x^i}\right) + \frac{\partial \lambda}{\partial x^i},$$

noting here that another term vanishes because n is orthogonal to all its derivatives. This allows us to substitute for $\bar{g}(\frac{\partial}{\partial t}n, \frac{\partial}{\partial x^i}F)$ in (7.3), giving

$$\frac{\partial n}{\partial t} = -g^{ij}\frac{\partial\lambda}{\partial x^i}\frac{\partial F}{\partial x^j},$$

and this is exactly $-F_*(\text{grad}^{M^n}\lambda)$.

Lemma 7.6 *The variation in the second fundamental form is given by the equations*

$$\frac{\partial}{\partial t}h_{ij} = -\nabla_i\nabla_j\lambda - \lambda(-h_{ik}h_j^k + \overline{\text{Riem}}_{injn})$$

$$\frac{\partial}{\partial t}h_i^j = -\nabla_i\nabla^j\lambda - \lambda(h_{ik}h^{kj} + \overline{\text{Riem}}_{in}{}^j{}_n)$$

$$\frac{\partial H}{\partial t} = -\Delta\lambda - (|A|^2 + \overline{\text{Ric}}(n,n))$$

$$\frac{\partial}{\partial t}|A|^2 = -2h^{ij}\nabla_i\nabla_j\lambda - 2\lambda(\text{tr } A^3 + h^{ij}\overline{\text{Riem}}_{injn})$$

Proof. The second fundamental form is, by definition,

$$h_{ij} = -\bar{g}\left(\bar{\nabla}_{\frac{\partial F}{\partial x^i}}\frac{\partial F}{\partial x^j}, n\right),$$

and so, differentiating,

$$\frac{\partial}{\partial t}h_{ij} = -\bar{g}\left(\frac{\partial}{\partial t}\left\{\bar{\nabla}_{\frac{\partial F}{\partial x^i}}\frac{\partial F}{\partial x^j}\right\}, n\right) - \bar{g}\left(\bar{\nabla}_{\frac{\partial F}{\partial x^i}}\frac{\partial F}{\partial x^j}, \frac{\partial n}{\partial t}\right).$$

Now we reinstate the assumptions of Lemma 7.4. In the normal co-ordinate system, the rightmost term vanishes altogether because $\frac{\partial}{\partial t}n$ is tangential and the spatial derivative normal. So, expressing the remaining covariant derivative in co-ordinates,

$$\frac{\partial}{\partial t}h_{ij} = -\bar{g}\left(\frac{\partial}{\partial t}\left\{\frac{\partial^2 F}{\partial x^i\partial x^j} + \bar{\Gamma}^\alpha_{\beta\gamma}\frac{\partial F^\beta}{\partial x^i}\frac{\partial F^\gamma}{\partial x^j}\frac{\partial}{\partial y^\alpha}\right\}, n\right).$$

Expanding the time derivative, and noting that the terms containing $\bar{\Gamma}^\alpha_{\beta\gamma}$ all vanish, this becomes

$$\frac{\partial}{\partial t}h_{ij} = -\bar{g}\left(\frac{\partial^2(\lambda n)}{\partial x^i\partial x^j} + \frac{\partial}{\partial t}\bar{\Gamma}^\alpha_{\beta\gamma}\cdot\frac{\partial F^\beta}{\partial x^i}\frac{\partial F^\gamma}{\partial x^j}\frac{\partial}{\partial y^\alpha}, n\right)$$

$$= -\bar{g}\left(\frac{\partial^2(\lambda n)}{\partial x^i\partial x^j} + \lambda\bar{\nabla}_n\bar{\Gamma}^\alpha_{\beta\gamma}\cdot\frac{\partial F^\beta}{\partial x^i}\frac{\partial F^\gamma}{\partial x^j}\frac{\partial}{\partial y^\alpha}, n\right).$$

At the point (x^*, t^*), the Weingarten equation for the derivative of the normal gives

$$\frac{\partial n^\alpha}{\partial x^i} = h_i^j\frac{\partial F^\alpha}{\partial x^j} \quad \text{and} \quad \frac{\partial^2 n^\alpha}{\partial x^i\partial x^j} = \frac{\partial h_j^k}{\partial x^i}\frac{\partial F^\alpha}{\partial x^k} + h_j^k\frac{\partial^2 F^\alpha}{\partial x^i\partial x^j} - \frac{\partial\bar{\Gamma}^\alpha_{\beta\gamma}}{\partial x^i}\frac{\partial F^\beta}{\partial x^j}n^\gamma.$$

It follows that

$$\bar{g}\left(\frac{\partial^2 n}{\partial x^i \partial x^j}, n\right) = -h_j^k h_{ki} - \bar{g}\left(\bar{\nabla}_{\frac{\partial}{\partial x^i}}\Gamma_{\beta\gamma}^\alpha \frac{\partial F^\beta}{\partial x^i} n^\gamma \frac{\partial}{\partial y^\alpha}, n\right).$$

This enables us to expand the product derivative in (7.4), which gives

$$\frac{\partial}{\partial t}h_{ij} = -\frac{\partial^2\lambda}{\partial x^i \partial x^j} - \lambda h_{ik}h_j^k + \lambda\bar{g}\left(\left(\bar{\nabla}_n\Gamma_{\beta\gamma}^\alpha \frac{\partial F^\beta}{\partial x^i}\frac{\partial F^\gamma}{\partial x^j} - \bar{\nabla}_{\frac{\partial F}{\partial x^i}}\Gamma_{\beta\gamma}^\alpha\frac{\partial F^\beta}{\partial x^j}n^\gamma\right)\frac{\partial}{\partial x^\alpha}, n\right).$$

However, in our normal co-ordinates, the second partial derivative of λ coincides with the covariant derivative, and the final term is simply the Riemann tensor of N^{n+1}. Permutation of the indices in one or other of the summands allows this term to be rewritten as

$$\bar{g}\left(\left(\bar{\nabla}_{\frac{\partial}{\partial y^\delta}}\Gamma_{\beta\gamma}^\alpha - \bar{\nabla}_{\frac{\partial}{\partial y^\beta}}\Gamma_{\gamma\delta}^\alpha\right)\frac{\partial}{\partial y^\alpha}, n\right) \cdot \frac{\partial F^\beta}{\partial x^i}\frac{\partial F^\gamma}{\partial x^j}n^\delta$$

and now the first factor matches the definition of the Riemann tensor; the product simplifies therefore to

$$\overline{\text{Riem}}_{\alpha\beta\delta\gamma}\, n^\alpha \frac{\partial F^\beta}{\partial x^i}\frac{\partial F^\gamma}{\partial x^j}n^\delta = \overline{\text{Riem}}_{injn}$$

In view of all this, the evolution equation for h_{ij} may be rewritten as

$$\frac{\partial}{\partial t}h_{ij} = -\nabla_i\nabla_j\lambda - \lambda(-h_{ik}h_j^k + \overline{\text{Riem}}_{injn}) \qquad (7.4)$$

which settles the first claim of the lemma. The equation for h_i^j follows quickly:

$$\frac{\partial}{\partial t}h_i^j = \frac{\partial}{\partial t}(g^{jk}h_{ik}) = -2\lambda \cdot h^{jk}h_{ik} - g^{jk}\left(\nabla_i\nabla_k\lambda - \lambda(-h_{il}h_k^l + \overline{\text{Riem}}_{inkn})\right)$$
$$= -\nabla_i\nabla^j\lambda - \lambda(h_i^k h_k^j + \overline{\text{Riem}}_{in}{}^j{}_n),$$

and tracing over i and j gives the equation for H at once. Lastly,

$$\frac{\partial}{\partial t}|A|^2 = \frac{\partial}{\partial t}(h_i^j h_j^i) = -2h^{ij}\nabla_i\nabla_j\lambda - 2\lambda(h_i^j h_j^k h_k^i + h^{ij}\overline{\text{Riem}}_{inij})$$

which is the final claim. The curvature equations hint at the structure concealed in (7.2) and (7.1). In the examples mentioned earlier, where the speed λ was the curvature itself, they are clearly parabolic.

The speeds which interest us in this work feature the laplacian of curvature as their leading term. These too will lead to parabolic equations.

The assumption of orientability demanded in order to make sense of (7.2) is heavy-handed. In real-life examples, it can typically be avoided. The mean curvature flow, for instance, may be rewritten using the Weingarten equation simply as

$$\frac{\partial F}{\partial t} = \Delta F,$$

where the laplacian is computed in the induced metric; and this is now perfectly meaningful even when the manifold is *not* orientable. The hypersurface flows we shall consider later can be redefined in the same way.

The first step towards understanding these problems is to prove that solutions can be found at all. The strategy here is inherited from the second-order theory. The geometric equation is shown to be equivalent to a quasilinear scalar equation, which may be solved for some short interval of time using linearisation techniques; this short-time solution may then be continued as long as it does not become singular.

In the remainder of this chapter, we construct the short-time existence theory. In the problems of higher order we consider, even the linear theory is incomplete, and we have to develop it for ourselves. This is the goal of the following two sections.

2.2 The linear problem

In this section, we prove the existence of solutions to the linear parabolic equation of $2p$-th order on a closed manifold. There are numerous proofs of the corresponding result in a euclidean domain — see, for instance, [18] or [50]. But these typically rely on the construction of a fundamental solution to the equation, a technique which is not easily adapted to the manifold setting. Friedman [21] describes an abstract approach based on a variant of the Lax-Milgram lemma, developing ideas originally due to J. L. Lions [44] and F. Trèves [51].

However, Friedman's account of the result is inaccurate. He proves the existence of a time-$W^{k,2}$ solution to the linear equation $D_t u + A_t u = g, u_0 \equiv 0$ under the unacceptably strong assumption that g vanishes at time zero along with all its derivatives of order up to $k - 1$. Considering the equation as a physical process, this is tantamount to assuming there are no external forces at time $t = 0$. This is clearly an undesirable condition, and in no way a natural one.

This is in itself not a mistake, though it limits the usefulness of the theorem. Friedman goes on to claim in a remark, however, that one may prescribe the initial values u_0 freely by considering the equation for $u - u_0$. But this gives an equation whose forcing term no longer satisfies the vanishing condition; the theorem as Friedman states it does not apply. In this section, we use techniques related to Friedman's to prove a minimal existence result. This will be strengthened later when we prove the natural a priori estimates for the linear problem.

Let (M^n, g) be a smooth, compact Riemannian manifold. Let A be a linear differential operator of order $2p$ on M^n; that is, for a $2p$-times differentiable function $u : M^n \to \mathbb{R}$, we set

$$Au(x) = \sum_{k \leq 2p} A_k^{i_1 i_2 \cdots i_k}(x) \nabla_{i_1 i_2 \cdots i_k} u(x),$$

in local co-ordinates, where the A_k are smooth tensor fields on M^n of type $(0, k)$. Let A be *elliptic* in the very strong sense that the leading term can be factorised as

$$A_{2p}^{i_1 j_1 i_2 j_2 \cdots i_p j_p} = E^{i_1 j_1} E^{i_2 j_2} \cdots E^{i_p j_p},$$

where the 2-form E is strictly positive: $E \geq \lambda g$ for some $\lambda > 0$. In words, the leading part of A should simply be the p-th power of some second-order elliptic operator.

It is possible — easy, even — to define much weaker notions of ellipticity. However, the operators which arise in our geometric problems do turn out to have the structure above; moreover, it is a much simpler matter to prove Gårding's inequality in this class.

Consider such an operator in the usual Sobolev-space way as a bilinear form defined on $W^{p,2}(M^n)$, which we shall also denote by A. Then:

Lemma 7.7 *(Gårding's Inequality for A)* *For any $\phi \in W^{p,2}(M^n)$,*

$$A(\phi, \phi) \geq \frac{\lambda^p}{2} \|\phi\|^2_{W^{p,2}(M^n)} - Q \|\phi\|^2_{L^2(M^n)},$$

where the constant Q depends on n, λ, and the C^{p-1} norms of the of the tensors A_k and $\mathrm{Riem}^{M^n}_{ijkl}$.

Proof. This result is easy; the only point that requires any explanation is the appearance of the Riemann tensor in the calculation. That arises through perhaps having to permute derivatives in order to ensure that the leading term is given by

$$A(\phi, \phi) = \int_{M^n} E^{i_1 j_1} E^{i_2 j_2} \cdots E^{i_p j_p} \nabla_{i_1 i_2 \cdots i_p} \phi \cdot \nabla_{j_1 j_2 \cdots j_p} \phi + (\text{errorterms}) \, d\mu.$$

The very strong ellipticity condition makes it clear at once that the leading term is at least $\lambda^p \int_{M^n} \|\nabla^p \phi\|^2_g \, d\mu$, and the usual interpolation argument can then be used to estimate each of the terms of lesser order between a fraction of this and a large multiple of the L^2-norm.

Now we consider the parabolic problem. Let A_t be a smooth family of elliptic operators of order $2p$. 'Smooth' means simply that the component tensor fields should vary smoothly over $M^n \times [0, \infty)$.

To prove the existence of a solution to $D_t u + A_t u = g$, we recast the problem in the natural Hilbert space setting, and solve the resulting operator equation using the following refinement of the Lax-Milgram lemma, which relaxes the continuity assumptions on the bilinear form:

Lemma 7.8 *Let $(H, \|\cdot\|_H)$ be a Hilbert space and $(\Phi, \|\cdot\|_\Phi)$ an inner-product space continuously embedded in H. Φ is not assumed to be complete. Let $F : H \times \Phi \to \mathbb{R}$ be a bilinear form with the properties that*

- *the mapping $h \to F(h, \phi)$ is continuous for each fixed $\phi \in \Phi$, and*
- *F is coercive on Φ: $F(\phi, \phi) \geq \lambda \|\phi\|^2_\Phi$, for some $\lambda > 0$.*

Then any smooth functional $L \in \Phi^*$ can be realised as a slice through F: there exists $u_L \in H$ such that $L(\phi) = F(u_L, \phi)$ for each $\phi \in \Phi$.

Proof. See [21], Chapter 10, Theorem 16.

For smooth functions $f, g : M^n \times [0, \infty) \to \mathbb{R}$, we introduce the weighted inner products:

$$\langle f, g \rangle_{LL_a} = \int_0^\infty e^{-2at} \langle f(\cdot, t), g(\cdot, t) \rangle_{L^2(M^n)} \, dt$$

$$\langle f, g \rangle_{LW_a} = \int_0^\infty e^{-2at} \langle f(\cdot, t), g(\cdot, t) \rangle_{W^{p,2}(M^n)} \, dt$$

$$\langle f, g \rangle_{WW_a} = \langle f, g \rangle_{LW_a} + \langle D_t f, D_t g \rangle_{LL_a};$$

we define LL_a, LW_a and WW_a to be the Hilbert spaces formed by completion of $C^\infty(M^n \times [0, \infty))$ in the corresponding norms. Further, let $\Phi = C_C^\infty(M^n \times (0, \infty))$ be the space of smooth functions which vanish for very large and very small times, and let WW_a^0 denote the completion of Φ in WW_a.

Theorem 7.9 *Let A_t be a smooth and uniformly elliptic family of operators of order $2p$. Then, for sufficiently large a (which depends only on A_t), the equation*

$$D_t u + A_t u = g, \qquad u(\bullet, 0) \equiv 0 \qquad\qquad (7.5)$$

has a unique weak solution in WW_a^0.

Proof. Note first that u is a solution to (7.5) if and only if $\exp(-Mt) \cdot u$ solves the equation $D_t w + (A_t + M \cdot id)w = g \exp(-Mt)$. Choosing $M = Q$, the weight of the error term in the Gårding inequality above, we see that it suffices to solve equations in which the elliptic operator is *strictly* coercive. From here on, we assume this is the case.
Define a bilinear form on $WW_a^0 \times \Phi$ by the formula

$$P(w, \phi) = \langle D_t w, \ D_t \phi \rangle_{LL_a} + \int_0^\infty e^{-2at} A_t(w, D_t \phi) \, dt,$$

and a linear functional on Φ by

$$L(\phi) = \langle g, \ D_t \phi \rangle_{LL_a}.$$

These are simply the results of testing the left and right hand sides of (7.5) with the function $e^{-2at} \cdot D_t \phi$. Fixing ϕ, P is easily seen to be continuous in w. It is just as obvious that L too is continuous with respect to the WW_a-norm. It remains only to show that P is coercive, and Lemma 7.8 will apply.
This is a simple but technical matter. Let $\phi \in \Phi$; then

$$P(\phi, \phi) = \|D_t \phi\|_{LL_a}^2 + \int_0^\infty e^{-2at} A_t(\phi, D_t \phi) \, dt.$$

Let I denote the second term on the right. Partial integration in time shows that:

$$I = \int_0^\infty e^{-2at} \left(a \cdot A_t(\phi, \phi) - \frac{1}{2}(D_t A_t)(\phi, \phi) \right) dt$$

$$\geq \left(a \frac{\lambda^p}{2} - \frac{1}{2} \sup |D_t A_t| \right) \cdot \|\phi\|_{LW_a}^2.$$

If a is chosen large enough, then, this ensures that P is coercive on $\Phi \times \Phi$ with respect to the WW_a-norm; thus, by Lemma 7.7, one can find a $w^* \in WW_a^0$ for which $P(w^*, \phi) = L(\phi)$ for any $\phi \in \Phi$.
It might seem at first that this is insufficient to deliver a weak solution of (7.5), as our test function space is still too small. We are restricted to those functions whose average over time is zero, which would normally mean only that w^* differs from a solution to (7.5) by a time-constant function.

In fact, this problem does not arise because of the weighting given to the measure. Fix some $\psi \in \Phi$, and consider the function

$$\Psi(x,t) = \psi(x,t) - \psi(x,t+B).$$

For B large enough (so large that the support of the second term does not overlap with that of the first), this averages over time to zero, and so it can be represented as $D_t\phi$ for some $\phi \in \Phi$. However, the contributions to $P(w^*, \phi)$ and $L(\phi)$ made by the second term are easily seen to diminish to zero as $B \to \infty$ because of the exponential factor; we therefore have

$$\langle D_t w^*, \ \psi \rangle_{LL_a} + \int_0^\infty e^{-2at} A_t(w^*, \psi)\, dt = \langle g, \ \psi \rangle_{LL_a} \qquad \text{for any } \psi \in \Phi,$$

and with that, w^* is a weak solution to the original equation.

Several points remain open. The solution above has the minimum of regularity needed to make sense of the equation; $W^{1,2}$ in time and $W^{p,2}$ in space. Such a function takes on the zero boundary data continuously only in L^2.

It is possible to accommodate sufficiently smooth nonzero initial data u_0 by considering the equation for $u - u_0$. To apply the result above, this means that $A_t u_0$ needs to be in LL_a, which in turn implies that u_0 has to be a $W^{2p,2}$ function.

This last result is less than optimal. In fact, the natural class for the initial values is $W^{p,2}$; this will follow from the estimates proved in the coming section.

The solution above is unique. This too will follow from the estimates.

7.3 A Priori Estimates for the Linear Equation

This section is concerned with the regularity of the solution obtained above. Crudely, the principal results are first, that the solution is as smooth as the forcing term g allows it to be, and second, that the correspondence between solution u and forcing term is an isomorphism of appropriately defined Banach spaces. The importance of the second of these will become clear in the next section, where we discuss the quasilinear problem.

First, we define the appropriate Hilbert spaces. Let

$$LW_a^s = \left\{ f : M^n \times [0,\infty) \to \mathbb{R} \ \Big| \ \int_0^\infty e^{-2at} \|f\|^2_{W^{s,2}(M^n)}\, dt < \infty \right\}$$

with inner product $\langle f, \ g \rangle_{LW_a^s} = \int_0^\infty e^{-2at} \langle f, \ g \rangle_{W^{s,2}(M^n)}\, dt$, and let

$$P_a^m = \{ f : M^n \times [0,\infty) \to \mathbb{R} \,|\, D_t^i f \text{ exists and is in } LW_a^{2(m-i)p} \text{ for each } i \le m \},$$

where the inner product is the obvious choice:

$$\langle f, \ g \rangle_{P_a^m} = \sum_{i \le m} \langle D_t^i f, \ D_t^i g \rangle_{LW_a^{2(m-i)p}}.$$

The description may appear unwieldy, but P_a^m is the natural space forced upon us by the scaling properties of the problem. Accepting the parabolic mantra that one time derivative corresponds to $2p$ derivatives in space, this somehow describes the space of all functions which are in total $2mp$ times differentiable.

The natural regularity of the boundary values of such a function (and, because M^n has no boundary of its own, we can use the word without ambiguity to refer to the time boundary $t = 0$) is $W^{2p(m-\frac{1}{2}),2}$. The principal result of this section, which we prove below, is this: for any m, the map

$$F(u) = (u_0, g) = (u_0, D_t u - Au) \tag{7.6}$$

is an isomorphism from P_a^m onto $W^{2p(m-\frac{1}{2}),2} \times P_a^{m-1}$.

The technique for proving regularity is standard — we prove Caccioppoli-esque energy estimates first for u and then for its difference quotients. So, suppose u is in some LW_a and solves the equation

$$\langle D_t u, \phi \rangle_{LL_a} + \int_0^\infty e^{-2at} A_t(u, \phi) \, dt = \langle g, \phi \rangle_{LL_a} \tag{7.7}$$

for any smooth, compactly supported ϕ, and hence by completion for any ϕ in the broader class LW_a. Suppose further that u has initial values u_0, which are taken on only in L^2.

Lemma 7.10 u *satisfies the energy estimate*

$$\|u\|_{LW_a^p}^2 \leq C \left(\|u_0\|_{L^2(M^n)}^2 + \|g\|_{LL_a}^2 \right).$$

Proof. Choose u itself as the test function in (7.7). Then,

$$\langle D_t u, u \rangle_{LL_a} + \int_0^\infty e^{-2at} A_t(u, u) \, dt = \langle g, u \rangle_{LL_a}. \tag{7.8}$$

For any f which is $W^{1,2}$ in time, partial integration gives

$$\langle D_t f, f \rangle_{LL_a} = a \|f\|_{LL_a}^2 - \frac{1}{2} \|f_0\|_{L^2(M^n)}^2. \tag{7.9}$$

Thus, returning to (7.6), and using the ellipticity of A_t,

$$\frac{\lambda^p}{2} \|u\|_{LW_a^p}^2 + (a - Q) \|u\|_{LL_a}^2 \leq \frac{1}{2} \|u_0\|_{L^2(M^n)}^2 + \|g\|_{LL_a} \cdot \|u\|_{LL_a}.$$

Now choosing $a > Q$ and handling the forcing term with Young's inequality, the lemma is proved.

To establish that u has derivatives of higher order, we prove estimates for its difference quotients. Defining the difference quotients requires a continuous co-ordinate system, so we have to focus on a single co-ordinate patch. This means multiplying u with a cut-off function in space.

Lemma 7.11 *If $u \in WW_a$ is a solution to (7.7) with initial values $u_0 \in W^{p,2}(M^n)$, then $u \in LW_a^{2p}$, with the estimate*

$$\|u\|_{LW_a^{2p}}^2 \leq C \left(\|u_0\|_{W^{p,2}(M^n)}^2 + \|g\|_{LL_a}^2 \right).$$

Proof. Let $\psi_i : D_1^n \to M^n$, $i = 1, \ldots, N$ be a collection of smooth co-ordinate patches, so chosen that the images $\psi_i(D_{1/2})$ between them cover all of M^n. Let η be a fixed C^∞ cut-off function between $D_{1/2}$ and $D_{3/4}$. Let B_i be the image $\psi_i(D_1)$ in M^n; let B_i^* denote the image of $D_{1/2}$.

Now assume that we already have estimates for u in LW_a^{l+p}, for each $l < k$, and set $\phi = \Delta_{-h}^k(\eta^{2p+2}\Delta_h^k u)$ in B_i, where η and the finite differencing operator $\Delta_h f(x) = h^{-1}(f(x + he_j) - f(x))$ are lifted to M^n with the co-ordinate map ψ_i. Outside B_i, we simply extend ϕ to be zero. Notice that, although we don't have uniform estimates in h for the norm of ϕ, the function is at least as regular in space as u itself.

Then, substituting for ϕ in (7.7):

$$\left\langle D_t u, \Delta_{-h}^k(\eta^{2p+2}\Delta_h^k u) \right\rangle_{LL_a} + \int_0^\infty e^{-2at} A_t(u, \Delta_{-h}^k(\eta^{2p+2}\Delta_h^k u))\, dt = \left\langle g, \Delta_{-h}^k(\eta^{2p+2}\Delta_h^k u) \right\rangle_{LL_a}$$

and so, shifting difference operators with the discrete analogue of partial integration in space,

$$\left\langle D_t(\eta^{p+1}\Delta_h^k u),\ \eta^{p+1}\Delta_h^k u \right\rangle_{LL_a} + \int_0^\infty e^{-2at} A_t(\Delta_h^k u, \eta^{2p+2}\Delta_h^k u)\, dt =$$
$$(-1)^k \left\langle g, \Delta_{-h}^k(\eta^{2p+2}\Delta_h^k u) \right\rangle_{LL_a} - (-1)^k \sum_{j=1}^k \binom{k}{j} \int_0^\infty e^{-2at}(\Delta_h^j A_t)(\Delta_h^{k-j}u, \eta^{2p+2}\Delta_h^k u)\, dt,$$

where the error term on the right arises through applying the product rule for difference quotients to the elliptic term. The discretised product rule is not completely clear-cut; it reads $\Delta_h(fg)(x) = f(x + he_j)\Delta_h g(x) + g(x)\Delta_h f(x)$ — that is, one of the functions is shifted. In our case, this has no bearing; A is completely smooth, and we are concerned only with its pointwise properties. In applying the product rule, then, we always shift the factor in A, never the one in u.

Denote this error term E_1. Then,

$$|E_1| \leq C \int_0^\infty \int_{M^n} e^{-2at} \sum_{i,j \leq p; 1 \leq m \leq k} |\nabla^i \Delta_h^{k-m} u| \cdot |\nabla^j(\eta^{2p+2}\Delta_h^k u)|\, d\mu \otimes dt.$$

The first factor in the quadratic part is harmless, because by assumption we already have an estimate for u in LW_a^{k+p-1}. The second may be handled in the same way for $j < p$, while the case $j = p$ is no worse than the helpful term which will be won using the ellipticity of A_t. So, by Young's inequality,

$$|E_1| \leq C\|u\|_{LW_a^{k+p-1}}^2 + \epsilon \left\|\eta^{p+1}\nabla^p(\Delta_h^k u)\right\|_{LL_a}^2. \tag{7.10}$$

Notice here that, if we differentiate η^{2p+2} up to p times, there always remain at least $p+1$ powers of η in every term.

Now consider the remaining elliptic term; call it E_2. We have:

$$
\begin{aligned}
E_2 &= \int_0^\infty e^{-2at} A_t(\Delta_h^k u, \eta^{2p+2}\Delta_h^k u)\, dt \\
&= \int_0^\infty \int_{M^n} e^{-2at} \sum_{|I|,|J| \leq p} A^{IJ} \nabla_I(\Delta_h^k u)\nabla_J(\eta^{2p+2}\Delta_h^k u)\, d\mu \otimes dt,
\end{aligned}
$$

and whatever else may happen, every term in the sum must contain at least $p+1$ powers of η. Bearing this in mind, and isolating the term of highest order in u,

$$E_2 \geq \int_0^\infty \int_{M^n} e^{-2at} \left(A^{I_p J_p} \nabla_{I_p} f \nabla_{J_p} f \cdot \eta^{2p+2} - C\eta^{p+1} \sum_{i,j \leq p; i+j < 2p} |\nabla^i f| \cdot |\nabla^j f| \right) d\mu \otimes dt$$

where now $f = \Delta_h^k u$, and so using the ellipticity of Λ, Young's inequality and the estimates assumed from the outset for u,

$$E_2 \geq \frac{\lambda^p}{2} \left\| \eta^{p+1} \nabla^p (\Delta_h^k u) \right\|_{LL_a}^2 - C \|u\|_{LW^{k+p-1}}^2 . \tag{7.11}$$

This controls the terms in A_t in (7.10); now consider those which remain. The first term is of the form handled in (7.9):

$$\left\langle D_t(\eta^{p+1} \Delta_h^k u), \, \eta^{p+1} \Delta_h^k u \right\rangle = a \left\| \eta^{p+1} \Delta_h^k u \right\|_{LL_a}^2 - \frac{1}{2} \left\| \eta^{p+1} \Delta_h^k u_0 \right\|_{L^2(M^n)}^2 \tag{7.12}$$

while the forcing term is easily estimated if we now assume $k \leq p$:

$$\left\langle g, \, \Delta_{-h}^k(\eta^{2p+2} \Delta_h^k u) \right\rangle_{LL_a} \leq \|g\|_{LL_a} \cdot \left\| \Delta_{-h}^k(\eta^{2p+2} \Delta_h^k u) \right\|_{LL_a} \tag{7.13}$$

$$\leq C \|g\|_{LL_a} \cdot \left\| \eta^{2p+2} \Delta_h^k u \right\|_{LW_a^k} \tag{7.14}$$

$$\leq C \|g\|_{LL_a} \cdot \left\| \eta^{2p+2} \Delta_h^k u \right\|_{LW_a^p} . \tag{7.15}$$

Combining (7.10), (7.10), (7.11), (7.12) and (7.15), we have

$$\left\| \eta^{p+1} \nabla^p (\Delta_h^k u) \right\|_{LL_a}^2 \leq C \left(\left\| \eta^{p+1} \Delta_h^k u_0 \right\|_{L^2(M^n)}^2 + \|u\|_{LW_a^{k+p-1}}^2 + \|g\|_{LL_a}^2 \right).$$

Now assume that u_0 is in the class $W^{p,2}(M^n)$. This isn't the same as assuming the initial data are taken on in $W^{p,2}$ — indeed, in the above calculation, we only make use of the fact that u, and hence $\Delta_h^k u$, assume their initial values in L^2. Nonetheless, if u_0 is (by chance) $W^{p,2}$, then we have

$$\left\| \eta^{p+1} \nabla^p (\Delta_h^k u) \right\|_{LL_a}^2 \leq C \left(\|u_0\|_{W^{p,2}(M^n)}^2 + \|u\|_{LW_a^{k+p-1}}^2 + \|g\|_{LL_a}^2 \right).$$

To keep the notation from becoming needlessly complicated, this estimate was written for the 'pure' k-th difference quotients; though it is clearly just as true for the mixed difference quotients as well. Thus, we have uniform $W^{p,2}$-bounds in h for all the difference quotients of k-th order in the ball B_i^*, which therefore converge weakly to genuine weak derivatives satisfying the estimate

$$\left\| \nabla^{k+p} u \right\|_{LL_a(B_i^*)}^2 \leq C \left(\|u_0\|_{W^{p,2}(M^n)}^2 + \|u\|_{LW_a^{k+p-1}}^2 + \|g\|_{LL_a}^2 \right),$$

and summing now over all co-ordinate charts, we have

$$\left\| \nabla^{k+p} u \right\|_{LL_a}^2 \leq C \left(\|u_0\|_{W^{p,2}(M^n)}^2 + \|u\|_{LW_a^{k+p-1}}^2 + \|g\|_{LL_a}^2 \right). \tag{7.16}$$

Starting from the energy estimate (7.10), then, and iterating (7.16) over $k = 0, 1, 2, \ldots, p$, this proves the lemma.

Once this much is settled, u is smooth enough that we no longer need represent $A_t u$ as an operator — indeed, we have $D_t u = g - A_t u$ pointwise almost everywhere in $M^n \times [0, \infty)$. This means in turn that $D_t u \in LL_a$ — well, this much we knew already — but with a concrete estimate. Combining this bound with that of the previous lemma, we have then

Lemma 7.12 *If u is a solution to (7.7), where $g \in P_a^0$ and $u_0 \in W^{p,2}(M^n)$, and if a is chosen larger than some constant which depends only on A_t, then $u \in P_a^1$, with the estimate*

$$\|u\|_{P_a^1}^2 \leq Q \left(\|u_0\|_{W^{p,2}(M^n)}^2 + \|g\|_{P_a^0}^2 \right).$$

The constant Q depends only on A_t and M^n.

Higher Regularity: The above proof made use only of the minimum of regularity needed to make sense of the defining equation. Now suppose that $g \in P_a^m$ and that $u_0 \in W^{2p(m+\frac{1}{2}),2}(M^n)$. In this case, we choose $\phi = \Delta_{-h}^{2mp+k}(\eta^{2p+2}\Delta_h^{2mp+k}u)$ as test function in (7.7) and proceed almost as before; the one difference is in the bound for the integral which comes from the forcing term. In this case, we exploit the greater smoothness of g:

$$\left\langle g, \, \Delta_{-h}^{2mp+k}(\eta^{2p+2}\Delta_h^{2mp+k}u) \right\rangle_{LL_a} \leq \|g\|_{LW_a^{2mp}} \cdot \|\eta^{2p+2}\Delta_h^{2mp+k}u\|_{LW_a^k}^2$$

and again, the term in u is subsumed by the ellipticity term $(\frac{1}{2}\lambda^p\|\eta^{2p+2}\nabla^p(\Delta_h^{2mp+k})\|_{LL_a}^2)$ provided $k \leq p$. In this case, then, we can iterate (7.16) as far as

$$\|u\|_{LW_a^{2p(m+1)}}^2 \leq Q \left(\|u_0\|_{W^{2p(m+\frac{1}{2}),2}(M^n)}^2 + \|g\|_{LW_a^{2mp}}^2 \right). \tag{7.17}$$

So much for the spatial regularity; now consider the time derivatives. Here again we argue incrementally. Assume that, for each $s \leq j$, $D_t^s u$ exists and is in the class $LW_a^{2p(m-s)}$. Since $D_t u = g - A_t u$ pointwise almost everywhere, this means that u is $(j+1)$ times differentiable in time; more to this, this gives us the estimate

$$\|D_t^{j+1}u\|_{LW_a^{2p(m-j)}}^2 \leq Q \left(\|D_t^j g\|_{LW_a^{2p(m-j)}}^2 + \sum_{l \leq j} \|D_t^l u\|_{LW_a^{2p(m-l)}}^2 \right).$$

Starting from (7.17) then, and iterating up to $j = m$, this proves

Lemma 7.13 *If u is a solution to (7.7), where now $g \in P_a^m$ and $u_0 \in W^{2p(m+\frac{1}{2}),2}(M^n)$, then $u \in P_a^{m+1}$, with the estimate*

$$\|u\|_{P_a^{m+1}}^2 \leq Q \left(\|u_0\|_{W^{2p(m+\frac{1}{2}),2}(M^n)}^2 + \|g\|_{P_a^m}^2 \right).$$

These estimates are the foundations for the theorem foreshadowed at the start of this section:

Theorem 7.14 *The map* $F : P_a^{m+1} \to W^{2p(m+\frac{1}{2}),2}(M^n) \times P_a^m$ *defined by*

$$F(u) = (u_0, g(u)) = (u_0, D_t u - A_t u) \tag{7.18}$$

is a Banach space isomorphism.

Proof. First we show that F is continuous. This much is clear for the second component; the interesting part is the continuity of the initial values.

Let $u \in P_a^{m+1}$. Then u has, at the very least, a weak time derivative which is in LL_a. This ensures that $\|u_t\|_{L^2(M^n)}$ is a Lipschitz function of time; in particular, u_t converges in L^2 to u_0 as $t \to 0$.

To prove u_0 is better than merely an L^2 function, we argue again with difference quotients. Recycling the notation used to derive the estimates above, we have

$$\left\| \eta^2 \Delta_h^{(2m+1)p} u_0 \right\|_{L^2(M^n)}^2 = 2a \left\| \eta^2 \Delta_h^{(2m+1)p} u \right\|_{LL_a}^2 - 2 \left\langle D_t \Delta_h^{(2m+1)p} u,\ \eta^2 \Delta_h^{(2m+1)p} u \right\rangle_{LL_a}$$

(which is simply equation (7.9) rearranged). The first term on the right is clearly controlled by the LW_a^{2mp+p}-norm of u, and in turn by the P_a^{m+1}-norm. To handle the second, we shift p difference operators from one factor in the inner product to the other, giving

$$\left\| \eta^2 \Delta_h^{(2m+1)p} u_0 \right\|_{L^2(M^n)}^2 \leq C \|u\|_{P_a^{m+1}}^2 - 2(-1)^p \left\langle D_t \Delta_h^{2mp} u,\ \Delta_{-h}^p(\eta^2 \Delta_h^{(2m+1)p} u) \right\rangle_{LL_a}.$$

The second term is now easily estimated using the Cauchy-Schwarz inequality: the first factor by the LW_a^{2mp}-norm of $D_t u$, the second by the LW_a^{2mp+2p}-norm of u itself. Again, both of these are contained within the P_a^{m+1} norm, so we have

$$\left\| \eta^2 \Delta_h^{(2m+1)p} u_0 \right\|_{L^2(M^n)}^2 \leq C \|u\|_{P_a^{m+1}}^2$$

Since this is independent of h, we may pass to the weak limit, and infer that u_0 has weak derivatives of order $(2m+1)p$, with the estimate

$$\|u_0\|_{W^{2p(m+\frac{1}{2}),2}(M^n)}^2 \leq C \|u\|_{P_a^{m+1}}^2,$$

and with this, the mapping $u \to u_0$ is continuous between the given spaces.

The next step is to show that F has an inverse. This means showing that the equation $Fu = (u_0, g)$ for given u_0 and g is uniquely solvable in the appropriate class. This follows from the existence theorem 7.9 together with the regularity theory above. The one remaining loophole is that the initial data in the existence theorem were assumed to be at least $W^{2p,2}$. We now close this.

Let $u_0 \in W^{p,2}(M^n)$, and let u_0^i be a sequence of smooth functions on M^n converging to u_0 in the $W^{p,2}$ norm. For each i, the existence result returns a function u^i in P_a^1 which solves the equation $Fu^i = (u_0^i, g)$. Because of the P_a^1 estimate above, these converge in the P_a^1 norm to a limit u. By the continuity of F, $Fu = (u_0, g)$ and we have our solution. The uniqueness claim is an immediate consequence of the P_a^1 estimate.

Lastly, we wish to show that the inverse of F is continuous. But this is exactly the content of the estimates above. ∎

7.4 The Quasilinear Equation

The isomorphism estimates of the previous section are the means by which we proceed from the linear equation to the quasilinear.

Consider the quasilinear problem of $2p$-th order

$$D_t u = Q(u) = A^{i_1 j_1 i_2 j_2 \ldots i_p j_p} D_{i_1 i_2 \ldots i_p j_1 \ldots j_p} u + b, \qquad u(\bullet, 0) = u_0, \qquad (7.19)$$

where the functions $A = A(x, u, \nabla u, \nabla^2 u, \ldots, \nabla^{2p-1} u)$ and $b = b(x, u, \nabla u, \ldots, \nabla^{2p-1} u)$ are smooth in all their arguments and where A is elliptic, at least in a neighbourhood of the initial data. By 'elliptic', we mean still that A factors as a power of a positive tensor of second rank.

Theorem 7.15 *For any smooth initial data u_0 for which $A(u_0)$ is ellipic, the quasilinear problem (7.19) has a smooth solution defined on some finite time interval $[0, T)$. The solution is unique and depends continuously on u_0.*

Proof. Define an operator $F : P_a^{m+1} \to W^{2p(m+\frac{1}{2}),2}(M^n) \times P_a^m$ thus:

$$F(u) = (u_0, D_t u - Q(u)).$$

F is Fréchet differentiable, and, by the results of the previous section, the linearised operator $DF(z)$ is an isomorphism provided $A(z_t)$ is elliptic for each t.

Let w be the solution to the 'frozen' linear problem,

$$D_t w - A(u_0) \cdot D^{2p} w = b(u_0), \qquad w_0 = u_0.$$

Since u_0 is smooth, the same is true of w, and in particular, w_t converges smoothly to u_0 as $t \to 0$. It follows that $A(w_t)$ is elliptic for sufficiently small times.

Now linearise F around w. To simplify the following argument, we shall assume that $A(w_t)$ is in fact elliptic for *all* times; if this is not the case by nature, we simply tamper with w outside some interval $0 \le t < \delta$. In the following analysis, we are in any case only concerned with the properties of w in a neighbourhood of $t = 0$.

The Implicit Function Theorem for mappings between Banach spaces (see, for instance, [15]) ensures that F is locally a diffeomorphism from a neighbourhood of w to a neighbourhood U of $F(w)$. However, $F(w) = (u_0, D_t w - Q(w))$, and hence, bearing in mind the definition of w,

$$F(w) = (u_0, (b(u_0) - b(w)) - (A(u_0) - A(w)) \cdot D^{2p} w),$$

and the second component here converges smoothly to zero for $t \to 0$. In particular, then, we can choose $q \in W^{(2m+1)p,2}(M^n) \times P_a^m$ approximating $F(w)$ with second component vanishing in some whole interval of time $[0, \epsilon)$. For appropriately small ϵ, this q will fall inside the neigbourhood U.

Since F is a local diffeomorphism, this means there is some $u \in P_a^{m+1}$ such that $F(u) = (u_0, q)$. In particular, then, the second component of $F(u)$ vanishes for small time: $D_t u = Q(u)$ for $t < \epsilon$. With that, we have a solution to the quasilinear problem in P_a^{m+1}.

However, we may conclude more. The initial values here need not be precisely u_0; they may be taken from some neighbourhood of u_0 in $W^{(2m+1)p,2}(M^n)$. So we have a solution for all nearby initial data; and more to this, the Implicit Function Theorem guarantees that the correspondence between solution and initial data is continuous in the appropriate norms.

All this does not yet suffice to prove the claim of the theorem. The argument above gives solutions of increasing smoothness, but possibly defined on intervals of decreasing duration. To prove the existence of a C^∞ solution on a definite time interval, we need a further property from the regularity theory.

Let u be a solution now to (7.19) in $[0, \epsilon)$ which is uniformly bounded in C^m. For m sufficiently large, this will imply that u is in fact absolutely smooth.

Precisely, we consider u as a fixed function now, and derive the evolution equation for $w_j = \nabla^j u$. If u is uniformly bounded in $C^{2p-1+\alpha}$, then this may be considered as a *linear* equation for w with coefficients which depend on u and its derivatives of up to $(2p-1)$-th order. These are therefore uniformly bounded in C^α. This limits the growth of w_j to exponential in time; in particular, each w_j remains bounded in the interval $[0, \epsilon)$, and this means that u is in fact C^∞ for $t < \epsilon$. This detail completes the proof of the theorem.

7.5 Short-time Existence for Geometric Equations

Now at last we are in a position to prove the existence of short-time solutions for a class of geometric evolution equations. Consider the deformation processes

$$\frac{\partial g}{\partial t} = \left(-(-\Delta)^p R + \phi(g, \text{Riem}, \nabla \text{Riem}, \nabla^2 \text{Riem}, \ldots, \nabla^{2p-1} \text{Riem})\right) g \qquad (7.20)$$

for metrics, and

$$\frac{\partial F}{\partial t} = \left(-(-\Delta)^p H + \phi(F, n, A, \nabla A, \nabla^2 A, \ldots, \nabla^{2p-1} A)\right) n \qquad (7.21)$$

for immersions, where ϕ is in each case smooth in all its arguments, but otherwise arbitrary. This is the structure of the total-curvature problems considered in subsequent chapters; it also appears in a number of equations unrelated to variational problems. The cases $n = 1, \phi = 0$ in the above equations fall into this last category; they arise in general relativity [13] and crystal formation [10].

The derived equations of lemmas 7.3 and 2.1.6 describe the change in the curvature which follows from the above deformations. To highest order,

$$\frac{\partial R}{\partial t} = -(-\Delta)^{p+1} R + \cdots, \qquad \frac{\partial H}{\partial t} = -(-\Delta)^{p+1} H + \cdots$$

and these are parabolic, in the sense that the linearisations are parabolic, even using the very restrictive notion of ellipticity introduced in section 2.2.

It is not surprising, then, that (7.20) and (7.21) themselves are, properly interpreted, also parabolic equations. In this section, we show how they may be reduced to strictly parabolic, quasilinear scalar equations, and hence that they have short-time solutions under appropriate initial conditions.

Theorem 7.16 *Let M^n ($n \geq 2$) be a smooth manifold with metric g_0. There exists a solution to (7.20) defined for some period of time $0 \leq t < T$ which takes g_0 as its initial values. It is unique.*

Proof. If there is a solution at all, it may be represented as in Lemma 7.1 in the form

$$g_t(x) = \exp 2u(x,t) \cdot g_0(x)$$

where the conformal factor u evolves according to

$$\frac{\partial u}{\partial t} = \frac{1}{2} \left(-(-\Delta)^p R + \phi(g, \text{Riem}, \nabla\text{Riem}, \nabla^2\text{Riem}, \dots, \nabla^{2p-1}\text{Riem}) \right). \tag{7.22}$$

Lemma 7.2 shows how the Laplace operator and Ricci curvature of g_t may be expressed in terms of those of g_0 and the conformal factor. The same is clearly true of the Riemann curvature of g and its covariant derivatives. So one can rewrite (7) in the form

$$\frac{\partial u}{\partial t} = -(p-1)\exp(-2(p+1)u) \cdot (-\Delta^0)^{p+1}u + \phi\left(x, u, \nabla^0 u, (\nabla^0)^2 u, \dots, (\nabla^0)^{2p+1}u\right),$$

where ϕ is smooth in all its arguments. This, however, is now a quasilinear scalar equation on (M^n, g_0); it is parabolic in the sense of Theorem 7.15, and it therefore has a unique solution with initial values $u_0 \equiv 0$ for some period of time $0 \leq t < T$. Starting from this solution, we now take (7.22) as the *definition* of g_t and this gives a solution to the geometric problem.

When it comes to hypersurfaces, life is not quite so simple. In this case, the reduction to a quasilinear problem is not as clear-cut. The idea is to represent the the hypersurface at time t as a graph in Fermi co-ordinates over the initial hypersurface and consider the deformation process as a quasilinear scalar equation for the height function.

Consider again an isometric immersion $F_0 : (M^n, g) \hookrightarrow (N^{n+1}, \bar{g})$. Let $M = M^n \times (-\epsilon, \epsilon)$, where ϵ is chosen so small that the map

$$F : M \to N^{n+1} : (x, h) \to \exp_{F_0(x)}(hn(x))$$

is itself an immersion. If F_0 is an embedding, then F is simply the map which generates Fermi co-ordinates on a tubular neighbourhood of $F_0(M^n)$. The tangent space to M is spanned by the vectors $\frac{\partial}{\partial x^i}$, $i = 1, 2, \dots n$ and $\frac{\partial}{\partial h}$.

Let G be the metric induced on M by F. It follows from the Gauß Lemma that the geodesics $\{x = \text{constant}\}$ are orthogonal to the parallel surfaces $\{h = \text{constant}\}$, and hence that G may be broken into a sum:

$$G(x, h) = G_h(x) + dh \otimes dh,$$

where G_h is the metric on the surface $M^n \times \{h\}$. In particular, $G_0 = F_0^* \bar{g} = g$.

Now let $u : M^n \to (-\epsilon, \epsilon)$ be a smooth function, and consider the graph of u in M, parametrised by $\psi : M^n \to M : x \to (x, u(x))$. The tangent space to graph u, considered as a submanifold of M, is spanned at a point $\psi(x)$ by the vectors

$$\psi_* \frac{\partial}{\partial x^i} = \frac{\partial \psi}{\partial x^i} = \frac{\partial}{\partial x^i} + \frac{\partial u}{\partial x^i} \frac{\partial}{\partial h},$$

and the induced metric M^n inherits from graph u is given by

$$\gamma_{ij}(x) = \psi^* G\left(\frac{\partial}{\partial x^i}, \frac{\partial}{\partial x^j}\right) = G\left(\frac{\partial \psi}{\partial x^i}, \frac{\partial \psi}{\partial x^j}\right) = (G_{u(x)})_{ij} + \frac{\partial u}{\partial x^i} \frac{\partial u}{\partial x^j}.$$

The normal to graph u — one sees this simply by computing the product with each of the basis vectors above — is given by

$$n_{\text{graph } u}(x) = \frac{1}{N} \left\{ \frac{\partial}{\partial h} - G_{u(x)}^{pq} \frac{\partial u}{\partial x^p} \frac{\partial}{\partial x^q} \right\},$$

where $N = \left\{1 + G_{u(x)}(\nabla^{G_u} u, \nabla^{G_u} u)\right\}^{\frac{1}{2}}$. From this, we derive an expression for the second fundamental form of the graph:

$$
\begin{aligned}
h_{ij}(x) &= G\left(\nabla^M_{\frac{\partial}{\partial x^i} \psi} n_{\text{graph } u}, \frac{\partial \psi}{\partial x^j}\right) = G\left(\nabla^M_{\frac{\partial}{\partial x^i} \psi}\left(\frac{1}{N}\left\{\frac{\partial}{\partial h} - G_{u(x)}^{pq} \frac{\partial u}{\partial x^p} \frac{\partial}{\partial x^q}\right\}\right), \frac{\partial \psi}{\partial x^j}\right) \\
&= -\frac{1}{N} G\left(\nabla^M_{\frac{\partial}{\partial x^i} \psi}\left\{G_{u(x)}^{pq} \frac{\partial u}{\partial x^p} \frac{\partial}{\partial x^q}\right\}, \frac{\partial \psi}{\partial x^j}\right) \\
&= -\frac{1}{N} \frac{\partial^2 u}{\partial x^i \partial x^j} + \text{terms in } x, u \text{ and } \nabla u \\
&= -\frac{1}{N} \nabla_i^{G_0} \nabla_j^{G_0} u + \sigma_{ij}(x, u, \nabla u)
\end{aligned}
$$

and in the last equation, σ is a smooth tensor field. Its exact form need not concern us. Tracing with the metric on graph u, this yields the mean curvature:

$$H = -\frac{1}{N} \gamma^{ij} \nabla_i^{G_0} \nabla_j^{G_0} u + (\gamma^{ij} \sigma_{ij})(x, u, \nabla u), \qquad (7.23)$$

nothing here that γ^{ij} too depends only on x, and u and its first derivatives.
Since γ depends on ∇u, its Christoffel symbols depend in turn upon derivatives of u of second order. The same is then true of covariant derivatives on graph u; while q-th covariant derivatives depend on derivatives of u up to and including the $(q+1)$-th.
In particular, if f is a function graph $u \to \mathbb{R}$, then

$$\Delta^\gamma f = \gamma^{ij} \nabla_i^{G_0} \nabla_j^{G_0} f + \text{terms in } x, u, \cdots, \nabla^3 u, f \text{ and } \nabla f. \qquad (7.24)$$

Thus, the leading term in the mean curvature of graph u could be rewritten (and more naturally so) in terms of the laplacian on graph u; this, however, is precisely what we *don't* want for the purposes of the short-time existence theory. We wish to repackage our equations for a fixed manifold with a fixed metric.

It follows from (7.23), (7.23) and (7.24), together with the preceding remarks about covariant derivatives, that the instantaneous speed of graph u under (7.21) is given by

$$(-1)^p \frac{1}{N} \gamma^{i_1 j_1} \gamma^{i_2 j_2} \cdots \gamma^{i_p+1 j_p+1} \nabla^{Go}_{i_1} \nabla^{Go}_{j_1} \cdots \nabla^{Go}_{i_{p+1}} \nabla^{Go}_{j_{p+1}} u + \tilde{\phi}(x, u, \nabla u, \ldots, \nabla^{2p+1} u),$$

where $\tilde{\phi}$ is another function which is smooth in all its arguments but whose precise form is left unstated.

Theorem 7.17 *For any smooth hypersurface immersion $F_0 : M^n \to N^{n+1}$, there exists a unique solution to the flow problem (7.21) defined on some interval $0 \le t < T$ and taking F_0 as its initial values.*

Proof. The quasilinear equation

$$\frac{\partial u}{\partial t} = (-1)^n \gamma^{i_1 j_1} \gamma^{i_2 j_2} \cdots \gamma^{i_{n+1} j_{n+1}} \nabla^{Go}_{i_1} \nabla^{Go}_{j_1} \cdots \nabla^{Go}_{i_{n+1}} \nabla^{Go}_{j_{n+1}} u + N \tilde{\phi}(x, u, \nabla u, \ldots, \nabla^{2n+1} u)$$

is parabolic at $u = 0$, and therefore has a unique smooth solution with zero initial values defined for some period of time $0 \le t < T$.

Consider now the family of surfaces graph u_t, parametrised by $\psi_t(x) = (x, u(x, t))$. The initial surface $u \equiv 0$ is simply M^n itself; and the family develops according to the equation

$$\frac{\partial \psi}{\partial t} = \frac{\partial u}{\partial t} \cdot \frac{\partial}{\partial h}.$$

In view of the preceding discussion, $\frac{\partial}{\partial t} u$ is the speed graph u_t would have under the flow (7.21), rescaled by a factor of N. So, ψ is a solution to the geometric equation

$$\frac{\partial \psi}{\partial t}(x) = N \omega \cdot \frac{\partial}{\partial h},$$

where ω is the speed in (7.21).

However, decomposing $\frac{\partial}{\partial h}$ into tangential and normal parts,

$$\frac{\partial}{\partial h} = \frac{1}{N} n_{\text{graph}\, u_t} + \lambda$$

where $\lambda(x)$ is a tangential vector field. Since ψ is smooth, so is λ. Thus, the evolution of ψ can be rewritten,

$$\frac{\partial \psi}{\partial t}(x) = \omega \cdot n_{\text{graph}\, u_t}(x) + N\lambda,$$

and so ψ is a solution to the original geometric problem, give or take a tangential motion. Tangential motion has no effect on the solution hypersurfaces, considered as sets; it affects only their parametrisations. In crude terms, it is the movement of a bicycle chain: each individual link is in motion, but the shape of the chain taken as a whole does not change.

Let $\alpha : M^n \times [0, T) \to M^n$ be a smoothly varying family of diffeomorphisms of M^n, with α_0 being the identity map and

$$\frac{\partial \alpha}{\partial t} = -\psi_i^* \lambda.$$

Since λ and ψ are completely smooth up to time T, this may be integrated directly to a unique solution for α. Now define

$$\Psi(x, t) = \psi(\alpha(x, t), t).$$

Then,

$$
\begin{aligned}
\frac{\partial \Psi}{\partial t}(x, t) &= \frac{\partial \psi}{\partial t}(\alpha(x, t), t) + \psi_*(\frac{\partial \alpha}{\partial t}(x, t), t) \\
&= (\omega \cdot n(\Psi(x, t)) + \lambda(\Psi(x, t))) - \lambda(\Psi(x, t)),
\end{aligned}
$$

and so Ψ now solves the geometric problem, at least in M, and because F is an isometry, $F \circ \Psi$ now satisfies the original geometric equation in N^{n+1}.

These two theorems provide the short-time existence results for the flows which appear in later chapters; as another special case, Theorem 7.17 incorporates the L^2-gradient flow for the Willmore energy of a surface immersed in a Riemannian manifold.

References

[1] B. Andrews, *Contraction of convex hypersurfaces in Euclidean space*, Calc. Var. **2** (1994), 151–171.

[2] B. Andrews, *Gauss curvature flow: The fate of the rolling stones*, preprint ANU Canberra (1998), pp10.

[3] B. Andrews, *Contraction of convex hypersurfaces by their affine normal*, J. Diff. Geom. **43** (1996), 207–230.

[4] B. Andrews, *Contraction of convex hypersurfaces in Riemannian spaces*, J. Diff. Geom. **39** (1994), 407–431.

[5] B. Andrews, *Monotone quantities and unique limits for evolving convex hypersurfaces*, IMRN **20** (1997), 1001–1031.

[6] B. Andrews, private communication.

[7] S.B. Angenent, J.J.L. Velazques, *Degenerate neckpinches in mean curvature flow*, J. Reine Angew. Math. **482** (1997), 15–66.

[8] T. Aubin, *Nonlinear Analysis on Manifolds. Monge-Ampère Equations*, Springer-Verlag, New York.

[9] K. A. Brakke, *The Motion of a Surface by its Mean Curvature*, Mathematical Notes, Princeton University Press, Princeton.

[10] J. W. Cahn, W. C. Carter, A. R. Roosen and J. E. Taylor, *Shape Evolution by Surface Diffusion and Surface Attachment Limited Kinetics on Completely Faceted Surfaces* available over *http://www.ctcms.nist.gov/~roosen/SD_SALK/*.

[11] Y.G. Chen, Y. Giga, and S. Goto, *Uniqueness and Existence of Viscosity Solutions of Generalized Mean Curvature Flow Equations*, J Diff.Geom. **33** (1991), 749–786.

[12] B. Chow, *Deforming convex hypersurfaces by the nth root of the Guassian curvature*, J.Diff.Geom. **23** (1985), 117–138.

[13] P. T. Chruściel, *On Robinson-Trautman Space-Times,* Centre for Mathematical Analysis Research Report CMA-R23-90, The Australian National University, Canberra.

[14] D.M.DeTurck, *Deforming metrics in direction of their Ricci tensors*, J.Diff.Geom. **18** (1983), 157–162.

[15] J. Dieudonné, *Foundations of Modern Analysis*, Academic Press, New York and London.

[16] K. Ecker, G. Huisken, *Interior estimates for hypersurfaces moving by mean curvature*, Invent.Math. **105** (1991), 547–569.

[17] K. Ecker, G. Huisken, *Mean curvature evolution of entire graphs*, Ann.of Math. **130** (1989), 453–471.

[18] S. D. Eidel'man, *Parabolic Equations*, in *Partial Differential Equations VI*, Encyclopaedia of Mathematical Sciences, Volume 63, editors Yu. V. Egerov and M. A. Shubin, Springer-Verlag, Berlin, Heidelberg, New York.

[19] L. C. Evans, J. Spruck, *Motion of Level Sets by Mean Curvature I*, J. Diff.Geom. **33** (1991), 635–681.

[20] W.J. Firey, *Shapes of worn stones*, Mathematica **21** (1974), 1–11.

[21] Avner F. Friedman, *Partial Differential Equations of Parabolic Type,* Prentice-Hall, Englewood Cliffs, N.J.

[22] M. E. Gage and R. S. Hamilton, *The Heat Equation Shrinking Convex Plane Curves*, J. Diff. Geom. **23**, 285–314.

[23] C. Gerhardt, *Flow of nonconvex hypersurfaces into spheres*, J.Diff.Geom. **32** (1990), 299–314.

[24] R. Geroch, *Energy Extraction*, Ann. New York Acad. Sci. **224** (1973), 108–17.

[25] M. Grayson, *The heat equation shrinks embedded plane curves to points*, J. Diff. Geom. **26** (1987), 285–314.

[26] M. Grayson, *Shortening embedded curves*, Annals Math. **129** (1989), 71–111.

[27] R. S. Hamilton, *The Ricci Flow on Surfaces*, Contemporary Mathematics **71**, 237–262.

[28] R. S. Hamilton, *The formation of singularities in the Ricci Flow*, Surveys in Differential Geometry Vol. II, International Press, Cambridge MA (1993), 7–136.

[29] R. S. Hamilton, *Harnack estimate for the mean curvature flow*, J. Diff. Geom. **41** (1995), 215–226.

[30] R. S. Hamilton, *Three–manifolds with positive Ricci curvature*, J. Diff. Geom. **17** (1982), 255–306.

[31] R. S. Hamilton, *Four manifolds with positive isotropic curvature*, Comm. Anal. Geom. **5** (1997), 1–92.

[32] R. S. Hamilton, *Isoperimetric estimates for the curve shrinking flow in the plane*, Modern Methods in Compl. Anal., Princeton Univ.Press (1992), 201-222.

[33] R. S. Hamilton, *CBMS Conference Notes*, Hawaii.

[34] G. Huisken, *Flow by mean curvature of convex surfaces into spheres*, J. Diff. Geometry **20** (1984), 237–266.

[35] G. Huisken, *Contracting convex hypersurfaces in Riemannian manifolds by their mean curvature*, Invent. Math. **84** (1986), 463–480.

[36] G. Huisken, *Asymptotic behaviour for singularities of the mean curvature flow*, J. Diff. Geometry **31** (1990), 285–299.

[37] G. Huisken, *Local and global behaviour of hypersurfaces moving by mean curvature*, Proceedings of Symposia in Pure Mathematics **54** (1993), 175–191.

[38] G. Huisken, *A distance comparison principle for evolving curves*, Asian J. Math. **2** (1998), 127–134.

[39] G. Huisken, T. Ilmanen, *The inverse mean curvature flow and the Riemannian Penrose inequality*, preprint http://poincare.mathematik.uni-tuebingen.de, to appear.

[40] G. Huisken, T. Ilmanen, *The Riemannian Penrose inequality*, IMRN **20** (1997), 1045-1058.

[41] G. Huisken, C. Sinestrari, *Mean curvature flow singularities for mean convex surfaces*, Calc. Variations, to appear.

[42] G. Huisken, C. Sinestrari, *Convexity estimates for mean curvature flow and singularities for mean convex surfaces*, preprint, to appear.

[43] T. Ilmanen, *Elliptic regularisation and partial regularity for motion by mean curvature*, Memoirs AMS **108** (1994), pp90.

[44] J. L. Lions, *Sur les problèmes mixtes pour certaines systèmes paraboliques dans des ouverts non cylindriques*, Ann. l'Insitute Fourier **7**, 143–182.

[45] W. W. Mullins, *Two-dimensional Motion of Idealised Grain Boundaries*, J. Appl. Phys. **27**/8 (August 1956), 900–904.

[46] G. Sapiro, A. Tannenbaum, *On affine plane curve evolution*, J. Funct. Anal. **119** (1994), 79–120.

[47] R. Schoen, L. Simon, S. T. Yau, *Curvature estimates for minimal hypersurfaces*, Acta Math. **134** (1975), 276–288.

[48] J. Simons, *Minimal varieties in Riemannian manifolds*, Ann.of Math. **88** (1968), 62–105.

[49] K. Smoczyk, *Starshaped hypersurfaces and the mean curvature flow*, Preprint (1997), 13pp.

[50] V. A. Solonnikov, *Green matrices for parabolic boundary value problems*, Zap. Nauch. Sem. Leningr. Otd. Math. Inst. Steklova **14** 256–287, translated in Semin. Math. Steklova Math. Inst. Leningrad (1972), 109–121.

[51] F. Trèves, *Relations de domination entre opérateurs différentiels*, Acta. Math. **101**, 1–139.

[52] K. Tso, *Deforming a hypersurface by its Gauss–Kronecker curvature*, Comm. Pure Appl. Math. **38** (1985), 867–882.

[53] J. Urbas, *On the Expansion of Starshaped Hypersurfaces by Symmetric Functions of Their Principal Curvatures*, Math. Z. **205** (1990), 355–372.

[54] B. White, *Partial Regularity of Mean Convex Hypersurfaces Flowing by Mean Curvature*, IMRN **4** (1994), 185–192.

Variational models for microstructure and phase transitions

Stefan Müller

Contents

1 Setting of the problem

1.1 What are microstructures?

For the purpose of these lectures, a microstructure is any structure on a scale between the macroscopic scale (on which we usually make observations) and the atomic scale. Such structures are abundant in nature: the fine hierarchical structures in a leaf and many other biological materials, the complex arrangements of fissures, cracks, voids and inclusions in rock or soil, fine scale mixing patterns in turbulent or multiphase flow, man-made layered or fibre-reinforced materials and fine phase mixtures in solid-solid phase transformations, to quote but a few examples. The microstructure influences in a crucial way the macroscopic behaviour of the material or system and is often chosen (or spontaneously generated) to optimize its performance (maximum strength at given weight, minimal energy, maximal entropy, maximal or minimal permeability, ...). Microstructures often develop on many different scales in space and time, and to understand the formation, interaction, and overall effect of these structures is a great scientific challenge, weather modelling providing an illustrative example. In the applied literature the passage from microscales to macroscales is frequently achieved by clever ad hoc "averaging" or "renormalization". A good mathematical framework in which these procedures could be justified and systematically improved is often lacking, and its development would be a difficult, but very rewarding, task.

The mathematical analysis of microstructures usually neglects the atomic scale by considering a continuum model from the outset. The issue is then to understand scales that are small (or converge to zero) compared to the fixed macroscopic scale. Research has mostly focused on three areas: homogenization, variational models of microstructure and optimal design which lies between the two first areas as the optimal structure often corresponds to a homogenization limit. The basic problem in homogenization is to determine the macroscopic behaviour (or at least bounds on it) induced by a given microstructure (given for example by a periodic mixture of two heat conductors in the limit of vanishing period, by a weakly convergent sequence of conductivity tensors or by statistical information). Variational models of microstructures try to model systems which spontaneously form internal microstructure by assuming that the structure formed has a certain optimality property. The reason for the formation of such microstructure is typically that no exact optimum exists and optimizing sequences have to develop finer and finer oscillations (which may only be limited by effects neglected in the model, such as the underlying atomic structure). An important task is to extract the relevant features of minimizing sequences. Young measures, which are discussed in Section 3 below, are one possibility to do this, but by no means the only one.

In these lectures I will focus on variational models for microstructures that arise from solid-solid phase transition in certain elastic crystals (usually alloys, such as In-Th, Cu-Al-Ni, Ni-Ti). These materials display a fascinating variety of microstructures (see Fig. 1) which is closely linked to unusual and technologically interesting material behaviour (shape memory effect, pseudoelasticity). A mathematical model for elastic crystals will be introduced in Section 1.3 below. Before doing this let us briefly review the relation between microstructure and energy minimization in more detail in some simple examples.

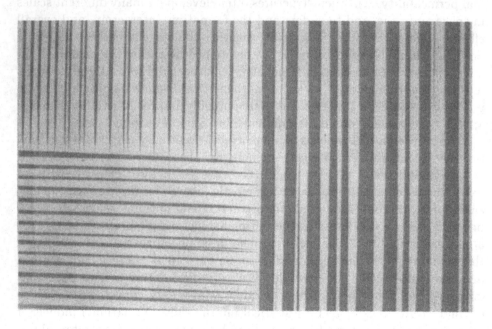

Figure 1: Microstructure in a Cu-Al-Ni single crystal; the imaged area is approximately 2 mm × 3 mm (courtesy of C. Chu and R.D. James, University of Minnesota)

1.2 Microstructures as energy minimizers

Example 1: Consider the problem:

$$\text{Minimize} \quad \int_0^1 (u_x^2 - 1)^2 \, dx$$

subject to

$$u(0) = u(1) = 0.$$

The minimum is attained but the set of minimizers is highly degenerate. Every Lipschitz function whose slopes are ± 1 almost everywhere and that attains the boundary values is a minimizer. In particular the weak$*$ closure in $W^{1,\infty}$ of the set of minimizers consists of all functions with Lipschitz constant less then or equal to one that are bounded by $\pm \min(x, 1 - x)$.

Example 2 (Bolza, L.C. Young): Consider the problem:

$$\text{Minimize} \quad I(u) := \int_0^1 (u_x^2 - 1)^2 + u^2 \, dx$$

subject to

$$u(0) = u(1) = 0.$$

The infimum of the functional is zero since there exist rapidly oscillating functions with slope ± 1 whose supremum is arbitrarily small. Indeed if s denotes the periodic extension of the sawtooth function

$$s(x) = \begin{cases} x & \text{on } [0, 1/4) \\ 1/2 - x & \text{on } [1/4, 3/4) \\ x - 1 & \text{on } [3/4, 1) \end{cases} \tag{1.1}$$

then $u_j(x) := j^{-1} s(jx)$ satisfy $I(u_j) \to 0$ as $j \to \infty$. The infimum cannot be attained since there is no function that satisfies simultaneously $u \equiv 0$ and $u_x = \pm 1$ almost everywhere. Minimizing sequences must oscillate and converge weakly (in the Sobolev space $W^{1,4}(0, 1)$), but not strongly, to zero.

This provides a first example how minimization can lead to fine scale oscillation or microstructure. The failure of classical minimization was investigated by L.C. Young in the 1930's in the context of optimal control. It led him to the introduction of generalized measure-valued solutions (see Section 3 below on Young measures). His book [Yo 69] describes various interesting situations where generalized solutions naturally arise, including applications to sailing and the construction of railway tracks.

Example 3: Let $\Omega = [0, L] \times [0, 1]$ be a rectangle and consider the problem:

$$\text{Minimize} \quad J(u) = \int_\Omega u_x^2 + (u_y^2 - 1)^2 \, dx \, dy$$

subject to

$$u = 0 \quad \text{on} \quad \partial\Omega.$$

Clearly $J(u) > 0$ since otherwise $u_x = 0$ almost everywhere, whence $u \equiv 0$ on Ω and $(u_y^2 - 1)^2 = 1$. On the other hand the infimum of J is zero. One way to see this is to consider the sawtooth function s given by (1.1), to define

$$u(x, y) = j^{-1} s(jy) \quad \text{for } \delta < x < L - \delta,$$

and to use linear interpolation to achieve the boundary values at $x = 0$ and $x = L$. Considering first the limit $j \to \infty$ and then $\delta \to 0$ one obtains inf $J = 0$. As in the Example 2 no (classical) minimizers exist and minimizing sequences must develop rapid oscillations.

Two questions arise from the consideration of these examples.

Question 1: Are there special minimizers or minimizing sequences? Are, for example, the maximal solutions $\pm \min(x, 1-x)$ in Example 1 in a certain way preferred minimizers?

Question 2: Are there certain common features of all minimizing sequences?

1.3 Variational models for elastic crystals

The basic idea is to model the elastic crystal as a nonlinearly elastic continuum. The crystalline structure enters in this approach through the symmetry properties of the stored-energy function. The (usually stress-free) reference configuration of the crystal is identified with a bounded domain $\Omega \subset \mathbf{R}^3$. A deformation $u : \Omega \to \mathbf{R}^3$ of the crystal requires an elastic energy

$$I(u) = \int_\Omega W(Du)\, dx, \qquad (1.2)$$

where $W : M^{m \times m} \to \mathbf{R}$ is the stored-energy density function that describes the properties of the material. Under the Cauchy-Born rule $W(F)$ is given by the (free) energy per unit volume that is required for an affine deformation $x \mapsto Fx$ of the crystal lattice.

The stored energy is invariant under rotations in the ambient space and under the action of the isotropy group $\mathcal{P}$ of the crystal lattice which usually is a discrete subgroup of $SO(3)$. Thus

$$W(QF) = W(F) \qquad \forall Q \in SO(3), \qquad (1.3)$$
$$W(FP) = W(F) \qquad \forall P \in \mathcal{P} \subset SO(3). \qquad (1.4)$$

Instead of the compact group $\mathcal{P}$ one could also consider the larger noncompact group of all lattice invariant transformation which is conjugate to the group $GL(3, \mathbf{Z})$. This leads to a highly degenerate situation and in particular such an invariance implies (in connection with the consideration of global rather than local mimimizers) that the material has no macroscopic shear resistance. We will thus use the point group and refer to [BJ 87, BJ 92], [CK 88], [Er 77, Er 79, Er 80, Er 84, Er 89], [Fo 87], [Pa 77, Pa 81], [Pi 84], [Za 92] for further discussion of this point.

The stored-energy also depends on temperature but we will always assume that the temperature is constant throughout the crystal and thus surpress this dependence.

The basic assumption of the variational approach to microstructure is:

> The observed microstructures correspond to minimizers or almost minimizers of the elastic energy I.

It is convenient to normalize W so that $\min W = 0$. The set $K = W^{-1}(0)$ then corresponds to the zero energy affine deformations of the crystal lattices. Experimentally it is often observed that microstructures do not only minimize the integral I (subject to suitable boundary conditions) but in fact minimize the integrand pointwise. We are thus led to consider the simpler problem:

Determine (Lipschitz) maps that satisfy exactly
or approximately $Du \in K$.

The difference in behaviour of different materials is thus closely related to the set K which depends on the material and temperature. For ordinary materials K is (conjugate to) $SO(3)$ (the smallest set compatible with the rotation invariance) while for material forming microstructures K consists of several copies of $SO(3)$. The Cu-Al-Ni alloy for which the microstructures in Fig.1 were observed undergoes a solid-solid phase transition at a critical temperature T_c, i.e. the preferred crystal structure, and hence the set K changes at T_c.

	$T > T_c$		$T < T_c$
		phase transition	
crystal structure	cubic	$\longrightarrow$	orthorhombic
K	$SO(3)$		$SO(3)U_1 \cup \ldots SO(3)U_6$
			$U_1 = \begin{pmatrix} \alpha & 0 & 0 \\ 0 & \beta & \gamma \\ 0 & \gamma & \beta \end{pmatrix}$
micro-structure	none (see Section 2.3)		large variety observed (see Section 2.2)

Figure 2: A cubic to orthorhombic transition

In view of (1.4) the matrices U_i are related by conjugation under the cubic group.

1.4 The basic problems

We slightly generalize the setting of the previous section and consider maps $u : \Omega \subset \mathbf{R}^n \to \mathbf{R}^m$ on a bounded domain Ω (with Lipschitz boundary if needed). In particular the Sobolev space $W^{1,\infty}$ agrees with the class of Lipschitz maps. Let $K \subset M^{m \times n}$ be a compact set in the space $M^{m \times n}$ of $m \times n$ matrices.

Problem 1 (exact solutions): Characterize all Lipschitz maps u that satisfy

$$Du \in K \quad \text{a.e. in } \Omega.$$

Problem 2 (approximate solutions): Characterize all sequences u_j of Lipschitz functions with uniformly bounded Lipschitz constant such that

$$\text{dist}\,(Du_j, K) \to 0 \quad \text{a.e. in } \Omega.$$

Problem 3 (relaxation of K): Determine the sets K^{ex} and $K^{app} \subset M^{m \times n}$ of all affine maps $x \mapsto Fx$ such that Problem 1 and 2 have a solution that satisfies

$$u(x) = Fx \text{ on } \partial\Omega,$$
$$u_j(x) = Fx \text{ on } \partial\Omega,$$

respectively.

Problems 1-3 also arise in many other contexts, e.g. in the theory of isometric immersions. An important technical difference is that in geometric problems one is often interested in connected sets K (and hence C^1 solutions u) while we will usually consider sets with more than one component. For further information we refer to Gromov's treatise [Gr 86] and to Šverák's ICM lecture [Sv 95].

In the context of crystal microstructure discussed in the previous section the sets K^{ex} and K^{app} in Problem 3 have an important interpretation. They consist of the affine *macroscopic* deformations of the crystal with (almost) zero energy. They trivially contain the set K of microscopic zero energy deformations but can be much larger. For the set $K = SO(2)A \cup SO(2)B$ one obtains (see Section 4.5) that under suitable conditions on A and B the sets K^{app} and K^{ex} contain an open set (relative to the constraint $\det F = 1$), leading to fluid-like behaviour.

Problem 4: Find an efficient description of approximating sequences that eliminates nonuniqueness due to trivial modifications while keeping the relevant "macroscopic" features.

We saw in Section 1.2 how failure of minimization can lead to "infinitely fine" microstructure. In practice crystal microstructures always arise on some finite scale (albeit on a wide range from a few atomic distances to $10 - 100 \ \mu m$). Minimization of elastic energy alone may not be enough to explain this since there is no natural scale in the theory.

Problem 5: Explain the length scale and the fine geometry of the microstructure, possibly by including other contributions to the energy, such as interfacial energy.

Another possible explanation for limited fineness is that infinitely fine mixtures are (generalized) energy minimizers but not accessible by the natural dynamics of the system. This is a very important issue, but we can only touch briefly on it in these notes and refer to Section 7.2 and the references quoted there.

2   Examples

It is instructive to look at some examples before studying a general theory related to Problems 1-3. These simple examples already show a rich variety of phenomena and interesting connections with (nonlinear) elliptic regularity, functional analytic properties of minors and quasiconformal geometry. In the following K always denotes a subset of the space $M^{m \times n}$ of $m \times n$ matrices, $m, n \geq 2$.

2.1   The two-gradient problem

Exact solutions: Let $K = \{A, B\}$. The simplest solutions of the relation

$$Du \in K$$

are so called simple laminates, i.e. maps for which Du is constant in alternating bands that are bounded by hyperplanes $x \cdot n = \text{const}$ (see Fig. 3). Tangential continuity of u at these interfaces enforces that $A\tau = B\tau$ for vec-

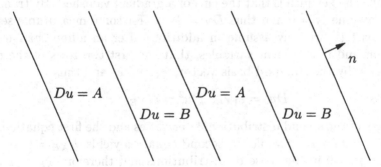

Figure 3: A simple laminate

tors τ perpendicular to n and thus $A - B$ has rank one and can be written as

$$B - A = a \otimes n.$$

In this case we say that A and B are rank-1 connected. We recall that the matrix $a \otimes n$ has entries $(a \otimes n)_{ij} = a_i n_j$. If one assumes that the interfaces between the regions $\{Du = A\}$ and $\{Du = B\}$ are smooth then a similar argument shows that they must be hyperplanes with fixed normal n. Moreover no such smooth arrangement is possible if $\mathrm{rk}(B - A) \geq 2$. The following proposition gives a much stronger statement because it shows that also among possibly very irregular maps there are no other solutions.

Proposition 2.1 *([BJ 87], Prop.1) Let Ω be a domain in $\mathbf{R}^n$ and let $u : \Omega \to \mathbf{R}^m$ be a Lipschitz map with $Du \in \{A, B\}$ a.e.*

(i) *If $\mathrm{rk}(B - A) \geq 2$, then $Du = A$ a.e. or $Du = B$ a.e.;*

(ii) *if $B - A = a \otimes n$ then u can locally be written in the form*

$$u(x) = Ax + ah(x \cdot n) + \mathrm{const}$$

where h is Lipschitz and $h' \in \{0, 1\}$ a.e. If Ω is convex this representation holds globally.

In particular, Du is constant if u satisfies an affine boundary condition $u(x) = Fx$ on $\partial\Omega$.

Proof. The key idea is that the curl of a gradient vanishes. By translation we may assume $A = 0$ and thus $Du = B\chi_E$, for some measurable set $E \subset \Omega$. For part (i) we may assume in addition, after an affine change of the dependent and independent variables, that the first two rows of the matrix B are given by the standard basis vectors e_1 and e_2 and thus

$$Du^1 = e_1\chi_E, \quad Du^2 = e_2\chi_E.$$

Symmetry of the second distributional derivaties and the first equation imply that $\partial_j\chi_E = 0$ for $j \neq 1$ while the second equation yields $\partial_k\chi_E = 0$ for $k \neq 2$. Hence $D\chi_E = 0$ in the sense of distributions and therefore $\chi_E = 1$ a.e. or $\chi_E = 0$ a.e. since Ω is connected.

To prove part (ii) we may assume $A = 0$, $a = n = e_1$ and thus $Du^1 = e_1\chi_B$, $Du^k = 0$, $k = 2, \ldots, m$. Hence $u^2, \ldots, u^m$ are constant and $\partial_k u^1 = 0$, for $k = 2, \ldots, m$. Therefore u^1 is locally only a function of x^1 as claimed. If Ω is convex then u^1 is constant on the hyperplanes $x^1 = \mathrm{const}$ that intersect Ω and thus globally of the desired form.

Finally if $u = Fx$ on $\partial\Omega$, then $F = (1 - \lambda)B$, $\lambda \in [0, 1]$ since by the Gauss-Green theorem

$$|E| \, B = \int_\Omega Du \, dx = \int_{\partial\Omega} u \otimes n \, d\mathcal{H}^{n-1} = \int_\Omega F \, dx,$$

where n is the outer normal of Ω. Extending u by Fx on $\mathbf{R}^n \setminus \Omega$ we can argue as in the proof of (ii) to deduce $u(x) = Ax + a\,\bar{h}(x \cdot n) + b$ on $\mathbf{R}^n$, where $\bar{h}' \in \{0, 1 - \lambda, 1\}$. Hence $u(x) \equiv Fx$ since each plane $x \cdot n = \text{const}$ intersects the set where $u = Fx$. $\qquad\square$

Approximate solutions: Consider again $K = \{A, B\}$ and suppose

$$B - A = a \otimes n, \quad F = \lambda A + (1 - \lambda)B, \lambda \in [0, 1].$$

We show that there exist sequences u_j with uniformly bounded Lipschitz constant such that in Ω

$$\text{dist}(Du_j, \{A, B\}) \to 0 \quad \text{in measure}, \tag{2.1}$$

and

$$u_j(x) = Fx \quad \partial\Omega. \tag{2.2}$$

Note that (2.1) and the bound on the Lipschitz constant imply that convergence also holds in L^p, $\forall p < \infty$. After translation we may assume

$$F = 0, A = -(1 - \lambda)a \otimes n, \quad B = \lambda a \otimes n.$$

Let h be the periodic extension of the function given by

$$h(t) = \begin{cases} -(1 - \lambda)t & t \in [0, \lambda), \\ \lambda(t - 1) & t \in [\lambda, 1], \end{cases}$$

and consider

$$v_j(x) = \frac{1}{j}\, a\, h(jx \cdot n).$$

Then $Dv_j \in \{A, B\}$ a.e. and $v_j \to 0$. To achieve the boundary conditions consider a cut-off function $\varphi \in C^\infty([0, \infty))$, $0 \le \varphi \le 1, \varphi = 0$ on $[0, 1/2]$, $\varphi = 1$ on $[1, \infty)$ and let

$$u_j(x) = \varphi(j\,\text{dist}(x, \partial\Omega))v_j(x).$$

Then $u_j = 0$ on $\partial\Omega$, Du_j is uniformly bounded and $Du_j = Dv_j$ except in a strip of thickness $1/j$ around $\partial\Omega$. If follows that u_j satisfies (2.1) and (2.2). Various modifications of this construction are possible, and we return in Section 3.2 to the question whether all approximating sequences are in a certain sense equivalent.

Note that due to the assumption $B - A = a \otimes n$, the problem (2.1), (2.2) essentially reduces to the scalar problem discussed in Example 3 of Section 1.2.

We now consider the case $\mathrm{rk}(B - A) \geq 2$. We have shown that in this case there are no nontrivial exact solutions. The argument used strongly the fact that Du only takes two values and that the curl of a gradient vanishes. It does not apply to approximating sequences. Nonetheless we have

Lemma 2.2 *([BJ 87], Prop.2) Suppose that* $\mathrm{rk}(B - A) \geq 2$ *and that* u_j *is a sequence with uniformly bounded Lipschitz constant such that*

$$\mathrm{dist}(Du_j, \{A, B\}) \to 0 \quad \text{in measure in } \Omega.$$

Then

$$Du_j \to A \text{ in measure or} \quad Du_j \to B \text{ in measure.}$$

In particular the problem (2.1), (2.2) has only the trivial solution, $F \in \{A, B\}$ *and* $Du_j \to F$ *in measure.*

The proof uses the following fundamental properties of minors. We recall that the semiarrow $\rightharpoonup$ denotes weak convergence.

Theorem 2.3 *[Ba 77, Mo 66, Re 67] Let M be an $r \times r$ minor (subdeterminant).*

(i) *If $p \geq r$ and $u, v \in W^{1,p}(\Omega), u - v \in W_0^{1,p}(\Omega)$ then*

$$\int_\Omega M(Du) = \int_\Omega M(Dv). \tag{2.3}$$

In particular

$$\int_\Omega M(Du) = \int_\Omega M(F) \quad \text{if } u = Fx \text{ on } \partial\Omega.$$

(ii) *If $p > r$ and if the sequence u_j satisfies*

$$u_j \rightharpoonup u \text{ in } W^{1,p}(\Omega, \mathbf{R}^m).$$

Then

$$M(Du_j) \rightharpoonup M(Du) \text{ in } L^{p/r}(\Omega).$$

Remark. Integrands f for which the integral $\int f(Du)$ only depends on the boundary values of u are called null Lagrangians, since the Euler-Lagrange equations are automatically satisfied for all functions u. Affine combinations of minors are the only null Lagrangians and the only functions that have the weak continuity property expressed in (ii) (see also Section 4.3).

Proof of Theorem 2.3. The main point is that minors can be written as divergences. For $n = m = 2$ one has

$$\det Du = \partial_1(u^1\partial_2 u^2) - \partial_2(u^1\partial_2 u^2), \tag{2.4}$$

for all $u \in C^2$ and hence for all $u \in W^{1,2}$ if the identity is understood in the sense of distributions. More generally for $n = m \geq 2$ the cofactor matrix that consists of the $(n-1) \times (n-1)$ minors of Du satisfies

$$\operatorname{div} \operatorname{cof} Du = 0, \quad \text{i.e. } \partial_j(\operatorname{cof} Du)_{ij} = 0 \tag{2.5}$$

and thus

$$\det Du = \frac{1}{n}\partial_j(u^i(\operatorname{cof} Du)_{ij}),$$

since $F(\operatorname{cof} F)^T = \operatorname{Id} \det F$. Similar formulae hold for general $r \times r$ minors, see [Mo 66, Da 89, GMS 96] for the detailed calculations. The multilinear algebra involved in these calculations can be expressed very concisely through the use of differential forms. In this setting one has for $n = m = 2$

$$\det Du\, dx^1 \wedge dx^2 = du^1 \wedge du^2 = d(u^1 \wedge du^2),$$

while for the $r \times r$ minor $M(Du)$ that involves the rows $1, \ldots, r$ and the columns $1, \ldots, r$ one has

$$\begin{aligned}
M(Du)dx^1 \wedge \ldots dx^n &= du^1 \wedge \ldots \wedge du^r \wedge dx^{r+1} \wedge \ldots \wedge dx^n \\
&= d(u^1 \wedge du^2 \wedge \ldots du^r \wedge dx^{r+1} \wedge \ldots \wedge dx^n).
\end{aligned}$$

In either formulation (i) follows from the Gauss-Green (or Stokes) theorem (and approximation by smooth functions) while (ii) follows from induction over the order r of minors and the fact that u_j converges strongly in L^p.

$\square$

Proof of Lemma 2.2. We may assume $A = 0$ and that there exists a 2×2 minor M such that $M(B) = 1$. By assumption there thus exist sets E_j such that

$$Du_j - B\chi_{E_j} \to 0 \text{ in measure} , \qquad (2.6)$$

and hence in L^p for all $p < \infty$. Moreover there exists a subsequence (not relabelled) such that

$$\chi_{E_j} \overset{*}{\rightharpoonup} \theta \text{ in } L^\infty(\Omega), \quad u_j \overset{*}{\rightharpoonup} u \text{ in } W^{1,\infty}(\Omega, \mathbf{R}^m). \qquad (2.7)$$

It follows from Theorem 2.3 and (2.6)

$$B\chi_{E_j} \overset{*}{\rightharpoonup} Du = B\theta$$

$$M(B)\chi_{E_j} \overset{*}{\rightharpoonup} M(Du) = M(B)\theta^2 \qquad (2.8)$$

Combining the first convergence in (2.7) and (2.8) we see that $\theta = \theta^2$ a.e. Thus θ must be a characteristic function χ_E. Hence (2.7) implies that (use e.g. the fact $\|\chi_{E_j}\|_{L^2} \to \|\chi_E\|_{L^2}$)

$$\chi_{E_j} \to \theta = \chi_E \quad \text{in measure.}$$

Therefore by (2.6)

$$Du_j \to Du = B\chi_E \quad \text{in measure.}$$

Finally Lemma 2.1 (i) implies that $Du = B$ a.e. or $Du = A = 0$ a.e. $\square$

2.2 Applications to crystal microstructures

Before proceeding with the mathematical discussion of the problem $Du \in K$ let us briefly review what can be learned about crystal microstructure so far. Which microstructures can form and why are they so fine?

First let us consider again the rôle of rank-1 connections. In the continuum theory discussed in the previous section they were related to continuity of the tangential derivatives or to the fact that the curl of a gradient vanishes (in Section 2.6 we still study the connections with the Fourier transform). The condition can also be understood in the discrete setting of crystal lattices. Two homogeneous lattices, obtained by affine deformations A and B of the same reference lattice can meet at a common plane S only if the deformations differ by a shear that leaves S invariant. Analytically we recover the condition $B - A = a \otimes n$, where n is the normal of S (see Figure 4).

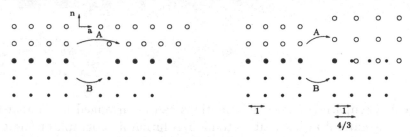

Figure 4: Compatible and incompatible lattice deformations. On the left the condition $B - A = a \otimes n$ is satisfied, on the right $B = \mathrm{Id}$, $A = 4/3\,\mathrm{Id}$, so the condition is violated. After deformation there is no interface on which the two lattices meet.

Under certain additional conditions the two sublattices are referred to as twins. There are different definitions what precisely constitutes a twin; a common requirement is that $B = QAH$, where $Q \in SO(3) \setminus \{\mathrm{Id}\}$, $Q^2 = \mathrm{Id}$ and where H belongs to the point group of the crystal, see [Ja 81] and [Za 92] for further discussion. Compatible lattice deformations can be arranged in alternating bands of different deformations, see Figure 5 (cf. also Fig.3).

If the set $K \in M^{m \times n}$ of minimizing affine deformations contains more rank-1 connections then more complicated patterns such as the double laminates (or 'twin crossings') in Figure 6 are possible.

In this way one can explain the observation of a number of microstructures through an analysis of rank-1 connections. The constructions based on rank-1 connections, however, involve no length scale. Why, then, are the observed structures often so fine?

For the situation of just two deformations A and B Proposition 2.1 (ii) and the discussion of approximate solutions provide an explanation. As soon as one imposes a nontrivial affine boundary condition $F = \lambda A + (1 - \lambda)B$ there are no exact solutions, and approximate solutions become the better the finer A and B are mixed (in a real crystal, additional contribution to the

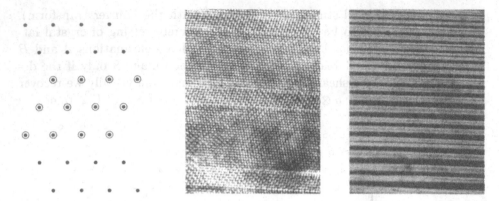

Figure 5: Compatible lattice deformations can be arranged in laminar patterns. Schematic drawing (left), atomic resolution micrograph of fine twinning in Ni-Al (middle; courtesy of D. Schryvers, RUCA, Antwerp), twinning in Cu-Al-Ni (right; courtesy of C. Chu and R. D. James), grey and black represent two different lattice deformations.

energy may eventually limit the fineness, see Section 6). In practise boundary conditions are often not so much imposed globally but by contact with other parts of the crystal where other deformation gradients prevail (e.g. because the phase transformation has not yet taken place there).

A typical example is the frequently observed austenite/finely-twinned martensite interface (see Figure 7). In an idealized situation this corresponds to a homogeneous affine deformation C on one side of the interface and a fine mixture of A and B on the other side. Neither A nor B are rank-1 connected to C but a suitable convex combination $\lambda A + (1 - \lambda)B$ is.

There is no deformation that uses all three gradients A, B and C and only these (see the end of the proof of Proposition 2.1). However, the volume fraction of gradients other than A, B and C can be made arbitrarily small by matching C to a fine mixture of layers of A and B in volume fractions λ and $1 - \lambda$.

The analysis of the rank-1 connections determines the volume fraction λ as well as the interface normals n and m, in very good agreement with experiment; see [BJ 87], Theorem 3 and [JK 89], Section 5 for a detailed discussion.

More complex patterns like the wedge microstructure in Figure 8 can be

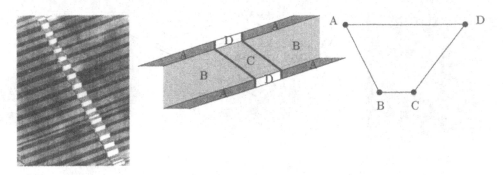

Figure 6: Twin crossings on Cu-Al-Ni (courtesy of C. Chu and R. D. James) and schematic drawings of the different deformation gradients and their rank-1 connections (indicated by solid lines).

understood in a similar vein. In this particular case so many rank-1 connections are required that the microstructure can only arise if the transformation strain satisfies a special relation; see [Bh 91], [Bh 92] for a comparison of theory and experiment.

The considerations in this subsection focused on constructions of microstructures based on rank-1 connections. Do these constructions cover (in a suitable sense) all possible microstructures? We return to this fundamental question in the remainder of this Section and in particular in Sections 4.3, 4.6 and 4.7.

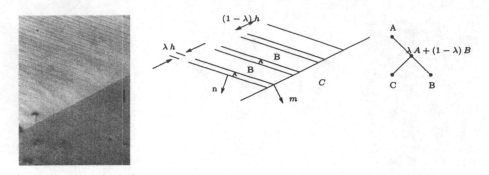

Figure 7: Austenite/finely twinned interface in Cu-Al-Ni (courtesy of C. Chu and R. D. James), schematic distribution of deformation gradients and rank-1 connections; a simple model for the refinement (branching) of the A/B twins towards the interface with C is discussed in section 6.2.

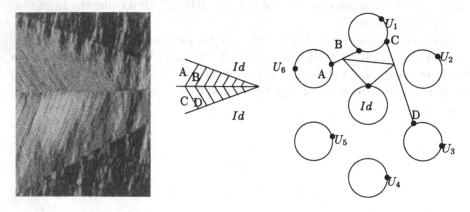

Figure 8: Wedge microstructure in Cu-Al-Ni (courtesy of C. Chu and R. D. James). The necessary rank-1 connections between the six orthorhombic wells $SO(3)U_i$ and the untransformed phase only exist for special transformation strains U_1.

2.3 The one-well problem

The simplest set K that is compatible with symmetry requirements (1.3) and (1.4) is $K = SO(n)$. In this case approximating sequences must converge strongly.

Theorem 2.4 *([Ki 88], p.231) Suppose that*

$$Du \in SO(n) \text{ a.e. in } \Omega.$$

Then Du is constant and $u(x) = Qx + b$, $Q \in SO(n)$. If u_j is a sequence of functions with uniformly bounded Lipschitz constant such that

$$\text{dist}(Du_j, SO(n)) \to 0 \quad \text{in measure}, \tag{2.9}$$

then

$$Du_j \to \text{const} \quad \text{in measure}.$$

Proof. To prove the first statement recall from (2.5) that

$$\text{div} \ \text{cof} Du = 0$$

for any Lipschitz map. Now $\text{cof } F = F$ for all $F \in SO(n)$ and thus u is harmonic and therefore smooth. Moreover $|Du|^2 = n$, where $|F|^2 = \text{tr} F^T F = \sum_{i,j} F_{ij}^2$, and therefore

$$2|D^2 u|^2 = \Delta |Du|^2 - 2\, Du \cdot D\Delta u = 0.$$

Thus Du is constant.

To prove the second assertion of the theorem we may assume that $u_j \overset{*}{\rightharpoonup} u$ in $W^{1,\infty}(\Omega; \mathbf{R}^m)$. Consider the function

$$f(F) = |F|^n - c_n \det F, \quad c_n = n^{n/2}.$$

One easily checks that $f \geq 0$ and that f vanishes exactly on matrices of the form λQ, $\lambda \geq 0$, $Q \in SO(n)$ (use polar decomposition, diagonalize and apply the arithmetic-geometric mean inequality). Hence (2.9), the weak continuity of minors (Theorem 2.3) and the weak lower semicontinuity of the L^n norm imply that

$$0 = \liminf_{n \to \infty} \int_\Omega f(Du_j)\, dx$$

$$= \liminf_{n \to \infty} \left(\int_\Omega |Du_j|^n\, dx - c_n \int_\Omega \det Du_j\, dx \right)$$

$$\geq \int_\Omega |Du|^n\, dx - c_n \int_\Omega \det Du\, dx = \int_\Omega f(Du) \geq 0.$$

Therefore all the inequalities must be equalities and in particular

$$f(Du) = 0 \text{ a.e.,} \quad \|Du_j\|_{L^n} \to \|Du\|_{L^n}.$$

It follows that

$$\begin{aligned} Du_j &\to Du \quad \text{in} \quad L^n(\Omega, M^{m \times n}) \text{ (hence in measure)} \\ Du(x) &= \lambda(x)Q(x), \quad \lambda \geq 0, \quad Q(x) \in SO(n) \text{ a.e.} \end{aligned}$$

Moreover $|Du_j|^2 = n$ a.e., whence $|Du|^2 = n$ a.e. Thus $Du \in SO(n)$ a.e. and, by the first part of the theorem $Du = \text{const}$. $\qquad\square$

The case $n = 2$ of the above result shows some interesting connections with the Cauchy-Riemann equations. Identify $\mathbf{C} \simeq \mathbf{R}^2$ as usual via $z = x + iy$ and let $\partial_z = 1/2(\partial_x - i\partial_y)$, $\partial_{\bar{z}} = 1/2(\partial_x + i\partial_y)$. Suppose that $1 < p < \infty$ and

$$\text{dist}(Du_j, SO(2)) \to 0 \quad \text{in } L^p(\Omega). \tag{2.10}$$

Then in particular $|\partial_z u_j| \to 1$ and

$$\partial_{\bar{z}} u_j \to 0 \quad \text{in } L^p(\Omega, \mathbf{C}),$$

and regularity for the Cauchy-Riemann operator implies that there exists a function u s.t.

$$u_j \to u \text{ in } W^{1,p}(\Omega, \mathbf{C}), \quad \partial_{\bar{z}} u = 0.$$

Thus u is (weakly) holomorphic and $|\partial_z u| = \lim_{j \to \infty} |\partial_z u_j| = 1$. Hence $\partial_z u = \text{const}$.

2.4 The three-gradient problem

Theorem 2.5 *([Sv 91b]). Let $K = \{A_1, A_2, A_3\}$ and suppose that $\mathrm{rk}(A_i - A_j) \neq 1$.*

(i) *If $Du \in K$ a.e. then Du is constant (a.e.).*

(ii) *If u_j is a sequence with uniformly bounded Lipschitz constant such that*

$$\mathrm{dist}(Du_j, K) \to 0 \quad \textit{in measure}$$

then

$$Du_j \to \mathrm{const} \quad \textit{in measure.}$$

Proof of part (i). For simplicity we only consider the case $n = m = 2$. The general case can be reduced to this if one considers separately the cases that the span E of $A_2 - A_1$ and $A_3 - A_1$ contains two, one or no rank-1 lines and uses Lemma 2.7 below, see also [Sv 91b].

We may assume that $A_1 = 0$ and thus $\det A_2 \neq 0$, $\det A_3 \neq 0$. Multiplying by A_2^{-1} we may further assume $A_2 = \mathrm{Id}$. Using the Jordan normal form we see that after a change of variables we have either

$$A_3 = \begin{pmatrix} \lambda & -\mu \\ \mu & \lambda \end{pmatrix}, \quad \lambda^2 + \mu^2 \neq 0$$

or

$$A_3 = \begin{pmatrix} \lambda & a \\ 0 & \mu \end{pmatrix}, \quad \lambda \neq 0, \mu \notin \{0, 1\}.$$

In the first case u satisfies the Cauchy-Riemann equations and is holomorphic and therefore smooth. Thus $Du \equiv A_i$ since K is discrete. In the second case $Du \in K$ implies that

$$\partial_1 u^2 = 0.$$

Hence $u^2(x) = h(x^2)$ (locally) and $\partial_2 u^2(x) = h'(x^2)$. Since $\mu \notin \{0, 1\}$ the value of $\partial_2 u^2$ uniquely determines one of the matrices A_i. Thus $Du(x) = g(x^2)$. In particular

$$\partial_1 \partial_1 u = 0, \quad \partial_2 \partial_1 u = \partial_1 \partial_2 u = 0$$

in the sense of distributions. Thus $\partial_1 u = \text{const}$ and $Du = \text{const} \otimes e_1 + \tilde{g}(x^2) \otimes e_2$. Therefore $\text{rk}(Du(x) - Du(\tilde{x})) \leq 1$ and thus $Du \equiv A_i$. $\qquad\square$

An alternative proof that features an interesting connection with the theory of quasiconformal (or more precisely quasiregular) maps proceeds as follows. After possible renumbering we may assume that $\det(A_2 - A_1)$ and $\det(A_3 - A_1)$ have the same sign. Taking $A_1 = 0$ and multiplying by $\text{diag}(1, -1)$ if needed we have $\det A_2 > 0, \det A_3 > 0$. Thus $Du \in K$ implies that

$$|Du|^2 \geq k \det Du$$

for a suitable constant k. Hence u is quasiregular and a deep result of Reshetnyak says that either $u = \text{const}$ or u is a local homeomorphism up to a discrete set B_u of branch points and that the (local) inverse u^{-1} preserves sets of measure zero (see [Re 89]). Hence either $Du = 0$ a.e. or $Du \neq 0$ a.e. In view of the results for the two-gradient problem this implies the assertion.

The proof of (ii) requires more subtle arguments (see [Sv 91b], [Sv 92b]). Šverák first shows that after suitable transformations (and elimination of some simpler special cases) one may assume

$$A_i = A_i{}^T, \quad \det A_i = 1.$$

Now a gradient Du is symmetric if and only if u is itself a gradient Dv. Thus assertion (ii) is essentially reduced to a study of approximate solutions of the Monge-Ampère equation

$$\det D^2 v_j \to 1, \quad v_j : \Omega \in \mathbf{R}^2 \to \mathbf{R}.$$

The difficulty is that, different from the usual literature on the Monge-Ampère equation, one cannot assume that $D^2 v_j$ is positive (semi-)definite. Indeed a crucial step in the proof that uses ideas from the theory of quasiregular maps is to show that $\det D^2 v > 0$ a.e. implies that v is locally convex or concave.

2.5 The four-gradient problem

The following example which was found independently by a number of authors (I am aware of [AH 86], [CT 93] and [Ta 93]; see [BFJK 94] for the adaptation of Tartar's construction for separately convex functions to diagonal matrices) shows that the absence of rank-1 connections does not guarantee absence of microstructures (i.e. strong convergence of approximating sequences).

Lemma 2.6 *Consider the 2×2 diagonal matrices $A_1 = -A_3 = \mathrm{diag}(-1, -3)$, $A_2 = -A_4 = \mathrm{diag}(-3, 1)$ and let $K = \{A_1, A_2, A_3, A_4\}$. Then $\mathrm{rk}(A_i - A_j) \neq 1$ but there exists a sequence u_j*

$$\mathrm{dist}(Du_j, K) \to 0 \quad \text{in measure,}$$

and Du_j does not converge in measure.

Exercise: Show that there is no nontrivial solution of $Du \in K$ for the above choice of K. Hint: consult the previous subsection.

It is not known whether there is another choice of four matrices with $\mathrm{rk}(A_i - A_j) \neq 1$ for which nontrivial solutions exist. It is known, but not trivial, that for each $\epsilon > 0$ there exist nontrivial maps such that $\mathrm{dist}(Du, K) < \epsilon$ (see the discussion after Theorem 5.4). Note that for small ϵ the set of admissible gradients still contains no rank-1 connections.

Proof. Since K contains no rank-1 connections the key idea is to 'borrow' four additional matrices J_i (see Fig. 9) and to successively remove the regions where Du assumes J_i. We will construct a sequence v_k that satisfies the affine boundary condition

$$v_k(x) = J_4 x \quad \text{on } \partial Q = \partial(0, 1)^2.$$

As a first approximation we may take $v^{(0)}(x) = J_4 x$. To increase the measure of the set where the gradients lie in K we observe that J_4 is a rank-1 convex combination of A_1 and J_1,

$$J_4 = \frac{1}{2}A_1 + \frac{1}{2}J_1.$$

As in Section 2.1 we can thus construct a map $v^{(1)}$ that agrees with $v^{(0)}$ on ∂Q and uses only gradients A_1 and J_1 (in layers of thickness $1/2k$) except for a boundary layer of thickness c/k where the gradient remains uniformly bounded. In the next step we replace the stripes where $Dv^{(1)} = J_1$ by fine

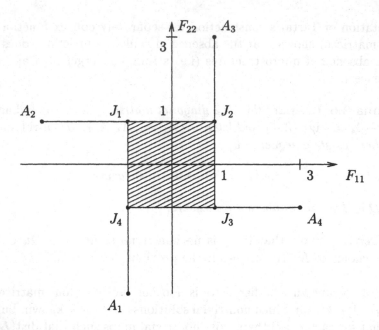

Figure 9: Four incompatible matrices that support a nontrivial minimizing sequence

layers of A_2 and J_2 and k new boundary layers of thickness c/k^2. This yields $v^{(2)}$ (see Fig. 10). The volume fraction of the J_i phases has been decreased to $\left(\frac{1}{2}\right)^2$ (up to small corrections due to the boundary layers). If we replace J_2 by fine layers of A_3 and J_3 (with k^2 boundary layers of thickness c/k^3) we obtain $v^{(3)}$ and replacing J_3 by A_4 and J_4 we obtain $v^{(4)}$. Up to the boundary layers $Dv^{(4)}$ only uses the values A_i and J_4. Compared to $v^{(0)}$ the volume fraction of the set where J_4 is taken has been reduced from one to (slightly less than) $(1/2)^4$. The volume fraction of the boundary layers is bounded by

$$\frac{c}{k} + k\frac{c}{k^2} + k^2\frac{c}{k^3} + k^3\frac{c}{k^4} = 4\frac{c}{k}.$$

Hence we have

$$|\{Dv^{(4)} \notin K\}| \leq \frac{4c}{k} + \frac{1}{16}.$$

To further reduce the volume fraction of the set $Dv \notin K$ we can now apply the same procedure to each of the small rectangles where $Dv^{(4)} = J_4$.

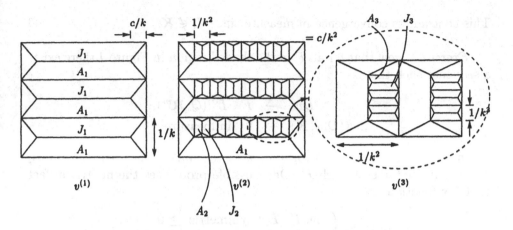

Figure 10: The first three stages in the construction of v_k.

After l iterations we obtain

$$|\{Dv^{(4l)} \notin K\}| \leq \sum_{m=0}^{l-1} \left(\frac{1}{16}\right)^m \frac{4c}{k} + \left(\frac{1}{16}\right)^l$$

$$\leq \frac{c}{k} + \left(\frac{1}{16}\right)^l.$$

With a suitable choice of l we thus find maps $v_k : Q \to \mathbf{R}^2$ such that $|Dv_k| \leq L$

$$|\{Dv_k \notin K\}| \leq \frac{C}{k} \to 0,$$
$$v_k(x) = J_4 x \text{ on } \partial Q.$$

In particular $\text{dist}(Dv_k, K) \to 0$ in measure.

We finally show that no subsequence of Dv_k can converge in measure. Indeed if $Dv_{k_j} \to Dv$ in measure then $Dv \in K$ in Q and $v = J_4 x$ on $\partial\Omega$. This is impossible since $Dv \in K$ implies $Dv \equiv \text{const}$ (see Exercise). Alternatively one can easily verify that

$$Dv_k \overset{*}{\rightharpoonup} J_4 \quad \text{in } L^\infty(Q; \mathbf{R}^m).$$

This contradicts convergence in measure since $J_4 \notin K$.   □

Exercise. Show that for all F in the shaded region in Figure 4 there exists a sequence such that

$$Dv_k \overset{*}{\rightharpoonup} F \quad L^\infty(Q; \mathbf{R}^m),$$
$$\text{dist}(Dv_k, K) \;\rightarrow\; 0 \quad \text{in measure.}$$

In fact the shaded region together with the rank-1 lines between the A_i and the J_i contains all such F. One possible proof uses the nontrivial fact that the function

$$f(F) = \begin{cases} \det F & F \quad \text{symmetric} \;\geq 0 \\ 0 & F \quad \text{symmetric} \;\not\geq 0 \\ +\infty & F \quad \text{not symmetric} \end{cases}$$

is quasiconvex (see [Sv 92b]). See Sections 4.3 and 4.4 for further information about the classification of weak limits.

2.6   Linear subspaces and elliptic systems

Lemma 2.7 *Let L be a linear subspace of $M^{m \times n}$ that contains no rank-1 line.*

(i) *If u is Lipschitz and $Du \in L$ a.e. then u is smooth*

(ii) *If u_j is a sequence that satisfies*

$$u_j \overset{*}{\rightharpoonup} u \quad \text{in } W^{1,\infty}(\Omega; \mathbf{R}^m),$$
$$\text{dist}(Du_j, L) \;\rightarrow\; 0 \quad \text{in measure.}$$

then $Du \in L$ a.e. and

$$Du_j \rightarrow Du \quad \text{in measure.}$$

Remark. In (i) it suffices to assume that $u \in W^{1,1}$, in (ii) it suffices that $u_j \rightarrow u$ in L^1_{loc} and that Du_j is bounded in L^1_{loc}.

Proof. Let $A : M^{m \times n} \to M^{m \times n}$ denote the projection onto the orthogonal complement of L. Then $Du \in L$ is equivalent to

$$A\,Du = 0. \tag{2.11}$$

The assumption that L contains no rank-1 lines essentially assures that (2.11) is a linear elliptic system and the assertions follow easily from the general theory of such systems. We sketch the proof for the convenience of the reader.

Suppose that v has compact support in Ω, f belongs to the Sobolev space $W^{k,2}(\Omega; \mathbf{R}^m)$ (i.e. all distributional derivatives up to order k belong to L^2) and v satisfies

$$A\,Dv = f. \tag{2.12}$$

We claim that $v \in W^{k+1,2}(\Omega; \mathbf{R}^m)$ and

$$\|D^{k+1}v\|_{L^2} \leq C\|D^k f\|_{L^2}. \tag{2.13}$$

To prove this consider the Fouriertransform

$$iA\,\hat{v}(\xi) \otimes \xi = \hat{f}(\xi)$$

of (2.11). Since L contains no rank-1 connections we have $A(a \otimes \xi) \neq 0$ if $a \neq 0$, $\xi \neq 0$, and by homogeneity

$$|A(a \otimes \xi)| \geq c|a|\,|\xi|$$

for some constant $c > 0$. The claim follows now from Plancherel's Theorem.

To prove (i) let $\varphi \in C_0^\infty(\Omega)$. Then

$$A\,D(\varphi u) \;=\; A(u \otimes D\varphi).$$

In view of (2.13) we have the implication $u \in W_{\text{loc}}^{k,2} \Rightarrow u \in W_{\text{loc}}^{k+1,2}$, and this yields (i).

To prove (ii) observe that the hypothesis and the linearity of (2.11) imply that $u_j \to u$ in L_{loc}^2, $ADu_j \to 0$ in L_{loc}^2, $ADu = 0$.

Application of (2.13) with $v = \varphi(u_j - u)$ yields the assertion. $\qquad\square$

To establish (i) for $u \in W^{1,1}$ it suffices to mollify u and to pass to the limit. To prove (ii) under the hypothesis in the remark one can use the weak L^1 estimates for elliptic systems.

Examples.

1. $L = \left\{ F \in M^{2\times 2} : F = \begin{pmatrix} a & b \\ -b & a \end{pmatrix} \right\}$; this corresponds to the Cauchy-Riemann equations

$$\partial_1 u^1 - \partial_2 u^2 = \partial_2 u^1 + \partial_1 u^2 = 0.$$

2. $L = \{ F \in M^{n\times n} : F^T = F, \operatorname{tr} F = 0 \}$; this corresponds to the Laplace equation $\Delta v = 0$, since Du symmetric implies $u = Dv$ (locally).

3. $L = \{ F \in M^{n\times n} : \operatorname{tr} F = 0, F_{ij}\xi_k - F_{ik}\xi_j = 0, \forall \xi \in \mathbf{R}^n \setminus \{0\} \}$; this corresponds to the system div $u = 0$, curl $u = 0$.

Problem. What is the largest dimension $d(m, n)$ of a subspace of $M^{m\times n}$ that contains no rank-1 line?

This is closely related to questions in algebraic geometry and K-theory, e.g. to the number of linear independent vector fields on S^{n-1}. For $m = n$ Example 2 provides the lower bound $d(n, n) \geq \frac{n(n+1)}{2}$. The upper bound $d(n, n) = n^2 - n$ is sharp exactly in dimension $n = 2, 4$ and 8. See [BFJK 94] for further information.

3 Efficient description of minimizing sequences - Young measures

3.1 The fundamental theorem on Young measures

We have seen in the examples in Section 1.2 and 2.1 that there are usually many minimizing sequences for a variational problem. We return now to the question whether all these sequences have some common features and whether one can describe the 'macroscopic' features of a sequence without paying attention to unnecessary details. Closely related is the issue of defining a notion of generalized solution for variational problems that do not admit classical solutions.

A reasonable condition for an object that describes the macroscopic behaviour of a sequence $z_j : E \to \mathbf{R}^d$ is that it should determine the limits of

$$\int_U f(z_j)$$

for continuous functions f (such as energy-, stress- or entropy density) and for all measurable subsets U of E. Such an object exists and was first introduced by L.C. Young in connection with generalized solutions of optimal control problems. By $C_0(\mathbf{R}^d)$ we denote the closure of continuous functions on $\mathbf{R}^d$ with compact support. The dual of $C_0(\mathbf{R}^d)$ can be identified with the space $\mathcal{M}(\mathbf{R}^d)$ of signed Radon measures with finite mass via the pairing

$$\langle \mu, f \rangle = \int_{\mathbf{R}^d} f \, d\mu.$$

A map $\mu : E \to \mathcal{M}(\mathbf{R}^d)$ is called weak∗ measurable if the functions $x \mapsto \langle \mu(x), f \rangle$ are measurable for all $f \in C_0(\mathbf{R}^d)$. We often write μ_x instead of $\mu(x)$.

Theorem 3.1 *(Fundamental theorem on Young measures)*

Let $E \subset \mathbf{R}^n$ be a measurable set of finite measure and let $z_j : E \to \mathbf{R}^d$ be a sequence of measurable functions. Then there exists a subsequence z_{j_k} and a weak∗ measurable map $\nu : E \to \mathcal{M}(\mathbf{R}^d)$ such that the following holds

(i) $\nu_x \geq 0, \quad \|\nu_x\|_{\mathcal{M}(\mathbf{R}^d)} = \int_{\mathbf{R}^d} d\nu_x \leq 1, \quad$ *for a.e. $x \in E$.*

(ii) *For all* $f \in C_0(\mathbf{R}^d)$

$$f(z_{j_k}) \overset{*}{\rightharpoonup} \bar{f} \quad in \ L^\infty(E),$$

where

$$\bar{f}(x) = \langle \nu_x, f \rangle = \int_{\mathbf{R}^d} f \, d\nu_x.$$

(iii) *Let* $K \subset \mathbf{R}^d$ *be compact. Then*

$$\mathrm{supp}\,\nu_x \subset K \quad if \ \mathrm{dist}(z_{j_k}, K) \to 0 \ in \ measure.$$

(iv) *Furthermore one has*

$$(i') \ \|\nu_x\|_{\mathcal{M}} = 1 \quad for \ a.e. \ x \in E$$

if and only if the sequence does not escape to infinity, i.e. if

$$\lim_{M \to \infty} \sup_k |\{|z_{j_k}| \ge M\}| = 0. \tag{3.1}$$

(v) *If (i') holds, if* $A \subset E$ *is measurable, if* $f \in C(\mathbf{R}^d)$ *and if*

$$f(z_{j_k}) \ is \ relatively \ weakly \ compact \ in \ L^1(A),$$

then

$$f(z_{j_k}) \rightharpoonup \bar{f} \ in \ L^1(A), \quad \bar{f}(x) = \langle \nu_x, f \rangle.$$

(vi) *If (i') holds, then in (iii) one can replace 'if' by 'if and only if'.*

Remarks. 1. The map $\nu : E \to \mathcal{M}(\mathbf{R}^d)$ is called the Young measure generated by (or: associated to) the sequence z_{j_k}. Every (weakly* measurable) map $\nu : E \to \mathcal{M}(\mathbf{R}^d)$ that satisfies (i) is generated by some sequence z_k.

2. The assumption $|E| < \infty$ was only introduced for notational convenience, cf. [Ba 89]. In fact $\mathbf{R}^d$ with Lebesgue measure can be replaced by a more general measure space $(\mathcal{S}, \Sigma, \mu)$, e.g. a locally compact space with a Radon measure. The converse statement in Remark 1 requires that μ be non-atomic.

3. The target $\mathbf{R}^d$ can be replaced e.g. by a compact metric space K. In this case one always has $\|\nu_x\| = 1$ a.e. The condition (3.1) has a simple interpretation if we replace $\mathbf{R}^d$ by its one-point compactification $K = \mathbf{R}^d \cup \{\infty\} \simeq S^d$ and consider the corresponding family of measures $\tilde{\nu}_x$ on K. Then $\|\tilde{\nu}_x\| = 1$ a.e., and (3.1) ensures that $\tilde{\nu}_x$ does not charge the point ∞.

4. If, for some $s > 0$ (!) and all $j \in \mathbf{N}$

$$\int_E |z_j|^s \leq C$$

then (3.1) holds.

5. Here is a typical application of (v): if $\{z_j\}$ is bounded in L^p and $|f(s)| \leq C(1+|s|^q), q < p$, then $f(z_{j_k}) \rightharpoonup \bar{f}$ in $L^{p/q}$. In particular, for $p > 1$ the choice $f = \mathrm{id}$ yields

$$z_{j_k} \rightharpoonup z, \quad z(x) = \langle \nu_x, \mathrm{id} \rangle. \tag{3.2}$$

Proof. The point is to pass from the functions z_j which take values in $\mathbf{R}^d$ to maps which take values in the space of $\mathcal{M}(\mathbf{R}^d)$ of measures in $\mathbf{R}^d$. Thus we allow new limiting objects which do not take a precise function value at every point but a probability distribution of values.

Let

$$Z_j(x) = \delta_{z_j(x)}.$$

Then $\|Z_j(x)\|_{\mathcal{M}(\mathbf{R}^d)} = 1$ and $\langle Z_j(x), f \rangle = f(z_j(x))$. Thus Z^j belongs to the space $L_w^\infty(E; \mathcal{M}(\mathbf{R}^d))$ of weak* measurable maps $\mu : E \to \mathcal{M}(\mathbf{R}^d)$ that are (essentially) bounded. Now it turns out $L_w^\infty(E, \mathcal{M}(\mathbf{R}^d))$ is the dual of the separable space $L^1(E, C_0(\mathbf{R}^d))$ (see e.g. [Ed 65, p.588], [IT 69, p.93], [Me 66, p.244]), where the duality pairing is given by

$$\langle \mu, g \rangle = \int_E \langle \mu(x), g(x) \rangle dx.$$

Hence the Banach-Alaoglu theorem yields a subsequence such that

$$Z_{j_k} = \delta_{z_{j_k}(\cdot)} \overset{*}{\rightharpoonup} \nu \quad \text{in } L_w^\infty(E, \mathcal{M}(\mathbf{R}^d)). \tag{3.3}$$

Lower semicontinuity of the norm implies that $\|\nu_x\| \leq 1$ for a.e. x. For $\varphi \in L^1(E)$ and $f \in C_0(\mathbf{R}^d)$ we denote by $\varphi \otimes f$ the element of $L^1(E, C_0(\mathbf{R}^d))$

given by $x \mapsto \varphi(x)f$. The definition of Z_j and (3.3) thus imply

$$\int_E \varphi(x)f(z_{j_k}(x))dx = \langle Z_{j_k}, \varphi \otimes f \rangle \to \int_E \varphi(x)\langle \nu_x, f \rangle dx.$$

Hence (ii) follows, and considering all functions $f \geq 0, \varphi \geq 0$ we also deduce $\nu_x \geq 0$.

To prove (iii) it suffices to show that

$$\langle \nu_x, f \rangle = 0 \qquad \forall f \in C_0(\mathbf{R}^d \setminus K). \tag{3.4}$$

Let $f \in C_0(\mathbf{R}^d \setminus K)$. Then for every $\epsilon > 0$ there exist C_ϵ such that $|f(y)| \leq \epsilon + C_\epsilon \operatorname{dist}(y, K)$. Hence the hypothesis $\operatorname{dist}(z_{j_k}, K) \to 0$ in measure implies that $(|f| - \epsilon)^+(z_{j_k}) \to 0$ in measure, and in view of (ii) we conclude that

$$\langle \nu_x, (|f| - \epsilon)^+ \rangle = 0 \qquad \text{for a.e. } x.$$

Now (3.4) follows since $\epsilon > 0$ was arbitrary.

The proof of (iv) and (v) is easily achieved by a careful truncation argument and the characterization of weakly compact sets in L^1 [Me 66], see [Ba 89] for the details. Finally the proof of (vi) follows by an application of (v) to the bounded function $f = \max(\operatorname{dist}(\cdot, K), 1)$. $\square$

Remark. Since the span of tensor products $\varphi \otimes f, \varphi \in L^1(\Omega), f \in C_0(\mathbf{R}^d)$, is dense in $L^1(\Omega; C_0(\mathbf{R}^d))$ assertion (ii) of the theorem is equivalent to $Z_{jk} \xrightarrow{*} \nu$.

The measure ν_{x_0} describes the probability of finding a certain value in the sequence $z_{j_k}(x)$ for x in a small neighbourhood $B_r(x_0)$ in the limits $j \to \infty$ and $r \to 0$. The following useful fact reflects this probabilistic interpretation.

Corollary 3.2 *Suppose that a sequence z_j of measurable functions from E to $\mathbf{R}^d$ generates the Young measure $\nu : E \to \mathcal{M}(\mathbf{R}^d)$. Then*

$$z_j \to z \text{ in measure} \quad \text{if and only if} \quad \nu_x = \delta_{z(x)} \text{ a.e.}$$

Proof. If $z_j \to z$ in measure then $f(z_j) \to f(z)$ in measure for all $f \in C_0(\mathbf{R}^d)$. Hence by Theorem 3.1 (ii) one has $\langle \nu_x, f \rangle = f(z(x))$ for all $f \in C_0(\mathbf{R}^d)$ and thus $\nu_x = \delta_{z(x)}$. If conversely $\nu_x = \delta_{z(x)}$ a.e. we claim that

$$\limsup_{j \to \infty} |\{|z_j - w| > \epsilon\}| \leq |\{|z - w| > \epsilon/2\}|,$$

for all piecewise constant measurable functions $w : E \to \mathbf{R}^d$. To see this it suffices to consider constant functions $w \equiv a$ and to apply (v) with $f(y) = \varphi(|y-a|)$ where φ is continuous $0 \leq \varphi \leq 1, \varphi = 1$ on $[\epsilon, \infty), \varphi = 0$ on $[0, \epsilon/2]$. Thus

$$\limsup_{j \to \infty} |\{|z_j - z| > \epsilon\}| \leq \limsup_{j \to \infty} |\{|z_j - w| > \epsilon/2\}| + |\{|w - z| > \epsilon/2\}|$$

$$\leq 2|\{|z - w| > \epsilon/4\}|.$$

The last term can be made arbitrarily small since measurable functions can be approximated by piecewise constant functions, and the assertion follows (note that z is measurable since $\{\nu_x\}_{x \in E}$ is weak* measurable). $\square$

An alternative approach to the 'if' part of the corollary is to apply Corollary 3.3 below to the Carathéodory function $f(x, y) = \min(|y - z(x)|, 1)$.

3.2 Examples

a) Let $h : \mathbf{R} \to \mathbf{R}$ be the periodic extension of the function given by

$$h(x) = \begin{cases} a & \text{if } 0 \leq x < \lambda, \\ b & \text{if } \lambda \leq x < 1, \end{cases}$$

and define $z_j : [0, 1] \to \mathbf{R}$ by

$$z_j(x) = h(jx). \tag{3.5}$$

Using the periodicity of h one easily checks that (see e.g. [Da 81], p.8),

$$z_j \stackrel{*}{\rightharpoonup} \int_0^1 h(y)dy = \lambda a + (1 - \lambda)b$$

and similarly

$$f(z_j) \stackrel{*}{\rightharpoonup} \lambda f(a) + (1 - \lambda)f(b).$$

Hence z_j generates a Young measure ν given by

$$\nu_x = \lambda \delta_a + (1 - \lambda)\delta_b.$$

In particular ν_x is independent of x. Such Young measures are called *homogeneous Young measures*.

Although there are many different minimizing sequences for I they all generate the same Young measure. The Young measure captures the essential feature of minimizing sequences. They have to use slopes (close to) ± 1 in equal proportion in a finer and finer mixture.

One may view the pair (u, ν) as a generalized solution of the problem $I \to$ min. The derivative u_x is replaced by a probability measure and the coupling between u and ν occurs through the centre of mass of ν (cf. (3.7), (3.8) and Theorem 4.9):

$$u_x = \langle \nu_x, \mathrm{id} \rangle.$$

c) (Approximate solutions of the two-well problem)

Let $A, B \in M^{m \times n}, B - A = a \otimes n, F = \lambda A + (1 - \lambda)B, \lambda \in (0, 1)$. Consider a sequence of maps $u_j : \Omega \subset \mathbf{R}^n \to \mathbf{R}^m$ with uniformly bounded Lipschitz constant that satisfies

$$\mathrm{dist}(Du_j, \{A, B\}) \to 0 \quad \text{in measure in } \Omega,$$

$$u_j(x) = Fx \quad \text{in } \partial \Omega.$$

Let ν be the Young measure generated by (a subsequence of) Du_j. Then $\|\nu_x\| = 1$ and Theorem 3.1 (iii) yields $\mathrm{supp}\,\nu_x \subset \{A, B\}$, i.e. $\nu_x = \mu(x)\delta_A + (1 - \mu(x))\delta_B$. Passing to a further subsequence we may assume $u_j \overset{*}{\rightharpoonup} u$ in $W^{1,\infty}(\Omega, \mathbf{R}^m)$, and in view of (3.2) we have

$$Du(x) = \mu(x)A + (1 - \mu(x))B = A + (1 - \mu(x))a \otimes n.$$

Extending u_j and u by Fx outside Ω we deduce that $v(x) = u(x) - Ax$ is constant on the planes $x \cdot n = \mathrm{const}$. Hence $u(x) = Fx$ and $\mu(x) = \lambda$. Thus $\{Du_j\}$ generates the unique (homogeneous) Young measure

$$\nu_x = \lambda \delta_A + (1 - \lambda)\delta_B.$$

d) (Four-gradient example)

The sequence Du^j constructed in Section 2.5 generates the unique homogeneous Young measure

$$\nu_x = \frac{8}{15}\delta_{A_1} + \frac{4}{15}\delta_{A_2} + \frac{2}{15}\delta_{A_3} + \frac{1}{15}\delta_{A_4}.$$

Proof. Exercise.

3.3 What the Young measure cannot detect

The Young measure describes the local phase proportions in an infinitesimally fine mixture (modelled mathematically by a sequence that develops finer and finer oscillations). This is exactly what is needed to compute limits of integrals $\int_U f(z_j)$. There are, however, other natural quantities that cannot be computed from the Young measure.

Example 1 (micromagnetism).
The energy of a large rigid magnetic body represented by a domain $\Omega \subset \mathbf{R}^3$, is given by

$$I(m) = \int_\Omega \varphi(m) + \int_{\mathbf{R}^3} |h_m|^2.$$

Here $m : \Omega \to \mathbf{R}^3$ is the magnetization and h_m is the Helmholtz projection of $-m$ (extended by zero outside Ω), i.e. the unique gradient field that satisfies $\mathrm{div} h_m = -\mathrm{div} m$ in the sense of distributions. In suitable units m satisfies the saturation condition $|m| = 1$. For simplicity we have neglected exchange energy (this is a good approximation for large bodies, see [DS 93]).

Let $m_j : \Omega \to S^2 \subset \mathbf{R}^3$ be a sequence of magnetizations that generates a Young measure ν. Then

$$\int_\Omega \varphi(m_j)dx \to \int_\Omega \langle \nu_x, \varphi \rangle dx.$$

The limit of $\int_{\mathbf{R}^3} |h_{m_j}|^2$, however, is in general not determined by the Young measure (see Fig. 11). Indeed let f be the periodic extension of the sign function on $[-1/2, 1/2]$, let $\Omega = [0,1]^3$ and let

$$m_j = f(jx^1)e^1\chi_\Omega; \qquad \tilde{m}_j = f(jx^2)e^1\chi_\Omega.$$

Both sequences generate the same (homogeneous) Young measure $\nu_x = \frac{1}{2}\delta_{e^1} + \frac{1}{2}\delta_{-e^1}$. On the other hand it is not difficult to verify that $\|h_{m_j}\|_2 \to 1$ while $\|h_{\tilde{m}_j}\|_2 \to 0$. First replace χ_Ω by a smooth function φ and show that the resulting fields M^j and $\tilde{M}^j$ satisfy $\mathrm{curl}\, M_j \to 0$, $\mathrm{div} \tilde{M}_j \to 0$ in H^{-1}; then use $\|h_{m_j} - h_{M_j}\|_2 \leq \|m_j - M_j\|_2$ since the map $m \mapsto -h_m$ is an orthogonal projection. Alternatively one may use the representation of h_m in Fourier space.

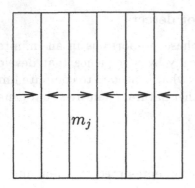 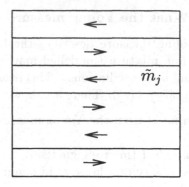

Figure 11: Both sequences generate the same Young measure but m_j is almost a gradient while $\tilde{m}_j$ is almost divergence free.

Example 2 (correlations).
The limit of

$$I_j(u_j) := \int\limits_0^1 u_j(x)u_j(x + \frac{1}{j})dx$$

is not determined by the Young measure of $\{u_j\}$. Indeed consider

$$u_j(x) = \sin j\pi x,$$
$$v_j(x) = \sin 2j^2\pi x.$$

Both sequences generate the same (homogeneous) Young measure $\nu_x = \frac{2}{\pi}(\sin^{-1})'(y)dy$ (cf. Section 3.2 a)), but

$$I_j(u_j) = \int\limits_0^1 \sin(j\pi x)\,\sin(j\pi x + \pi) \to -1/2,$$
$$I_j(v_j) = \int\limits_0^1 \sin(j^2\pi x)\,\sin(j^2\pi x + 2j\pi) \to 1/2,$$

3.4 More about Young measures and lower semicontinuity

We have seen that Young measures are useful as a concept since they allow one to give a precise meaning to the idea of 'infinitesimally fine phase mixture' and provide a framework for generalized solutions where no classical minimizers exist.

In this section, which may be omitted on first reading, we briefly discuss the advantages of Young measures as a technical tool. The following two results allow one, among other things, to extend lower semicontinuity results for integrals $\int f(Du(x))dx$ to integrals $\int f(x, u(x), Du(x))dx$ without additional effort. More generally Young measures are a rather efficient tool to eliminate all dependence on 'lower order' terms by soft general arguments. The first result shows that the Young measure suffices to compute limits of Carathéodory functions, the second extends the characterization of strong convergence in Corollary 3.2.

Corollary 3.3 *Suppose that the sequence of maps $z_k : E \to \mathbf{R}^d$ generates the Young measure ν. Let $f : E \times \mathbf{R}^d \to \mathbf{R}$ be a Carathéodory function (measurable in the first argument and continuous in the second) and assume that the negative part $f^-(x, z_k(x))$ is weakly relatively compact in $L^1(E)$. Then*

$$\liminf_{k \to \infty} \int_E f(x, z_k(x))dx \geq \int_E \int_{\mathbf{R}^d} f(x, \lambda)d\nu_x(\lambda)dx. \qquad (3.9)$$

If, in addition the sequence of functions $x \mapsto |f|(x, z_k(x))$ is weakly relatively compact in $L^1(E)$ then

$$f(\cdot, z_k(\cdot)) \rightharpoonup \bar{f} \text{ in } L^1(E), \quad \bar{f}(x) = \int_{\mathbf{R}^d} f(x, \lambda)d\nu_x(\lambda)dx. \qquad (3.10)$$

Remarks. 1. Assertion (3.9) still holds if f is (Borel) measurable on $E \times \mathbf{R}^d$ and lower semicontinuous in the second argument rather than a Carathéodory function (see [BL 73]).

2. The choice $f(x, p) = \min(|p - z(x)|, 1)$ in (3.10) can be used to prove the 'if' statement in Corollary 3.2.

Proof. It suffices to prove (3.9). The second assertion follows by application of this inequality to $\tilde{f}(x, p) = \pm\varphi(x)f(x, p)$ for all $\varphi \in L^\infty(E), \varphi \geq 0$.

To prove (3.9) first consider the case $f \geq 0$. Assume temporarily that, in addition,

$$f(x, \lambda) = 0 \quad \text{if } |\lambda| \geq R. \qquad (3.11)$$

By the Scorza-Dragoni theorem there exists an increasing sequence of compact sets E_j such that $|E \setminus E_j| \to 0$ and $f_{|E_j \times \mathbf{R}^d}$ is continuous. Define

$F_j : E \to C_0(\mathbf{R}^d)$ by $F_j(x) = \chi_{E_j}(x)f(x,\cdot)$. Then $F_j \in L^1(E; C_0(\mathbf{R}^d))$ and the convergence of $\delta_{z_k(\cdot)}$ to ν in the dual space yields

$$\int_E f(x, z_k(x))dx \geq \int_E \langle \delta_{z_k(x)}, F_j(x) \rangle$$
$$\to \int_E \langle \nu_x, F_j(x) \rangle dx = \int_{E_j} f(x, \lambda)d\nu_x(\lambda).$$

Letting $j \to \infty$ we obtain the assertion by the monotone convergence theorem. To remove the assumption (3.11) consider an increasing sequence $\{\eta_l\} \subset C_0^\infty(\mathbf{R}^d)$, that converges to 1, use the estimate for $f_l(x, \lambda) = f(x, \lambda)\eta_l(\lambda)$ and apply again the monotone convergence theorem. This finishes the proof if $f \geq 0$ or more generally if f is bounded from below.

For general f let

$$h_k(x) \quad = f(x, z_k(x)) = h_k^+(x) - h_k^-(x).$$
$$f_M(x, \lambda) \quad = \max(f(x, \lambda), -M).$$

By the equivalent characterizations of equiintegrability (see e.g. [Me 66]) for each $\epsilon > 0$ there exists an $M > 0$ such that

$$\sup_k \int_{h_k^- \geq M} h_k^-(x)dx < \epsilon.$$

Hence

$$\liminf_{k \to \infty} \int_E f(x, z_k(x))dx + \epsilon \geq \liminf_{k \to \infty} \int_E f_M(x, z_k(x))dx$$

$$\geq \int_E \int_{\mathbf{R}^d} f_M(x, \lambda)d\nu_x(\lambda)dx \geq \int_E f(x, \lambda)d\nu_x(\lambda)dx.$$

Since $\epsilon > 0$ was arbitrary the proof is finished. $\qquad\qquad\square$

Corollary 3.4 *Let $u_j : E \to \mathbf{R}^d, v_j : E \to \mathbf{R}^{d'}$ be measurable and suppose that $u_j \to u$ a.e. while v_j generates the Young measure ν. Then the sequence of pairs $(u_j, v_j) : E \to \mathbf{R}^{d+d'}$ generates the Young measure $x \mapsto \delta_{u(x)} \otimes \nu_x$.*

Proof. Let $\varphi \in C_0(\mathbf{R}^d), \psi \in C_0(\mathbf{R}^{d'}), \eta \in L^1(E)$. Then $\varphi(u_j) \to \varphi(u)$a.e. and $\eta\varphi(u_j) \to \eta\varphi(u)$ in $L^1(E)$ by the dominated convergence theorem. Moreover by assumption

$$\psi(v_j) \overset{*}{\rightharpoonup} \bar{\psi} \quad \text{in } L^\infty, \qquad \bar{\psi}(x) = \langle \nu_x, \psi \rangle.$$

Hence

$$\int_E \eta(\varphi \otimes \psi)(u_j, v_j) dx = \int_E \eta\varphi(u_j)\psi(v_j) dx \to \int_E \eta\varphi(u)\langle \nu_x, \psi\rangle dx$$

or

$$(\varphi \otimes \psi)(u_j, v_j) \overset{*}{\rightharpoonup} \langle \delta_{u(\cdot)} \otimes \nu_{\cdot}, \varphi \otimes \psi\rangle \text{ in } L^\infty(E).$$

The assertion follows since linear combinations of tensor products $\varphi \otimes \psi$ are dense in $C_0(\mathbf{R}^{d+d'})$. $\qquad \square$

A typical application of the corollaries is as follows. Let $f : \Omega \times (\mathbf{R}^m \times M^{m \times n}) \to \mathbf{R}$ be a Carathéodory function and suppose that $f \geq 0$. Suppose that $u_j \rightharpoonup u$ in $W^{1,p}(\Omega; \mathbf{R}^m)$ and that Du_j generates the Young measure ν. Taking $v_j = Du_j$, $z_j = (u_j, v_j)$ we obtain.

$$\lim_{j\to\infty} \int_\Omega f(x, u_j(x), Du_j(x)) dx$$
$$\geq \int_\Omega \int_{\mathbf{R}^m \times M^{m \times n}} f(x, \lambda, \mu) d\delta_{u(x)}(\lambda) \otimes d\nu_x(\mu) dx$$
$$= \int_\Omega \int_{M^{m \times n}} f(x, u(x), \lambda) d\nu_x(\lambda) dx.$$

The proof of the lower semicontinuity is thus reduced to the verification of the inequality

$$\int_{M^{m \times n}} g(\lambda) \nu_x(\lambda) \geq g(Du(x)) = g(\langle \nu_x, \text{id}\rangle) \qquad (3.12)$$

for the function

$$g(\lambda) = f(x, u(x), \lambda)$$

with 'frozen' first and second argument. To see when (3.12) holds we need to understand which Young measures are generated by gradients. This is the topic of the next section.

4 Which Young measures arise from gradients?

To employ Young measures in the study of crystal microstructure we need to understand which Young measures arise from sequences of gradients $\{Du_j\}$. As before $\Omega \subset \mathbf{R}^n$ denotes a bounded domain with Lipschitz boundary.

Definition 4.1 *A (weakly* measurable) map* $\nu : \Omega \to \mathcal{M}(M^{m \times n})$ *is a* $W^{1,p}$ *gradient Young measure if there exists a sequence of maps* $u_j : \Omega \to \mathbf{R}^m$ *such that*

$$u_j \rightharpoonup u \quad in \quad W^{1,p}(\Omega; \mathbf{R}^m) \quad (\overset{*}{\rightharpoonup} \ if\ p = \infty),$$

$$\delta_{Du(\cdot)} \overset{*}{\rightharpoonup} \nu \quad L_w^\infty(\Omega; \mathcal{M}(M^{m \times n})).$$

Using this notion we may reformulate Problem 2 (approximate solutions) as follows.

    Problem 2'   Given a set $K \subset M^{m \times n}$, characterize all $W^{1,\infty}$- gradient Young measures ν such that

$$\operatorname{supp}\nu_x \subset K \quad for \quad a.e.\ x.$$

An abstract characterization of gradient Young measures due to Kinderlehrer and Pedregal will be derived in Section 4.3 below. It involves the notion of quasiconvexity. Quasiconvexity, first introduced by Morrey in 1952, is clearly the natural notion of convexity for vector-valued problems (see Section 4.2) but still remains largely mysterious since it is very hard to determine whether a given function is quasiconvex. Therefore further notions of convexity were introduced to obtain necessary or sufficient conditions for quasiconvexity. We begin by reviewing these notions and their relationship.

4.1 Notions of convexity

For a matrix $F \in M^{m \times n}$ let $M(F)$ denote the vector that consists of all minors of F and let $d(n,m) = \sum_{r=1}^{\min(n,m)} \binom{n}{r}\binom{m}{r}$ denote its length.

Definition 4.2 *A function* $f : M^{m \times n} \to \mathbf{R} \cup \{+\infty\} = (-\infty, \infty]$ *is*

(i) *convex if*

$$f(\lambda A + (1 - \lambda)B) \leq \lambda f(A) + (1 - \lambda)f(B)$$

$$\forall\, A, B \in M^{m \times n}, \lambda \in (0,1);$$

(ii) *polyconvex if there exists a convex function* $g : \mathbf{R}^{d(n,m)} \to \mathbf{R} \cup \{+\infty\}$
such that

$$f(F) = g(\boldsymbol{M}(F));$$

(iii) *quasiconvex if for every open and bounded set* U *with* $|\partial U| = 0$ *one has*

$$\int_U f(F + D\varphi)dx \geq \int_U f(F)dx = |U|f(F) \quad \forall \varphi \in W_0^{1,\infty}(U; \mathbf{R}^m), \quad (4.1)$$

whenever the integral on the left hand side exists;

(iv) *rank-1 convex, if* f *is convex along rank-1 lines, i.e. if*

$$f(\lambda A + (1 - \lambda)B) \leq \lambda f(A) + (1 - \lambda)f(B)$$

$$\forall A, B \in M^{m \times n} \text{ with } \mathrm{rk}(B - A) = 1, \quad \forall \lambda \in (0, 1).$$

Remarks. 1. If $f \in C^2$ then rank-1 convexity is equivalent to the Legendre-Hadamard condition

$$\frac{\partial^2 f}{\partial F^2}(F)(a \otimes b, a \otimes b) = \frac{\partial^2 f}{\partial F_\alpha^i \partial F_\beta^j}(F)a^i b_\alpha a^j b_\beta \geq 0.$$

2. Quasiconvexity is independent of the set U, i.e. if (4.1) holds for one open and bounded set with $|\partial U| = 0$ then it holds for all such sets. If f takes values in $\mathbf{R}$ it suffices to extend φ by zero outside U and to translate and scale U. For general f one can use the Vitali covering theorem.

3. If f takes values in $\mathbf{R}$ and is quasiconvex then it is rank-1 convex (see Lemma 4.3 below) and thus locally Lipschitz continuous (use that f is convex and thus locally Lipschitz in each coordinate direction in $M^{m \times n}$; see [Da 89], Chapter 2, Thm. 2.3, or [MP 98], Observation 2.3 for the details). In this case the integral on the left hand side of (4.1) always exists.

It is sometimes convenient to consider quasiconvex functions that take values in $[-\infty, \infty)$. The argument below shows that such functions are rank-1 convex and thus either take values in $\mathbf{R}$ or are identically $-\infty$.

If $n = 1$ or $m = 1$ then convexity, polyconvexity and rank-1 convexity are equivalent and they are equivalent to quasiconvexity if, in addition, f takes values in $\mathbf{R}$.

Lemma 4.3 *If $n \geq 2, m \geq 2$ then the following implications hold.*

$$
\begin{array}{ll}
f & convex \\
\Downarrow & \qquad\qquad \not\Uparrow \\
f & polyconvex \\
\Downarrow & \qquad\qquad \not\Uparrow \\
f & quasiconvex \\
\Downarrow f < \infty & \qquad \not\Uparrow \text{ if } m \geq 3 \\
f & rank\text{-}1 \ convex
\end{array}
$$

The most difficult question is whether rank-1 convexity implies quasiconvexity. Šverák's [Sv 92a] ingenious counterexample solved this long standing problem in the negative if $m \geq 3$; the case $m = 2, n \geq 2$ is completely open.

Proof. The first implication is obvious, the second follows from the fact that minors are null Lagrangians (see Theorem 2.3) and Jensen's inequality. To prove the last implication let f be quasiconvex, consider $A, B \in M^{m \times n}$, with $\mathrm{rk}(B - A) = 1$, and a convex combination $F = \lambda A + (1 - \lambda)B$. After translation and rotation we may assume that $F = 0, A = (1 - \lambda)a \otimes e_1, B = -\lambda a \otimes e_1$. Let h be a 1-periodic sawtooth function which satisfies $h(0) = 0$, $h' = (1 - \lambda)$ on $(0, \lambda)$ and $h' = -\lambda$ on $(\lambda, 1)$. Define for $x \in Q = (0, 1)^n$

$$
\begin{aligned}
u_k &= ak^{-1}h(kx^1), \\
v_k &= a \min\{k^{-1}h(kx^1), \mathrm{dist}_\infty(x, Q)\},
\end{aligned}
$$

where

$$
\begin{aligned}
\mathrm{dist}_\infty(x, Q) &= \inf\{\|x - y\|_\infty : y \in Q\}, \\
\|x\|_\infty &= \sup\{|x^i|, i = 1, \dots, n\}.
\end{aligned}
$$

Then $Dv_k \in \{A, B\} \cup \{\pm a \otimes e_i\}, v_k = 0$ on ∂Q, and $|\{Dv_k \neq Du_k\}| \to 0$ as $k \to 0$ (see Fig. 12).

It follows from the definition of quasiconvexity that

$$
\lambda f(A) + (1 - \lambda)f(B) = \lim_{k \to \infty} \int_Q f(Du_k)\,dx = \lim_{k \to \infty} \int_Q f(Dv_k)\,dx \geq f(0),
$$

as desired. Note that the inequality $\lambda f(A) + (1 - \lambda)f(B) \geq f(0)$ still holds if f takes values in $[-\infty, \infty)$.

As for the reverse implications, the minors (subdeterminants) of order greater than one are trivially polyconvex but not convex. An example of a

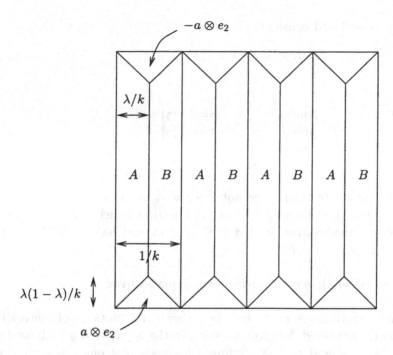

Figure 12: The gradients of v_k, for $n = 2$.

quasiconvex but not polyconvex function is given below. Šverák's counterexample of a rank-1 convex function that is not quasiconvex will be discussed in Section 4.7. □

Remark. The proof that quasiconvexity implies rank-1 convexity is similar to Fonseca's ([Fo 88], Theorem 2.4). In fact her method yields a slightly stronger result: if $f : M^{m \times n} \to [-\infty, \infty]$ is finite in a neighbourhood of F and quasiconvex then f does not take the value $-\infty$ on any rank-1 line through F and f is rank-1 convex at F, i.e. $f(F) \le \lambda f(F - (1 - \lambda)a \otimes b)$ $+(1 - \lambda)f(F + \lambda a \otimes b), \forall a \in \mathbf{R}^n, b \in \mathbf{R}^m, \lambda \in (0, 1)$. To obtain this refinement it suffices to replace $\text{dist}_\infty(x, Q)$ in the definition of v_k by $\epsilon \, \text{dist}_\infty(x, Q)$ for small enough $\epsilon > 0$.

The following example, due to Dacorogna and Marcellini [AD 92], [DM 88], [Da 89], may serve as a simple illustration of the different notions of convexity. Let $n = m = 2$ and consider

$$f(F) = |F|^4 - \gamma |F|^2 \det F. \tag{4.2}$$

Then

f	convex	$\Longleftrightarrow$	$|\gamma| \le \frac{4}{3}\sqrt{2},$
f	polyconvex	$\Longleftrightarrow$	$|\gamma| \le 2,$
f	quasiconvex	$\Longleftrightarrow$	$|\gamma| \le 2 + \epsilon,$
f	rank-1 convex	$\Longleftrightarrow$	$|\gamma| \le \frac{4}{\sqrt{3}}.$

It is known that $\epsilon > 0$; whether or not $2 + \epsilon = \frac{4}{\sqrt{3}}$ is open.

Alberti raised the following interesting question which shows how little we know about quasiconvexity. Let $2 \le n \le m$ and let $g : M^{m \times n} \to \mathbf{R}$, $\tilde{g} : M^{n \times m} \to \mathbf{R}, \tilde{g}(F) = g(F^T)$.

Question (Alberti): g quasiconvex $\overset{?}{\Longleftrightarrow}$ $\tilde{g}$ quasiconvex.

Obviously equivalence holds for the other three notions of convexity. Kružik recently answered Alberti's question in the negative if g is allowed to take the value $+\infty$ and $m \ge 3$. Refining his argument one can show that Šverák's quartic polynomial provides a finite-valued counterexample (see the end of section 4.7).

Ball, Kirchheim and Kristensen [BKK 98] recently solved a long-standing problem by proving that the quasiconvex hull of a C^1 function f (i.e. the largest quasiconvex function below f) is again C^1, provided that f satisfies polynomial growth conditions. The representation of the quasiconvex hull through gradient Young measures (see Section 4.3) plays a crucial rôle in their argument.

4.2 Properties of quasiconvexity

Quasiconvexity is the fundamental notion of convexity for vector-valued variational problems. It is closely related to lower semicontinuity of integral functionals, existence and regularity of minimizers and the passage from microscopic and macroscopic energies. Quasiconvex functions are the natural dual objects to gradient Young measures (see Section 4.3).

In the following Ω always denotes a bounded (Lipschitz) domain in $\mathbf{R}^n$ and we consider maps $u : \Omega \to \mathbf{R}^m$ and the functional

$$I(u) = \int_\Omega f(Du)dx.$$

In this section we merely summarize the results. Some of the proofs for $p = \infty$ are given in Sections 4.8 and 4.9 below. Further comments and references can be found at the end of these notes.

Theorem 4.4 *Suppose that* $f : M^{m \times n} \to \mathbf{R}$ *is continuous.*

(i) *The functional I is weak$*$ sequentially lower semicontinuous (w$*$slsc) on $W^{1,\infty}(\Omega; \mathbf{R}^m)$ if and only if f is quasiconvex.*

(ii) *Suppose, in addition, that*

$$0 \le f(F) \le C(|F|^p + 1) \tag{4.3}$$

for some $p \in [1, \infty)$. If f is quasiconvex then I is wslsc on $W^{1,p}(\Omega; \mathbf{R}^m)$.

Remarks. If $f \ge 0$ it can be shown that I is finite and wslsc on $W^{1,p}$ if and only if f satisfies (4.3) and is quasiconvex (see [Kr 94]). Part (i) is an essential ingredient in the classification of gradient Young measures. Using this classification and simple general facts about Young measures (see Section 3.4) one easily obtains similar lower semicontinuity results for integrands $f(x, u(x), Du(x))$.

Theorem 4.5 *(existence and relaxation)*
Suppose that $p \in (1, \infty), c > 0$ and that f satisfies

$$c|F|^p \le f(F) \le C(|F|^p + 1).$$

(i) *If f is quasiconvex and $v \in W^{1,p}(\Omega; \mathbf{R}^m)$ then I attains its minimum in the class*

$$W_v^{1,p}(\Omega; \mathbf{R}^m) := \{u \in W^{1,p}(\Omega; \mathbf{R}^m) : u - v \in W_0^{1,p}(\Omega; \mathbf{R}^m)\}.$$

(ii) *If f^{qc} denotes the quasiconvex envelope of f, i.e. the largest quasiconvex function below f, then*

$$\inf_{W_v^{1,p}} I = \min_{W_v^{1,p}} \bar{I},$$

where

$$\bar{I}(u) = \int_\Omega f^{qc}(Du). \tag{4.4}$$

Moreover a function $\bar{u}$ is a minimizer of $\bar{I}$ in $W_v^{1,p}$ if and only if it is a cluster point (with respect to weak convergence in $W^{1,p}$) of a minimizing sequence for I.

(iii) *For any $f : M^{m \times n} \to [-\infty, \infty)$ and every bounded domain U with $|\partial U| = 0$ one has*

$$f^{qc}(F) = \inf_{\varphi \in W_0^{1,\infty}} \frac{1}{|U|} \int_U f(F + D\varphi) dx. \tag{4.5}$$

The passage from I to $\bar{I}$ is called relaxation. It replaces a variational problem which may have no solution by one which has a solution. This sounds almost too good to be true and indeed there is a price to pay. The minimizers of $\bar{I}$ are in general only weak limits of a minimizing sequence of I, and important features of the sequence may be lost. If, for example, $\bar{I}$ has a homogeneous minimizer it is not clear whether minimizing sequences of I are (nearly) homogeneous or whether they involve an increasingly finer mixture of several states. A different approach, that keeps more information about the minimizing sequence is to derive a (relaxed) problem for the gradient Young measures generated by minimizing sequences (see Theorem 4.9 of the next section).

Physically, relaxation corresponds to the passage from a microscopic energy I to a macroscopic energy $\bar{I}$, which is obtained by averaging over fine scale oscillations; cf. the representation (4.5).

Theorem 4.6 *(regularity). Suppose that f is smooth, satisfies*

$$0 \leq f(F) \leq C(|F|^2 + 1)$$

and is uniformly quasiconvex, i.e. there exists $c > 0$ such that

$$\int_U [f(F + D\varphi) - f(F)] dx \geq c \int_U |D\varphi|^2, \quad \forall \varphi \in W_0^{1,\infty}(\Omega; \mathbf{R}^m).$$

Let $\bar{u} \in W^{1,2}(\Omega; \mathbf{R}^m)$ be a local minimizer of I, i.e.

$$I(\bar{u} + \varphi) \geq I(\bar{u}) \quad \forall \varphi \in C_0^\infty(\Omega).$$

Then there exists an open set Ω_0 of full measure such that

$$u \in C^\infty(\Omega_0).$$

4.3 Classification of gradient Young measures

Recall that a map $\nu : \Omega \to \mathcal{M}(M^{m \times n})$ is a $W^{1,\infty}$ gradient Young measure if it is the Young measure generated by a sequence of gradients Du_j, where $u_j \overset{*}{\rightharpoonup} u$ $W^{1,\infty}$ (see Definition 4.1).

Theorem 4.7 *([KP 91]) A (weak* measurable) map $\nu : \Omega \to \mathcal{M}(M^{m \times n})$ is a $W^{1,\infty}$ gradient Young measure if and only if $\nu_x \geq 0$ a.e. and there exists a compact set $K \subset M^{m \times n}$ and $u \in W^{1,\infty}(\Omega; \mathbf{R}^m)$ such that the following three conditions hold.*

(i) $\operatorname{supp}\nu_x \subset K$ *for a.e. x*

(ii) $\langle \nu_x, \operatorname{id} \rangle = Du$ *for a.e. x*

(iii) $\langle \nu_x, f \rangle \geq f(\langle \nu_x, \operatorname{id} \rangle)$ *for a.e. x and all quasiconvex $f : M^{m \times n} \to \mathbf{R}$.*

Remarks. 1. The key point is (iii) which is in nice duality with the definition of quasiconvexity. Roughly speaking, quasiconvex functions satisfy Jensen's inequality for gradients, while gradient Young measures must satisfy Jensen's inequality for all quasiconvex functions.

2. Let $K \subset M^{m \times n}$ be compact. For future reference we define the set of nonnegative measures that satify condition (iii) of the theorem as

$$\begin{aligned} \mathcal{M}^{qc}(K) = \{\nu \in \mathcal{M}(M^{m \times n}) : \nu \geq 0, \operatorname{supp}\nu \subset K, \\ \langle \nu, f \rangle \geq f(\langle \nu, \operatorname{id} \rangle) \quad \forall f : M^{m \times n} \to \mathbf{R} \quad \text{quasiconvex}\}. \end{aligned} \tag{4.6}$$

By the theorem $\mathcal{M}^{qc}(K)$ consist exactly of the homogeneous gradient Young measures supported in K. Similarly one defines $\mathcal{M}^{rc}(x)$ and $\mathcal{M}^{pc}(x)$ using rank-one convex and polyconvex functions, respectively.

3. Every minor M is a quasiaffine function (i.e. (4.1) holds with equality; see Theorem 2.3(i)). Hence application of (iii) with $\pm M$ yields the minors relations

$$\langle \nu_x, M \rangle = M(\langle \nu_x, \mathrm{id} \rangle) \tag{4.7}$$

as a necessary condition for gradient Young measures. This condition in fact follows directly from Theorem 2.3 (ii) and does not require Theorem 4.7. The minors relations often prove very useful for problems with large symmetries that arise e.g. in models of microstructure in crystals (see e.g. [Bh 92]). They are, however, far from being sufficient in general.

Exercise. Find a nontrivial measure supported on three diagonal 2×2 matrices without rank-1 connections that satisfies (4.7) and compare with Theorem 2.5.

Hint: Look for matrices on the two hyperbolae given by $\{\det = 1\}$.

Proof of Theorem 4.7 (necessity). Conditions (i) and (ii) follow from basic facts about Young measures (see Theorem 3.1 (ii) and (iii)) while (iii) follows form Morrey's lower semicontinuity result (Theorem 4.3(ii)), applied to all open subsets U of Ω. Sufficiency is discussed in Section 4.9. □

To finish the section we briefly mention the analogous result for $p < \infty$ and its relation to relaxation and generalized solutions. This may be omitted on first reading.

Theorem 4.8 *([KP 94]) Let $p \in [1, \infty)$. A (weakly measurable) map $\nu : \Omega \to \mathcal{M}(M^{m \times n})$ is a $W^{1,p}$ gradient Young measure if and only if $\nu_x \geq 0$ a.e. and the following three conditions hold*

(i) $\int\limits_{\Omega} \int\limits_{M^{m \times n}} |F|^p d\nu_x(F) dx < \infty;$

(ii) $\langle \nu_x, \mathrm{id} \rangle = Du, \quad u \in W^{1,p}(\Omega; \mathbf{R}^m);$

(iii) $\langle \nu_x, f \rangle \geq f(\langle \nu_x, \mathrm{id} \rangle)$ *for a.e. x and all quasiconvex f with $|f|(F) \leq C(|F|^p + 1).$*

Young measures arise naturally as generalized solutions of variational problems that have no classical solution. To this end extend the functional

$$I(u) = \int\limits_{\Omega} f(Du) dx$$

on functions to a functional

$$J(\nu) = \int_\Omega \langle \nu_x, f \rangle dx$$

on Young measures. For $v \in W^{1,p}(\Omega; \mathbf{R}^m)$ consider the admissible classes

$$\begin{aligned}
\mathcal{A} &= \{u \in W^{1,p}(\Omega; \mathbf{R}^m) : u - v \in W_0^{1,p}(\Omega; \mathbf{R}^m), \\
\mathcal{G} &= \{\nu : \Omega \to \mathcal{M}(R^m) : \nu \quad W^{1,p} \text{ gradient Young measure}, \\
&\qquad \langle \nu_x, \mathrm{id} \rangle = Du(x), u \in \mathcal{A}\}.
\end{aligned}$$

Theorem 4.9 *Suppose that f is continuous and satisfies $c|F|^p \leq f(F) \leq C(|F|^p + 1), c > 0, p > 1$. Then*

$$\inf_{\mathcal{A}} I = \min_{\mathcal{G}} J(\nu).$$

Moreover the minimizers of J are Young measures that are generated by gradients of minimizing sequences of I.

In particular, I has a minimizer in $\mathcal{A}$ if and only if there exists a minimizer ν of J such that ν_x is a Dirac mass for a.e. *x.*

4.4 Convex hulls and resolution of Problem 3

To return to the setting of Sections 1 and 2 we extend the different notions of convexity from functions to sets. We first recall that the quasiconvex (convex, polyconvex, rank-1 convex) envelope or hull of a function $f : M^{m \times n} \to \mathbf{R}$ is the largest quasiconvex (convex, polyconvex, rank-1 convex) function below f and is denoted by f^{qc} ($f^c = f^{**}, f^{pc}, f^{rc}$). Similarly the quasiconvex hull of a set $K \subset M^{m \times n}$ is defined via sublevel sets as

$$K^{qc} = \{F \in M^{m \times n} : f(F) \leq \inf_K f, \quad \forall f : M^{m \times n} \to \mathbf{R} \text{ quasiconvex}\},$$

with similar definitions for K^c, K^{pc}, K^{rc}. Note that K^c is the *closed* convex hull. A set is called quasiconvex if $K = K^{qc}$. In the case of rank-1 convexity one can also define a hull by pointwise operations rather than by sublevel sets. A set K is called *lamination convex* if the conditions $A, B \in K$ and $\mathrm{rk}(B - A) = 1$ imply that convex combinations of A and B are in K. The lamination convex hull K^{lc} of K is the smallest lamination convex set containing K. It is easy to verify that K^{lc} can equivalently be defined by inductively adding rank-1 segments:

$$K^{lc} := \bigcup_{i=1}^{\infty} K^{(i)}, \qquad K^{(1)} := K,$$

$$K^{(i+1)} := K^{(i)} \cup \{\lambda A + (1 - \lambda)B : A, B \in K^{(i)}, \mathrm{rk}(B - A) = 1, \lambda \in (0, 1)\}.$$

One has the following inclusions (see Lemma 4.3):

$$K^{lc} \subset K^{rc} \subset K^{qc} \subset K^{pc} \subset K^{c}. \tag{4.8}$$

The example in Section 2.5 shows that in general $K^{lc} \neq K^{rc}$. In this example $K^{lc} = K, K^{rc} \supset K \cup \{\mathrm{diag}(\lambda, \mu) : |\lambda| \leq 1, |\mu| \leq 1\}$. The characterization of laminates (see Section 4.6) as well as recent work of Matoušek and Plecháč [MP 98] suggest that K^{rc} is the more natural object, but more difficult to handle (Matoušek and Plecháč use the terms set-theoretic rank-1 convex hull and functional rank-1 convex hull to distinguish K^{lc} and K^{rc}).

The polyconvex hull is closest to the ordinary (closed) convex hull and is in fact the intersection of a convex set with a nonconvex constraint. Let $\mathbf{M}(F)$ denote the vector of all minors of F (see Section 4.1) and let

$$\hat{K} = \{\mathbf{M}(F) : F \in K\}.$$

Exercise. Show that

$$K^{pc} = \{F : M(F) \in (\hat{K})^c\} \tag{4.9}$$

and moreover

$$K^{pc} = \{\langle \nu, \mathrm{id} \rangle : \nu \in \mathcal{M}^{pc}(K)\}.$$

With this notation in place we have the following abstract resolution of Problem 3 (see Section 1.4). Recall that the set K^{app} (interpreted as the macroscopically stress free affine deformations) was the set of all matrices F such that there exists a sequence u_j bounded in $W^{1,\infty}(\Omega; \mathbf{R}^m)$, such that

$$\mathrm{dist}(Du_j, K) \to 0 \qquad \text{in measure in } \Omega, \tag{4.10}$$

$$u_j = Fx \qquad \text{on } \partial\Omega, \tag{4.11}$$

and that $\mathcal{M}^{qc}(K)$ denotes the set of homogeneous gradient Young measures (see (4.6)).

Theorem 4.10 *Suppose that K is compact and denote by* dist_K *the distance function from K. Then*

(i) $K^{app} = K^{qc}$,

(ii) $K^{qc} = \{\text{dist}_K^{qc} = 0\}$,

(iii) K^{qc} *is the set of barycentres of homogeneous gradient Young measures:*

$$K^{qc} = \{\langle \nu, \text{id} \rangle : \nu \in \mathcal{M}^{qc}(K)\}.$$

Proof. (i) To show that $K^{app} \subset K^{qc}$ let $F \in K^{app}$ and let $f : M^{m \times n} \to \mathbf{R}$ be quasiconvex and suppose that u_j is bounded in $W^{1,\infty}$ and satisfies (4.10) and (4.11). We may assume that $\inf_K f = 0$ and we need to show $f(F) \leq 0$. Since f is continuous (see Remark 3 after Definition 4.2) we have $f^+(Du_j) \to 0$ in measure, and $|f^+(Du_j)| \leq C$ since Du_j is bounded in L^∞. Quasiconvexity, (4.11) and dominated convergence yield

$$|\Omega| f(F) \leq \liminf_{j \to \infty} \int_\Omega f(Du_j) dx \leq \liminf_{j \to \infty} \int_\Omega f^+(Du_j) dx = 0.$$

To prove the converse inclusion $K^{qc} \subset K^{app}$ let $F \in K^{qc}$. Then $\text{dist}_K^{qc}(F) = 0$ by definition of K^{qc}. In view of the representation formula for dist_K^{qc} (Theorem 4.5 (iii)) there exist $\varphi_j \in W_0^{1,\infty}(\Omega; \mathbf{R}^m)$ such that

$$0 = \text{dist}_K^{qc}(F) = \lim_{j \to \infty} \int_\Omega \text{dist}_K(F + D\varphi_j) dx.$$

The functions $u_j(x) = Fx + \varphi_j$ thus satisfy (4.10) and (4.11). The problem is that a priori Du_j only needs to be bounded in L^1 (in fact weakly relatively compact in L^1) and may not be bounded in L^∞. Zhang's lemma (see Lemma 4.21 (ii) below) assures that u_j can be modified on small sets such that (4.10) and (4.11) hold and Du_j is bounded in L^∞.

(ii) The inclusion $\subset$ follows from the definition of K^{qc}. On the other hand we have just shown that $\text{dist}_K^{qc}(F) = 0$ implies $F \in K^{app} = K^{qc}$.

(iii) Let $\nu \in \mathcal{M}^{qc}(K)$. By definition $f(\langle \nu, \text{id} \rangle) \leq \langle \nu, f \rangle \leq \sup_K f$ and hence $\langle \nu, \text{id} \rangle \in K^{qc}$. Suppose conversely that $F \in K^{qc}$. We need to show that there exists $\nu \in \mathcal{M}^{qc}(K)$ with $\langle \nu, \mathrm{id} \rangle = F$. After an affine transformation we may assume $F = 0$. By part (i) there exists a sequence u_j (bounded in $W^{1,\infty}$) that satisfies (4.10) and (4.11). Passing to a subsequence we may assume that $\{Du_j\}$ generates the Young measure ν and $Du_j \overset{*}{\rightharpoonup} Du$ in $L^\infty(\Omega; M^{m \times n})$. By the divergence theorem

$$\int_\Omega \langle \nu_x, \mathrm{id} \rangle dx = \int_\Omega Du\, dx = 0. \qquad (4.12)$$

To obtain a homogeneous Young measure we define the average $\mathrm{Av}\,\nu$ by duality as the unique Radon measure that satisfies

$$\langle \mathrm{Av}\,\nu, f \rangle = \frac{1}{|\Omega|} \int_\Omega \langle \nu_x, f \rangle dx \quad \forall f \in C_0(M^{m \times n}).$$

By Theorem 4.7 we have $\nu_x \in \mathcal{M}^{qc}(K)$ for a.e. x and hence $\mathrm{Av}\,\nu \in \mathcal{M}^{qc}(K)$. Moreover (4.12) yields $\langle \mathrm{Av}\nu, \mathrm{id} \rangle = 0$ as desired. $\qquad \square$

4.5 The two-well problem

To see what the various convexity notions can do to understand microstructure in crystals we consider the two-well problem in two dimensions. This is the simplest multiphase problem consistent with the rotational symmetry and was analyzed completely in a beautiful paper by Šverák [Sv 93a]. Let

$$K = SO(2)A \cup SO(2)B \subset M^{2 \times 2}, \quad \det B \geq \det A > 0. \qquad (4.13)$$

Various normalizations are possible. Multiplication by A^{-1}, polar decomposition and diagonalization show, for example, that it suffices to consider

$$A = \begin{pmatrix} 1 & 0 \\ 0 & 1 \end{pmatrix}, \quad B = \begin{pmatrix} \lambda & 0 \\ 0 & \mu \end{pmatrix}, \quad 0 < \lambda \leq \mu, \quad \lambda\mu \geq 1. \qquad (4.14)$$

The first step towards the resolution of Problems 1-3 is to look for rank-1 connections in K.

Exercise. Prove the following classification.

(i) If $\lambda > 1$ then there are no rank-1 connections in K;

(ii) if $\lambda = 1$ (and $A \neq B$) each matrix in K is rank-1 connected to exactly one other matrix in K;

(iii) if $\lambda < 1$ each matrix in K is rank-1 connected to exactly two other matrices in K.

Theorem 4.11 *Suppose that K given by (4.13) contains no rank-1 connections. Then every Young measure $\nu : \Omega \to \mathcal{M}(M^{2\times 2})$ with $\mathrm{supp}\nu_x \in K$ is a constant Dirac mass. Moreover*

$$K^{lc} = K^{rc} = K^{qc} = K^{pc} = K \qquad (4.15)$$

Remark. It is not known whether the same result holds for $K = SO(3)A \cup SO(3)B \subset M^{3\times 3}$; some special cases are known ([Sv 93a], [Ma 92]).

Proof. The crucial observation is that

$$\det(F - G) > 0 \quad \forall F, G \in K, F \neq G. \qquad (4.16)$$

By symmetry and $SO(2)$ invariance it suffices to verify this for $G = \mathrm{Id}$. The inequality clearly holds for $F = B$ (by the above exercise) and hence by connectedness and the absence of rank-1 connections for $G \in SO(2)B$. Similarly $\det(\mathrm{Id} - (-\mathrm{Id})) > 0$ and hence by connectedness (4.16) holds also for all other $G \in SO(2)$.

To determine K^{qc} consider first a homogeneous gradient Young measure ν supported in K and let $\bar{\nu} = \langle \nu, \mathrm{id} \rangle$ denote its barycentre. We have for $F, G \in M^{2\times 2}$

$$\det(F - G) = \det F - \mathrm{cof}F : G + \det G,$$

where $F : G = trF^t G = \sum_{i,j} F_{ij}G_{ij}$. The minors relations yield

$$
\begin{aligned}
0 &\leq \int_{M^{2\times 2} \times M^{2\times 2}} \det(F - G)d\nu(F)d\nu(G) \\
&= \int_{M^{2\times 2} \times M^{2\times 2}} (\det F - \mathrm{cof}F : G + \det G)d\nu(F)d\nu(G) \\
&= \int_{M^{2\times 2}} (\det \bar{\nu} - \mathrm{cof}\,\bar{\nu} : G + \det G)d\nu(G) \\
&= \det \bar{\nu} - \mathrm{cof}\,\bar{\nu} : \bar{\nu} + \det \bar{\nu} = \det(\bar{\nu} - \bar{\nu}) = 0.
\end{aligned}
$$

Hence the first inequality must be an equality, and (4.16) implies that the product measure $\nu \otimes \nu$ is supported on the diagonal of $M^{2\times 2} \times M^{2\times 2}$. Hence ν must be a Dirac mass. This implies $K^{qc} = K$ by Theorem 4.10. Since the argument used only the minors relations we even have $K^{pc} = K$.

Now let $\nu : \Omega \to \mathcal{M}(M^{m \times n})$ be an arbitrary gradient Young measure with $\text{supp}\nu_x \subset K$ a.e. By the above argument $\nu_x = \delta_{Du(x)}$ and $Du(x) \in K$ a.e. We show that $Du \equiv \text{const}$. To this end observe first that (4.16) can be strengthened to

$$\det(X - Y) \geq c|X - Y|^2, \quad c > 0, \quad \forall X, Y \in K. \qquad (4.17)$$

Indeed by compactness and $SO(2)$ invariance it suffices to verify that the tangent space of $SO(2)$ at the identity contains no rank-1 connections. This is obvious. Now let e be a unit vector in $\mathbf{R}^2$ and for $0 < h < 1$ consider the translates $v(x) = u(x + he)$ and a cut-off function $\varphi \in C_0^\infty(\Omega)$. Since the determinant is a null Lagrangian (see Theorem 2.3(i)) integration of (4.17) yields

$$c \int_\Omega \varphi^2 |Du - Dv|^2 \, dx \leq \int_\Omega \det[\varphi(Du - Dv)] \, dx$$

$$\leq \int_\Omega \det[D\varphi(u - v)] \, dx - \int_\Omega \text{cof} D\varphi(u - v) : (u - v) \otimes D\varphi \, dx$$

$$+ \int_\Omega \det[(u - v) \otimes D\varphi] \, dx$$

$$\leq \frac{c}{2} \int_\Omega \varphi^2 |Du - Dv|^2 \, dx + C \int_\Omega |D\varphi|^2 |u - v|^2 \, dx.$$

Hence the difference quotients $\varphi \frac{Du - Dv}{h}$ are uniformly bounded in L^2 and thus $Du \in W_{\text{loc}}^{1,2}(\Omega; M^{m \times n})$. Therfore Du can only take values in one connected component of K, and the assertion follows from Theorem 2.4. $\qquad \square$

To consider the case where K has rank-1 connections it is convenient to introduce new coordinates on $M^{2 \times 2}$. Since A and B are not conformally equivalent (as $\lambda < \mu$) for every matrix F there exist a unique pair $(y, z) \in \mathbf{R}^2 \times \mathbf{R}^2$ such that

$$F = \begin{pmatrix} y_1 & -y_2 \\ y_2 & y_1 \end{pmatrix} + \begin{pmatrix} z_1 & -z_2 \\ z_2 & z_1 \end{pmatrix} B.$$

Theorem 4.12 *Suppose that K is given by (4.13), (4.14) and that $\lambda < 1$. Then*

$$K^{lc} = K^{rc} = K^{qc} = K^{pc}$$

and

$$K^{pc} = K^c \cap \{\det = 1\}, \quad \text{if } \det B = 1,$$
$$K^{pc} = \{F = (y,z) : |y| \leq \tfrac{\det B - \det F}{\det B - 1}, \quad |z| \leq \tfrac{\det F - 1}{\det B - 1}\}$$
$$\text{if } \det B > 1.$$

To characterize the polyconvex hull we use the following

Proposition 4.13 *The convex hull of the set* $K \subset \mathbf{R}^n \times \mathbf{R}^n \times \mathbf{R}$ *given by* $K = \{(y,0,a) : |y| = 1\} \cup \{(0,z,b) : |z| = 1\}$ *is given by*

$$K^c = \begin{cases} \{(y,z,a), |y| + |z| \leq 1\} & \text{if } a = b, \\ \{(y,z,t), |y| \leq \tfrac{b-t}{b-a}, |z| \leq \tfrac{t-a}{b-a}\} & \text{if } a < b. \end{cases}$$

Proof. This is obvious for $n = 1$, and the general case follows by invariance under $(y,z,t) \to (Ry, Qz, t)$, $R, Q \in SO(n)$. $\qquad\square$

Proof of Theorem 4.12. The formula for K^{pc} follows from the characterization (4.9) and Proposition 4.13. In view of the general relation (4.8) between the different convex hulls it only remains to show that $K^{lc} = K^{pc}$.

First case: $\det B = 1$.

Let $F \in K^{pc} = K^c \cap \{\det = 1\}$. If $F = (y,z) \in \partial K^c$, then by Proposition 4.13 we have $|y| + |z| = 1$. If $y = 0$ or $z = 0$ then $F \in K$ and we are done. If $y \neq 0, z \neq 0$ then we can consider $G(t) = ty/|y| + (1-t)z/|z|$. Let $g(t) = \det G(t)$. Then g is quadratic in t and $g(0) = g(1) = g(|y|) = 1$. Hence $g \equiv 1$ and $t \to G(t)$ must be a rank-1 line. This shows that F is a rank-1 combination of $(y/|y|, 0) \in K$ and $(0, z/|z|) \in K$.

If $F \in \mathrm{int} K^c \cap \{\det = 1\}$ then there exist $a, n \in S^1$ such that $\mathrm{cof} F : a \otimes n = Fn \cdot a = 0$. Hence the determinant is constant on the line $F + ta \otimes n$ and this rank-1 line intersects ∂K^c for positive and negative values of t. Therefore every matrix in $\mathrm{int} K^c \cap \{\det = 1\}$ is a rank-1 combination of two matrices in $\partial K^c \cap \{\det = 1\}$ and thus belongs to K^{lc}. This finishes the proof for $\det B = 1$.

Second case: $\det B > 1$.

Since every rank-1 half-line through an interior point of K^{pc} intersects ∂K^{pc} it suffices to show $\partial K^{pc} \subset K^{lc}$. Let $\bar{F} = (\bar{y}, \bar{z}) \in \partial K^{pc}$ and define

$$f(y,z) = (\det B - 1)|y| - \det B + \det F,$$
$$g(y,z) = (\det B - 1)|z| + 1 - \det F.$$

The polyconvex hull is given by $\{f \leq 0\} \cap \{g \leq 0\}$ and thus $f(\bar{y}, \bar{z}) = 0$ or $g(\bar{y}, \bar{z}) = 0$. For convenience we assume that latter, the other case is analogous.

If $f(\bar{F}) = 0$ then $|y| + |z| = 1$. Moreover f and g are quadratic functions on the segment $ty/|y| + (1-t)z/|z|$, and vanish at $t = 0, 1, |y|$. Hence they vanish identically on the segment which therefore is a rank-1 segment. Thus $F \in K^{lc}$.

Now suppose that $f(\bar{F}) < 0$. Using the $SO(2)$ invariance we may assume that $z_2 = 0$. For definiteness we suppose $z_1 > 0$, the case $z_1 < 0$ is analogous. Note that the linear space $\{z_2 = 0\}$ agrees with $\{F_{12} + F_{21} = 0\}$. We claim that there exists a rank-1 line in $\{z_2 = 0\}$ through $\bar{F}$ on which g vanishes (as long as $z_1 > 0$). One way to see this is to consider $\tilde{g}(y, z) = (\det B - 1)z_1 + 1 - \det F$ and to note that $\tilde{g} = 0$ defines a one sheeted hyperboloid H in the three dimensional space $\{z_2 = 0\}$. Hence through each point in H there exist two lines that lie on H and thus must be rank-1 line since $\det F$ is an affine function on these lines. Alternatively one can consider $(y(t), z(t)) = F(t) = \bar{F} + t(\mu - \lambda) \, a \otimes Pa$, with $Pa = (-a_1, a_2)$. Then $z_2(t) = 0$, $\dot{z}_1 = |a|^2 > 0$ and $\tilde{g}$ is affine on the line $t \to F(t)$. A short calculation shows that $\frac{d}{dt}g(F(t))_{|t=0} = (Qa, a)$ and the quadratic form

$$Q = (\det B - 1)\mathrm{Id} + \frac{1}{2}(\mu - \lambda)[(\operatorname{cof}\bar{F})P + P(\operatorname{cof}\bar{F})^T]$$

is indefinite and hence has a nontrivial kernel.

Consider thus the rank-1 line $F(t) = \bar{F} + t(\mu - \lambda)a \otimes Pa$ on which z_2 and g vanish.

Let $t_0 < 0$ be defined by $z_1(t_0) = 0$. Since $g(F(t_0)) = 0$ we deduce that $F(t_0) = (y(0), 0) \in K$. On the other hand $f(F(0)) < 0$ and using the fact that g vanishes on $F(t)$ we have $f(F(t)) = (\det B - 1)(|y(t)| + |z(t)| - 1) \to \infty$ as $t \to \infty$. Hence there exist $t_1 > 0$ such that $f(F(t_1)) = g(F(t_1)) = 0$ and therefore $F(t_1) \in K^{lc}$ by the considerations above. Thus $\bar{F} = F(0) \in K^{lc}$ and the proof is finished. $\qquad\square$

4.6 Are all microstructures laminates?

Theorems 4.7 and 4.10 completely classify gradient Young measures $\mathcal{M}^{qc}(K)$ and quasiconvex hulls K^{qc} and thus lead to an abstract solution of problems 2 and 3 in Section 1.4. The catch is that very few quasiconvex functions are known and that the abstract results are therefore of limited use to understand specific sets K. A manageable necessary condition is given by the minors relations (4.7). In this section we discuss the issue of sufficient conditions, i.e. constructions of (homogeneous) gradient Young measures supported on

More generally (see e.g. [BM 84]) if $h : \mathbf{R}^n \to \mathbf{R}$ is locally integrable and periodic with unit cell $[0,1]^n$ and z_j is defined by (3.5), then z_j generates a homogeneous Young measure ν given by

$$\int_{\mathbf{R}} g d\nu = \int_{[0,1]^n} g(h(y)) dy.$$

For a Borel set $B \subset \mathbf{R}$ one has

$$\nu(B) = |(0,1)^n \cap h^{-1}(B)|.$$

b) Let

$$I(u) = \int_0^1 (u_x^2 - 1)^2 + u^2 dx,$$

let u_j be a sequence such that

$$I(u_j) \to 0, \qquad u_j(0) = u_j(1) = 0, \tag{3.6}$$

and let $z_j = (u_j)_x$ (cf. Example 2 in Section 1.2). Then z_j is bounded in L^4, a subsequence generates a Young measure ν and $\|\nu_x\| = 1$ a.e. If we let $g(p) = \min((p^2 - 1)^2, 1)$ we deduce from (3.6) that

$$\langle \nu_x, g \rangle = 0 \qquad \text{for a.e. } x.$$

Hence $\mathrm{supp}\,\nu_x \subset \{-1, 1\}$ and $\nu_x = \lambda(x)\delta_{-1} + (1 - \lambda(x))\delta_1$ a.e. By Remark 5 after Theorem 3.1

$$z_{j_k} \overset{*}{\rightharpoonup} \langle \nu_x, \mathrm{id} \rangle = 1 - 2\lambda(x) \tag{3.7}$$

and

$$u_{j_k}(a) = \int_0^a z_{j_k} dx \to \int_0^a (1 - 2\lambda(x)) dx. \tag{3.8}$$

By (3.6) $u_j \to 0$ in L^2 and thus $\lambda(x) = 1/2$ a.e. Hence z_{j_k} generates the unique (homogeneous) Young measure

$$\nu_x = \frac{1}{2}\delta_{-1} + \frac{1}{2}\delta_1.$$

By uniqueness the whole sequence z_j generates this Young measure.

a given set K. The simplest case is $K = \{A, B\}$. If A and B are rank-1 connected every convex combination

$$\nu = \lambda\delta_A + (1 - \lambda)\delta_B, \quad \lambda \in [0, 1],$$

is a (homogeneous) gradient Young measure. It arises as a limit of a sequence of gradients Du_j arranged in a fine lamellar pattern (see Fig. 13).

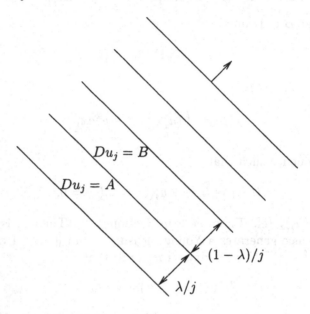

Figure 13: Fine layering of the rank-1 connected matrices A and B generates the homogeneous gradient Young measure $\lambda\delta_A + (1 - \lambda)\delta_B$.

We saw in Section 2.5 that this construction can be iterated for larger sets K. More precisely let C be a matrix that is rank-1 connected to $\lambda A + (1-\lambda)B$. Then every convex combination

$$\nu = \mu(\lambda\delta_A + (1 - \lambda)\delta_B) + (1 - \mu)\delta_C \tag{4.18}$$

is a (homogeneous) gradient Young measure (see Figure 8).

This construction can be iterated and motivates the following definition.

Definition 4.14 ([Da 89]) *For a finite family of pairs* $(\lambda_i, F_i) \in (0,1) \times M^{m \times n}$ *the condition* (H_l) *is defined inductively as follows.*

(i) *Two pairs* $(\lambda_1, F_1), (\lambda_2, F_2)$ *satisfy* (H_2) *if*

$$\text{rk}(F_2 - F_1) \le 1, \quad \lambda_1 + \lambda_2 = 1.$$

(ii) *A family* $\{(\lambda_i, F_i)\}_{i=1,\ldots,l}$ *satisfies* (H_l) *if, after possible renumbering*

$$\text{rk}(F_l - F_{l-1}) = 1 \tag{4.19}$$

and the new family $\{(\tilde{\lambda}_i, \tilde{F}_i)\}_{i=1,\ldots,l-1}$ *given by*

$$\tilde{F}_{l-1} = \frac{\lambda_{l-1}}{\lambda_{l-1} + \lambda_l} F_{l-1} + \frac{\lambda_l}{\lambda_{l-1} + \lambda_l} F_l, \quad \tilde{\lambda}_{l-1} = \lambda_{l-1} + \lambda_l, \tag{4.20}$$

$$(\tilde{\lambda}_i, \tilde{F}_i) = (\lambda_i, F_i) \text{ if } i \le l - 2, \tag{4.21}$$

satisfies (H_{l-1}).

If we call the process defined by (4.19), (4.20) and (4.21) contraction then the family $\{(\lambda_i, F_i)\}_{i=1,\ldots,l}$ satisfies (H_l) if it can be inductively contracted to $(1, \bar{F})$ where $\bar{F} = \sum \lambda_i F_i$ is the barycentre. Note that the F_i may take the same value for different i. To see that this can be useful consider the 8 matrices $\{A_1, \ldots A_4\}$ and $\{I_1, \ldots I_4\}$ in the four gradient example in Section 2.5. The family $(1/2, A_1)$, $(1/4, A_2)$, $(1/8, A_3)$, $(1/16, A_4)$, $(1/32, A_1)$, $(1/32, I_1)$ satisfies (H_6), but the family obtained by combining the two pairs involving A_1 to $(17/32, A_1)$ does not satisfy (H_5).

Definition 4.15 *A (probability) measure* ν *on* $M^{m \times n}$ *is called a laminate of finite order if there exists a family* $\{(\lambda_i, F_i)\}_{i=1,\ldots,l}$ *that satisfies* (H_l) *and*

$$\nu = \sum_{i=1}^{l} \lambda_i \delta_{F_i}.$$

A (probability) measure ν *is a laminate if there exists a sequence* ν_j *of laminates of finite order with support in a fixed compact set such that*

$$\nu_j \overset{*}{\rightharpoonup} \nu \text{ in } \mathcal{M}(M^{m \times n}).$$

Example. Again in the context of the four gradient example in Section 2.5 the measures

$$\frac{1}{2}\delta_{A_1} + \frac{1}{4}\delta_{A_2} + \frac{1}{8}\delta_{A_3} + \frac{1}{16}\delta_{A_4} + \frac{1}{16}\delta_{I_2}$$

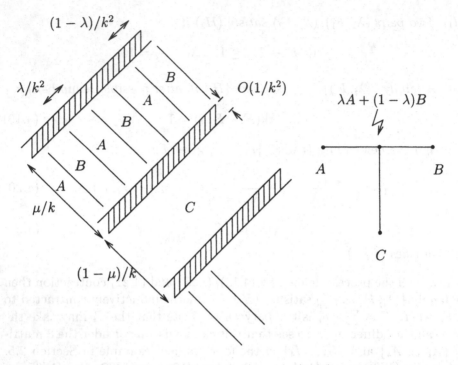

Figure 14: An order-2 laminate that generates (4.18) and the corresponding rank-1 connections.

or

$$(1 - (\frac{1}{16})^j)(\frac{8}{15}\delta_{A_1} + \frac{4}{15}\delta_{A_2} + \frac{2}{15}\delta_{A_3} + \frac{1}{15}\delta_{A_4}) + (\frac{1}{16})^j\delta_{I_2}$$

are laminates of finite order, while $\frac{8}{15}\delta_{A_1} + \frac{4}{15}\delta_{A_2} + \frac{2}{15}\delta_{A_3} + \frac{1}{15}\delta_{A_4}$ is a laminate but not a laminate of finite order.

Condition (H_l) implies that for every rank-1 convex function $f : M^{m \times n} \to \mathbf{R}$ one has

$$\langle \nu, f \rangle \geq f(\langle \nu, \mathrm{id} \rangle)$$

for all laminates of finite order ν. Since (finite) rank-1 convex functions are continuous the same inequality holds for all laminates ν. Pedregal [Pe 93] showed that this property characterizes laminates.

Theorem 4.16 *A compactly supported probability measure $\nu \in \mathcal{M}(M^{m \times n})$*

is a laminate if and only if

$$\langle \nu; f \rangle \geq f(\langle \nu, \mathrm{id} \rangle)$$

for all rank-1 convex functions $f : M^{m \times n} \to \mathbf{R}$. *In other words, the laminates supported on a compact set* K *are given exactly by* $\mathcal{M}^{rc}(K)$.

The question raised in the title of this subsection may now be stated more precisely:

Are all gradient Young measures laminates?

In view of Theorem 4.16 this may be concisely stated as

$$M^{rc} \stackrel{?}{=} \mathcal{M}^{qc}.$$

This would clearly be true if rank-1 convexity implied quasiconvexity. Conversely if $\mathcal{M}^{rc} = \mathcal{M}^{qc}$ then rank-1 convexity would imply quasiconvexity in view of the definition of $\mathcal{M}^{rc}$ and the fact that $f^{qc}(F) = \inf\{\langle \nu, f \rangle : \nu \in M^{qc},$ $\langle \nu, \mathrm{id} \rangle = F\}$ (one equality follows from the definition of $\mathcal{M}^{qc}$; for the other use Theorem 4.5 (iii) for $\Omega = (0,1)^n$, extend φ periodically, let $\varphi_k(x) = k^{-1}\varphi(kx)$ and note that $\{D\varphi_k\}$ generates a homogeneous gradient Young measure).

In the next section we discuss Šverák's example that shows that rank-1 convexity does not imply quasiconvexity if the target dimension satisfies $m \geq 3$.

4.7 Šverák's counterexample

Theorem 4.17 *(Šverák [Sv 92a]) Suppose that* $m \geq 3, n \geq 2$. *Then there exists a function* $f : M^{m \times n} \to \mathbf{R}$ *which is rank-1 convex but not quasiconvex.*

Using this result Kristensen recently showed that there is no local condition that implies quasiconvexity. This finally resolves, for $m \geq 3$, the conjecture carefully expressed by Morrey in his fundamental paper [Mo 52], p. 26: 'In fact, after a great deal of experimentation, the writer is inclined to think that there is no condition of the type discussed, which involves f and only a finite number of its derivatives, and which is both necessary and sufficient for quasi-convexity in the general case.'

To state Kristensen's result let us denote by $\mathcal{F}$ the space of extended real-valued functions $f : M^{m \times n} \to [-\infty, \infty]$. An operator $\mathcal{P} : C^\infty(M^{m \times n}) \to \mathcal{F}$

is called local if the implication

$f = g$ in a neighbourhood of $F \Longrightarrow \mathcal{P}(f) = \mathcal{P}(g)$ in a neighbourhood of F

holds.

Theorem 4.18 *([Kr 97a]) Suppose that $m \geq 3, n \geq 2$. There exists no local operator $\mathcal{P} : C^\infty(M^{m \times n}) \to \mathcal{F}$ such that*

$$\mathcal{P}(f) = 0 \Longleftrightarrow f \text{ is quasiconvex.}$$

By contrast, the local operator

$$\mathcal{P}_{rc}(f)(F) = \inf\{D^2 f(F)(a \otimes b, a \otimes b) : a \in \mathbf{R}^m, b \in \mathbf{R}^n\}$$

characterizes rank-1 convexity. At the end of this subsection we will give an argument of Šverák that proves Theorem 4.18 for $m \geq 6$.

Most research before Šverák's result focused on choosing a particular rank-1 convex integrand f (e.g. the Dacorogna-Marcellini example given by (4.2)) and trying to prove or disprove that there exists a function $u \in W_0^{1,\infty}(\Omega; \mathbf{R}^m)$ and $F \in M^{m \times n}$ such that

$$\int_\Omega f(F + Du)dx < \int_\Omega f(F)dx. \tag{4.22}$$

Šverák's key idea was to first fix a function u and to look for integrands f that satisfy (4.22) but are rank-1 convex. He made the crucial observation that the linear space spanned by gradients of trigonometric polynomials contains very few rank-1 direction and hence supports many rank-1 convex functions.

To proceed, it is useful to note that quasiconvexity can be defined using periodic test functions rather than functions that vanish on the boundary.

Proposition 4.19 *A continuous function $f : M^{m \times n} \to \mathbf{R}$ is quasiconvex if and only if*

$$\int_Q f(F + Du)dx \geq f(F)$$

for all Lipschitz functions u that are periodic on the unit cube Q and all $F \in M^{m \times n}$.

Proof. Sufficiency of the condition is clear since it suffices to verify condition (4.1) for Q (see Remark 2 after Definition 4.2). To establish necessity consider a periodic Lipschitz function u and cut-off functions $\varphi_k \in C_0^\infty((-k,k))^n$ such that $0 \le \varphi_k \le 1$, $\varphi_k = 1$ on $(-(k-1),(k-1))^n$ and $|D\varphi| \le C$. If we let $v_k = \varphi_k u$, $w_k(x) = \frac{1}{k}v_k(kx)$ then quasiconvexity implies that

$$(k-1)^n \int_Q f(F+Du)dy \ge \int_{(-k,k)^n} f(F+Dv_k)dx - Ck^{n-1}$$

$$= k^n \int_Q f(F+Dw_k)dx - Ck^{n-1} \ge k^n f(F) - Ck^{n-1}.$$

Division by k^n yields the assertion as $k \to \infty$. $\qquad\square$

Proof of Theorem 4.17. Consider the periodic function $u : \mathbf{R}^2 \to \mathbf{R}^3$

$$u(x) = \frac{1}{2\pi} \begin{pmatrix} \sin 2\pi x^1 \\ \sin 2\pi x^2 \\ \sin 2\pi(x^1+x^2) \end{pmatrix}.$$

Then

$$Du(x) = \begin{pmatrix} \cos 2\pi x^1 & 0 \\ 0 & \cos 2\pi x^2 \\ \cos 2\pi(x^1+x^2) & \cos 2\pi(x^1+x^2) \end{pmatrix}$$

and

$$L := \text{span}\{Du(x)\}_{x\in\mathbf{R}^2} = \left\{ \begin{pmatrix} r & 0 \\ 0 & s \\ t & t \end{pmatrix} : r,s,t \in \mathbf{R} \right\}.$$

The only rank-1 lines in L are lines parallel to the coordinate axes. In particular the function $g(F) = -rst$ is rank-1 convex (in fact rank-1 affine) on L. On the other hand

$$\int_{(0,1)^2} g(Du(x)) = -\frac{1}{4} < 0 = g(0). \tag{4.23}$$

To prove the theorem it only remains to show that g can be extended to a rank-1 convex function on $M^{3\times 2}$. Whether this is possible is unknown. There is, however, a rank-1 convex function that almost agrees with g in L and this is enough. Let P denote the orthogonal projection onto L and consider the quartic polynomial

$$f_{\epsilon,k}(F) = g(PF) + \epsilon(|F|^2 + |F|^4) + k|F - PF|^2.$$

We claim that for every $\epsilon > 0$ there exists a $k(\epsilon) > 0$ such that $f_{\epsilon,k(\epsilon)}$ is rank-1 convex. Suppose otherwise. Then there exists an $\epsilon > 0$ such that $f_{\epsilon,k}$ is not rank-1 convex for any $k > 0$. Hence there exist $F_k \in M^{m \times n}$, $a_k \in \mathbf{R}^m$, $b_k \in \mathbf{R}^n$, $|a_k| = |b_k| = 1$ such that

$$D^2 f_{\epsilon,k}(F_k)(a_k \otimes b_k, a_k \otimes b_k) \leq 0.$$

Now

$$D^2 f_{\epsilon,k}(f)(X, X) =$$
$$D^2 g(PF)(PX, PX) + 2\epsilon|X|^2 + \epsilon(4|F|^2|X|^2 + 8|F : X|^2) + k|X - PX|^2.$$

The term $D^2 g(PF)$ is linear in F while the third term on the right hand side is quadratic and positive definite. Hence F_k is bounded as $k \to \infty$, and passing to a subsequence if needed we may assume $F_k \to \bar{F}, a_k \to \bar{a}, b_k \to \bar{b}$. Since $D^2 f_{\epsilon,k} \geq D^2 f_{\epsilon,j}$ for $k \geq j$ we deduce

$$D^2 g(P\bar{F})(P\bar{a} \otimes \bar{b}, P\bar{a} \otimes \bar{b}) + 2\epsilon + j|\bar{a} \otimes \bar{b} - P\bar{a} \otimes \bar{b}|^2 \leq 0 \ \forall j. \qquad (4.24)$$

Thus $P(\bar{a} \otimes \bar{b}) = \bar{a} \otimes \bar{b}$, i.e. $\bar{a} \otimes \bar{b} \in L$. Therefore $t \mapsto g(P(\bar{F} + t\bar{a} \otimes \bar{b}))$ is affine, and the first term in (4.24) vanishes. This yields the contradiction $\epsilon \leq 0$.

Thus there exist $k(\epsilon)$ such that $f_\epsilon := f_{\epsilon,k(\epsilon)}$ is rank-1 convex. By (4.23) and the definition of u, the function f_ϵ is not quasiconvex as long as $\epsilon > 0$ is sufficiently small. $\qquad \square$

An immediate consequence of Šverák's result and the considerations in the previous section is that there exist gradient Young measures that are not laminates. In fact the measure ν defined by averaging $\delta_{Du(x)}$, i.e.

$$\langle \nu, h \rangle = \int_{(0,1)^2} h(Du(x)) dx, \quad \forall h \in C_0(M^{m \times n}),$$

provides an example, since $\langle \nu, f_\epsilon \rangle = \langle \nu, g \rangle + C\epsilon < f_\epsilon(\langle \nu, \mathrm{id} \rangle)$ (for small $\epsilon > 0$).

The following modification, due to James, provides an even simpler example and a nice illustration of the failure of quasiconvexity for g (or more

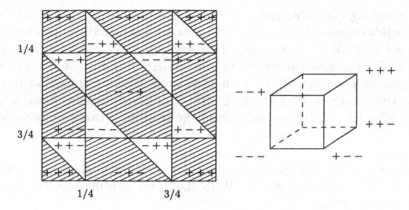

Figure 15: The gradients in James' modification of Šverák's example. Regions of positive parity are shaded. The picture on the right shows the rank-1 connections between the eight gradients.

precisely f_ϵ). Let $s : \mathbf{R} \to \mathbf{R}$ denote the periodic sawtooth function with mean zero and $s' = 1$ on $(0, 1/2)$, $s' = -1$ on $(1/2, 1)$ and define

$$\tilde{u}(x) = \begin{pmatrix} s(x^1) \\ s(x^2) \\ s(x^1 + x^2) \end{pmatrix}.$$

Then $D\tilde{u} \in L$ and $D\tilde{u}$ takes 8 values which can be denoted by $(+++), (++-), \ldots$ according to the signs of $\partial_1 \tilde{u}^1$, $\partial_2 \tilde{u}^2$ and $\partial_1 \tilde{u}^3 = \partial_2 \tilde{u}^3$. We say that $D\tilde{u}$ has positive parity if the number of $-$ signs is even. The analogue of (4.23) can be proved by inspecting Figure 15.

$$\int_{(0,1)^3} g(D\tilde{u}(x))dx = |\{\text{parity} D\tilde{u} = -\}| - |\{\text{parity} D\tilde{u} = +\}| = -\frac{1}{2}.$$

In particular

$$\nu = \frac{3}{16}(\delta_{+++} + \delta_{+--} + \delta_{-+-} + \delta_{--+}) + \frac{1}{16}(\delta_{++-} + \delta_{+-+} + \delta_{-++} + \delta_{---})$$

is a gradient Young measure that is not a laminate. Indeed for every laminate that only involves the eight matrices $(\pm \pm \pm)$ one has

$$\nu(\text{positive parity}) = \nu(\text{negative parity})$$

since g and $-g$ can be extended to rank-1 convex functions on $M^{3\times2}$.

Šverák's example leaves open the question whether there exists a (compact) set $K \subset M^{3\times2}$ that is rank-1 convex (i.e. $K^{rc} = K$) but not quasiconvex. For the gradient set above one has $K^{rc} = K^{qc} = K^c =$ unit cube in L. Using a variant of James' modification Milton [Mi 98] has recently shown that there exists a set $K \in M^{3\times12}$ of seven matrices that satisfies $K^{rc} \neq K^{qc}$. His motivation arose from the relation between quasiconvexification and optimal composites. His example shows that certain composites cannot be obtained by successive lamination.

Šverák showed that the complex version of the original example yields a set $K \subset M^{6\times2}$ with $K^{rc} \neq K^{qc}$. With the usual identification $\mathbf{R}^2 \simeq \mathbf{C}$ via $z = x + iy$ we define

$$K = \left\{ \begin{pmatrix} z_1 & 0 \\ 0 & z_2 \\ z_3 & z_3 \end{pmatrix} : z_i \in \mathbf{C}, |z_i| = 1, z_3 = z_1 z_2 \right\}, \qquad (4.25)$$

$L = \mathrm{span}K$, P orthogonal projection $M^{6\times2} \to L$. The periodic function $w : \mathbf{R}^2 \to \mathbf{C}^3$, given by

$$w(x) = \begin{pmatrix} e^{ix^1} \\ e^{ix^2} \\ e^{i(x^1+x^2)} \end{pmatrix}$$

satisfies $Dw \in K$. Hence $0 \in K^{qc}$ (use e.g. that $\mathrm{dist}_K^{qc}(0) = 0$ by Proposition 4.19).

We claim that

$$K^{qc} = K^{pc} = K \cup \{0\}.$$

To prove this consider on L the function

$$g(z_1, z_2, z_3) = |z_1 z_2 - z_3|^2 + |\bar{z}_2 z_3 - z_1|^2 + |z_3 \bar{z}_1 - z_2|^2$$

and note that g vanishes exactly on $K \cup \{0\}$. Now g can easily be extended to a polyconvex function f on $\mathbf{C}^{3\times2} \simeq M^{6\times2}$ with $f > 0$ outside L. If

$$F = \begin{pmatrix} F_{11} & F_{12} \\ F_{21} & F_{22} \\ F_{31} & F_{32} \end{pmatrix} = \begin{pmatrix} F_1 \\ F_2 \\ F_3 \end{pmatrix}$$

we may take

$$f(F) = |\det(F_1, F_2) - F_{31}|^2 + |\det(\bar{F}_2, F_3) + F_{11}|^2$$
$$+ |\det(F_3, \bar{F}_1) + F_{22}|^2 + |F - PF|^2.$$

Therefore $K^{pc} \subset K \cup \{0\}$ and equality holds since $0 \in K^{qc} \subset K^{pc}$. Moreover either $K^{rc} = K$ or $K^{rc} = K \cup \{0\}$. The following result shows that rank-1 convexity is a local condition and hence $K^{rc} = K \neq K^{qc}$.

Lemma 4.20 *Let K_1 and K_2 be disjoint compact sets and suppose that $K = K_1 \cup K_2$ is rank-1 convex. Then both K_1 and K_2 are rank-1 convex.*

Proof. See [Pe 93], Thm. 5.1 or [MP 98], Prop. 2.8.

There is also a simple direct proof that $K^{rc} = K$. Note that $f \in C^\infty$ and $Df^2(0) \geq c\,\mathrm{Id}$, $c > 0$. Indeed $f(0) = Df(0) = 0$ and thus

$$D^2 f(0)(F, F) = \lim_{t\to\infty} \frac{2}{t^2} f(tF) = |F_{31}|^2 + |F_{11}|^2 + |F_{22}|^2 + |F - PF|^2$$

and the right hand side vanishes only if $F = 0$. Hence there exists a neighbourhood $B(0, \epsilon)$ and a new function $\tilde{f}$ such that $\tilde{f} > 0$ in $B(0, \epsilon)$, $D^2\tilde{f} \geq \frac{c}{2}\mathrm{Id}$ in $B(0, 2\epsilon)$, $\tilde{f} = f$ outside $B(0, \epsilon)$. Then $\tilde{f}$ is locally polyconvex and hence rank-1 convex and $\{\tilde{f} \leq 0\} = K$. Thus $K^{rc} = K$.

Note that $\tilde{f}$ is in particular locally quasiconvex (i.e. for every point there is a neighbourhood in which $\tilde{f}$ agrees with a quasiconvex function) but not quasiconvex since $\tilde{f}(0) > 0$ and $\int_{T^2} \tilde{f}(Dw) = \int_{T^2} f(Dw) = 0$. This proves Kristensen's theorem for $m \geq 6$.

Kružik [Kr 97] used Šverák's counterexample to show that there exists an integrand $f : M^{3\times 2} \to \mathbf{R} \cup \{+\infty\}$ that is not quasiconvex such that $F \mapsto \tilde{f}(F) = f(F^T)$ is quasiconvex. Recall that

$$L = \left\{ \begin{pmatrix} r & 0 \\ 0 & s \\ t & t \end{pmatrix} \; : \; r, s, t \in \mathbf{R} \right\},$$

and let

$$f(F) = \begin{cases} -rst & \text{if } F \in L, \\ +\infty & \text{else.} \end{cases}$$

Then f, and hence $\tilde{F}$, is rank-1 convex since $\{f < \infty\} = L$ is convex and f is rank-1 convex on L. We have already seen that f is not quasiconvex and it only remains to show that

$$\int_{(0,1)^3} \tilde{f}(F + Du)\, dx \geq \tilde{f}(F)$$

for all periodic (Lipschitz) functions $u : \mathbf{R}^3 \to \mathbf{R}^2$. We may assume that $(F + Du)^T \in L$ a.e. Since $\int_{(0,1)^3} Du = 0$ by periodicity we deduce that $F^T \in L$ and $(Du)^T \in L$ a.e. Thus

$$\partial_2 u^1 = \partial_1 u^2 = 0, \quad \partial_3(u^1 - u^2) = 0.$$

Therefore u^1 is independent of x^2, while u^2 is independent of x^1, and differentiation of the second identity yields $\partial_1 \partial_3 u^1 = \partial_2 \partial_3 u^2 = 0$. Thus

$$u^1 = a(x^1) + b(x^3), \quad u^2 = c(x^2) + d(x^3),$$
$$Du = \begin{pmatrix} a'(x^1) & 0 & b'(x^3) \\ 0 & c'(x^2) & d'(x^3) \end{pmatrix},$$

and an application of Fubini's theorem in connection with the rank-1 convexity of $\tilde{f}$ yields the desired estimate.

By a more refined argument one can show that the function f

$$f_{\varepsilon,k}(F) = f(PF) + \varepsilon(|F|^2 + |F|^4) + k|F - PF|^2$$

considered above provides a finite-valued counterexample if $\varepsilon > 0$ is small enough and $k \geq k(\varepsilon)$. To show that

$$\int_{(0,1)^3} \tilde{f}_{\varepsilon,k}(F + Du) - \tilde{f}_{\varepsilon,k}(F) + D\tilde{f}_{\varepsilon,k}(F)Du\, dx \geq 0,$$

one introduces $v = (v^1, v^2, v^3)$ and $w = (w^1, w^2, w^3)$ by

$$P(D\varphi)^T = \begin{pmatrix} v^1 & 0 \\ 0 & v^2 \\ v^3 & v^3 \end{pmatrix}, \quad (D\varphi)^T - P(D\varphi)^T = \begin{pmatrix} 0 & w^2 \\ w^1 & 0 \\ w^3 & -w^3 \end{pmatrix}$$

and observes that the differential operator

$$A(Dv) = (\partial_2 v^1, \partial_3 v^1, \partial_1 v^2, \partial_3 v^2, \partial_1 v^3, \partial_2 v^3)$$

can be expressed as a linear combination of derivatives of w. Hence

$$\|A(Dv)\|_{W^{-1,2}(Q)} \leq C\|(D\varphi)^T - P(D\varphi)^T\|_{L^2(Q)}$$

and the crucial ingredient in the proof are the estimates

$$\left|\int_{(0,1)^3} v^1 v^2 v^3 \, dx\right| \leq C\|v\|_{L^4}^2 \|A(Dv)\|_{W^{-1,2}},$$

$$\left|\int_{(0,1)^3} v^i v^j\right| \leq C\|v\|_{L^2} \|A(Dv)\|_{W^{-1,2}}, \quad \text{for } i \neq j,$$

which are proved by a suitable decomposition of the (discrete) Fourier transforms $\mathcal{F}v^i$ into a part that is supported in a narrow cone near the i-th coordinates axis and a part that vanishes near that axis. The second part is then easely estimated in terms of $A(Dv)$.

4.8 Proofs: lower semicontinuity and relaxation

Proof of Theorem 4.4(i) ($W^{1,\infty}w^*slsc$ is equivalent to quasiconvexity of the integrand). To establish necessity of quasiconvexity let $Q = (0,1)^n$, $\varphi \in W_0^{1,\infty}(Q, \mathbf{R}^m)$, extend φ 1-periodically to $\mathbf{R}^n$ and let

$$u_j(x) = Fx + \frac{1}{j}\varphi(jx), \quad \text{for } x \in \Omega.$$

Then $u_j \overset{*}{\rightharpoonup} u$ in $W^{1,\infty}(\Omega, \mathbf{R}^m)$, $u = Fx$ and

$$f(Du_j) \overset{*}{\rightharpoonup} \text{const} = \int_Q f(F + D\varphi(y)) dy \quad \text{in } L^\infty(\Omega),$$

cf. Section 3.2 a). The necessity of quasiconvexity follows.

To prove sufficiency consider $u_j \overset{*}{\rightharpoonup} u$ in $W^{1,\infty}(\Omega, \mathbf{R}^m)$ and suppose first that $u(x) = Fx$. If $u_j - u$ was zero on $\partial\Omega$ the assertion would follow from the definition of quasiconvexity. For general u_j consider a compactly contained subdomain $\Omega' \subset\subset \Omega$, a cut-off function $\eta \in C_0^\infty(\Omega)$ with $\eta = 1$ on Ω' and let

$$v_j = u + \eta(u_j - u).$$

Since $u_j \to u$ locally uniformly in Ω by the Sobolev embedding theorem (or by the Arzela-Ascoli theorem) and since $|Du_j| \leq C$ we may assume that

$|Dv_j| \leq C'$ for $j \geq j_0(\eta)$. If we let $M = \sup\{|f(F)| : |F| \leq C + C'\}$ and use quasiconvexity we obtain

$$\liminf_{j \to \infty} I(u_j) \geq \liminf_{j \to \infty} \left[\int_\Omega f(Dv_j)\, dx + \int_{\Omega \setminus \Omega'} (f(Du_j) - f(Dv_j))\, dx \right]$$

$$\geq |\Omega| f(F) - 2M |\Omega \setminus \Omega'|.$$

Since $\Omega' \subset\subset \Omega$ was arbitrary the assertion follows for $u = Fx$ and similarly for piecewise affine u.

For arbitrary $u \in W^{1,\infty}(\Omega, \mathbf{R}^m)$ the result is established by approximation as follows. For compactly contained subdomains $\Omega' \subset\subset \Omega'' \subset\subset \Omega$ there exist v_k such that v_k is piecewise affine in Ω', $u = v_k$ in $\Omega \setminus \Omega''$, $|Dv_k| \leq C$, $Dv_k \to Du$ in measure (and hence in all $L^p, p < \infty$). To construct such v_k first approximate u in Ω'' by a C^1 function and then consider piecewise linear approximations on a sufficiently fine (regular) triangulation. Let $u_{j,k} = u_j + v_k - u$. Then

$$u_{j,k} \overset{*}{\rightharpoonup} v_k \quad \text{in } W^{1,\infty}(\Omega, \mathbf{R}^m) \text{ as } j \to \infty, \tag{4.26}$$

$$|Du_{j,k}| \leq C \tag{4.27}$$

Hence, by the previous result and the dominated convergence theorem

$$\lim_{k \to \infty} \liminf_{j \to \infty} \int_{\Omega'} f(Du_{j,k}) dx \geq \lim_{k \to \infty} \int_{\Omega'} f(Dv_k)\, dx$$
$$= \int_{\Omega'} f(Du) dx \geq \int_\Omega f(Du) - C|\Omega \setminus \Omega'|$$

On the other hand by (4.27), the uniform continuity of f on compact sets and the convergence of Dv_k in measure

$$\lim_{k \to \infty} \sup_j \int_{\Omega'} |f(Du_{j,k}) - f(Du_j)|\, dx = 0.$$

Hence

$$\liminf_{j \to \infty} \int_\Omega f(Du_j)\, dx \geq \int_\Omega f(Du)\, dx - 2C|\Omega \setminus \Omega'|,$$

and the assertion follows since Ω' was arbitrary. $\qquad\qquad$ □

Proof of Theorem 4.5(iii) (formula for f^{qc}).
Let

$$Qf(F,U) := \inf_{\varphi \in W_0^{1,\infty}} \frac{1}{|U|} \int\limits_U f(F + D\varphi)\,dx.$$

We have to show that $f^{qc}(F) = Qf(F,U)$. A simple scaling and covering argument shows that Qf is independent of U. By the definition of quasi-convexity $Qf \geq Qf^{qc} = f^{qc}$. To prove the converse inequality $Qf \leq f^{qc}$ it suffices to show that Qf is quasiconvex since $Qf \leq f$. We first claim that

$$\frac{1}{|U|} \int\limits_U Qf(F + D\psi)\,dx \geq Qf(F),$$

$$\tag{4.28}$$

$\forall \psi \in W_0^{1,\infty}(\Omega, \mathbf{R}^m)$, ψ piecewise affine.

Let U be a finite union (up to a null set) of disjoint open subsets U_i such that ψ is affine on U_i and let $\epsilon > 0$. By the definition of Qf (applied to U_i) there exist $\varphi_i \in W_0^{1,\infty}(U_i, \mathbf{R}^m)$ such that

$$Qf(F + D\psi) \geq \frac{1}{|U_i|} \int\limits_{U_i} f(F + D\psi + D\varphi_i)\,dx - \epsilon \quad \text{on } U_i.$$

Set $\varphi = \psi + \sum \varphi_i \in W_0^{1,\infty}(U, \mathbf{R}^m)$. Rearranging terms we find

$$\int\limits_U Qf(F + D\psi)\,dx \;\geq\; \int\limits_U f(F + D\varphi)\,dx - \epsilon|U|$$

$$\geq\; Qf(F) - \epsilon|U|,$$

and assertion (4.28) follows as $\epsilon > 0$ was arbitrary. Now (4.28) is enough to conclude that Qf is rank-1 convex and therefore locally Lipschitz continous (see Remark 3 after Definition 4.2). Hence Qf is quasiconvex by (4.28) and density arguments and therefore $f^{qc} = Qf$.

So far we have assumed that Qf does not take the value $-\infty$. If $Qf(F + D\psi) = -\infty$ on U_i then an obvious modification of the above argument shows that (4.28) still holds. Hence Qf is rank-1 convex (see the proof of Lemma 4.3) and one easily concludes that $f^{qc} \equiv Qf \equiv -\infty$ since the rank-1 directions span the space of all matrices. $\qquad\qquad$ □

4.9 Proofs: classification

The main point is to show that Jensen's inequality for quasiconvex functions characterizes homogeneous Young measures (see Lemma 4.23). The proof relies on the Hahn-Banach separation theorems and the representation (4.5) for f^{qc}. The extension to nonhomogeneous Young measures uses mainly generalities about measurable maps, in particular their approximation by piecewise constant ones.

An important technical tool of independent interest is a truncation result for sequences of gradients sometimes known as Zhang's lemma. (Closely related results were obtained previously by Acerbi and Fusco based on earlier work of Liu.) It implies that every gradient Young measure supported on a compact set $K \subset M^{m \times n}$ can be generated by a sequence $\{Dv_j\}$ whose L^∞ norm can be bounded in terms of K alone. For the rest of this section we adopt the following conventions:

$$K \quad \text{is a compact set in } M^{m \times n},$$

$$U, \Omega \quad \text{are bounded domains in } \mathbf{R}^n, |\partial \Omega| = |\partial U| = 0.$$

Lemma 4.21 *(Zhang's lemma). Let $|K|_\infty = \sup\{|F| : F \in K\}$.*

(i) *Let $u_j \in W^{1,1}_{\text{loc}}(\mathbf{R}^n; \mathbf{R}^m)$ and suppose that*

$$\text{dist}(Du_j, K) \to 0 \text{ in } L^1(\mathbf{R}^n). \tag{4.29}$$

Then there exists a sequence $v_j \in W^{1,1}_{\text{loc}}(\mathbf{R}^n; \mathbf{R}^m)$ such that

$$|Dv_j| \le c(n, m)|K|_\infty, \tag{4.30}$$

$$|\{u_j \ne v_j\}| \to 0. \tag{4.31}$$

(ii) *Let $U \in \mathbf{R}^n$ be a bounded domain and let $u_j \in W^{1,1}_{\text{loc}}(U; \mathbf{R}^m)$. Suppose that*

$$\text{dist}(Du_j, K) \to 0 \text{ in } L^1(U), \quad u_j \to u \text{ in } L^1_{\text{loc}}(U). \tag{4.32}$$

Then there exist $v_j \in W^{1,1}_{\text{loc}}(U; \mathbf{R}^m)$ such that

$$|Dv_j| \le c(n, m)|K|_\infty, \tag{4.33}$$

$$|\{(u_j \ne v_j\}| \to 0, \quad v_j = u \text{ near } \partial U.$$

Remarks. Estimates (4.29) and (4.32) can be replaced by the stronger assertion dist$(Dv_j, K^c) \to 0$ in L^∞, see [Mu 97b]. Note also that the assertion $|\{u_j \neq v_j\}| \to 0$ implies

$$|\{Du_j \neq Dv_j\}| \to 0,$$

since for any Sobolev function $Du = 0$ a.e. on $\{u = 0\}$.

Proof. Part (i) is essentially Lemma 3.1 in [Zh 92]. Alternatively it follows from (the proof of) Theorem 6.6.3 in [EG 92], pp. 254-255, with $\lambda = 3C|K|_\infty$. Part (ii) follows by a standard localization argument, see [Mu 97b] for the details. □

Now suppose that $\{u_j\}$ is bounded in $W^{1,\infty}(\Omega; \mathbf{R}^m)$ and $\{Du_j\}$ generates the (gradient) Young measure ν. Then $Du_j \overset{*}{\rightharpoonup} Du$ in L^∞, $Du(x) = \langle \nu_x, \mathrm{id} \rangle$ and $u_j \to u$ (locally) in L^∞. We call u the underlying deformation of ν. The Young measure $\nu : \Omega \to \mathcal{M}(M^{m \times n})$ is called homogeneous if it is constant in Ω (up to a null set). As usual we identify constant maps with their values and view the set $H(\Omega)$ of homogeneous gradient Young measures as a subset of $\mathcal{M}(M^{m \times n})$. By $H(\Omega, K)$ we denote the set of homogeneous gradient Young measures supported on K.

Lemma 4.22 *We have*

(i) *If $\nu \in H(\Omega, K)$ and $\langle \nu, \mathrm{id} \rangle = 0$ then there exists a sequence $u_j \in W_0^{1,\infty}(\Omega; \mathbf{R}^m)$ such that Du_j generates ν and satisfies $|Du_j| \leq C|K|_\infty$.*

(ii) *$H(\Omega, K)$ is weak* compact in $\mathcal{M}(M^{m \times n})$.*

(iii) *The set $H(\Omega, K)$ is independent of Ω. If ν is a gradient Young measure with $\mathrm{supp}\,\nu(x) \subset K$ a.e. whose underlying deformation u agrees with an affine map on $\partial\Omega$ (in the sence of $W_0^{1,\infty}$) then the average $\mathrm{Av}\nu$ defined by*

$$\langle \mathrm{Av}\nu, f \rangle = \frac{1}{|\Omega|} \int_\Omega \langle \nu_x, f \rangle dx$$

belongs to $H(K)$.

(iv) *The set $H_F(K) = \{\nu \in H(K) : \langle \nu, \mathrm{id} \rangle = F\}$ is weak* closed and convex.*

Proof. Assertion (i) follows from the definition of $H(\Omega, K)$ and part (ii) of Zhang's lemma. The proof of (ii) uses (i) and a diagonalization argument. Note that $H(\Omega, K)$ is contained in the weak* compact set $\mathcal{P}(K)$ of probability measures on K. Hence the weak* topology is metrizable on $\mathcal{P}(K)$ and can be described by sequences. Suppose that $\nu_k \in H(\Omega, K)$ and $\nu_k \overset{*}{\rightharpoonup} \nu$. After subtraction of affine functions in the generating sequences for ν_k we may assume that $\langle \nu_k, \text{id} \rangle = 0$. By (i) there exist $u_{k,j} \in W_0^{1,\infty}(\Omega; M^{m \times n})$ such that

$$\delta_{Du_{k,j}(\cdot)} \underset{j \to \infty}{\overset{*}{\rightharpoonup}} \nu_k \quad \text{in } L_w^\infty(\Omega; \mathcal{M}(M^{m \times n})), \quad |Du_{k,j}| \le C|K|_\infty.$$

Here we identified ν_k with the constant map $x \mapsto \nu_k$. Since the weak* topology is metrizable on $L_w^\infty(\Omega; P(B(0, C|K|_\infty)))$ we can apply a standard diagonalization argument to find $j_k \to \infty$ such that

$$\delta_{Du_{k,j_k}(\cdot)} \overset{*}{\rightharpoonup} \nu \text{ in } L_w^\infty(\Omega; \mathcal{M}(M^{m \times n})).$$

Since $|Du_{k,j_k}| \le C$ we have $\nu \in H(\Omega, K)$. Thus $H(\Omega, K)$ is weak* closed and therefore weak* compact as a subset of $\mathcal{P}(K)$.

To prove (iii) consider first $v \in W_0^{1,\infty}(U; \mathbf{R}^m)$ and the trivial Young measure given by $\mu(x) = \delta_{Dv(x)}$. We claim that $\text{Av}\mu$ is a homogeneous gradient Young measure (for all domains Ω). By the Vitali covering theorem there exist disjoint scaled copies $U_i = a_i + r_i U$ of U that are contained in the unit cube Q and cover it up to a null set. Define

$$w(x) = \begin{cases} r_i v \left(\dfrac{x - a_i}{r_i} \right) & \text{in } U_i, \\ 0 & \text{in } Q \setminus U_i, \end{cases}$$

extend w 1-periodically to $\mathbf{R}^n$, and let $w_k(x) = k^{-1} w(kx)$. Then for all continuous functions g (see Section 3.2 a))

$$g(Dw_k) \overset{*}{\rightharpoonup} \bar{g} \quad \text{in } L^\infty(\mathbf{R}^n),$$

where

$$\bar{g} = \int_Q g(Dw) dx = \frac{1}{|U|} \int_U g(Dv) dx = \langle \text{Av}\mu, g \rangle.$$

Thus, for all Ω, $\delta_{Dw_k(\cdot)}$ converges to the homogeneous Young measure $\text{Av}\mu$ in $L_w^\infty(\Omega; \mathcal{M}(M^{m \times n}))$ in the weak* topology. Hence $\text{Av}\mu \in H(\Omega)$ as claimed.

Now let ν satisfy the assumption of (iii). We may suppose that $u \in W_0^{1,\infty}(\Omega; \mathbf{R}^m)$. By the definition of gradient Young measures and part (ii) of

Zhang's lemma there exists a sequence $u_j \in W_0^{1,\infty}(\Omega; \mathbf{R}^m)$ such that $|Du_j| \leq R$ and

$$\delta_{Du_j(\cdot)} \overset{*}{\rightharpoonup} \nu \quad \text{in } L_w^\infty(\Omega; \mathcal{M}(M^{m \times n})).$$

Taking test functions of the form $1 \otimes g$ we see that

$$\mathrm{Av}\,\delta_{Du_j(\cdot)} \overset{*}{\rightharpoonup} \mathrm{Av}\nu.$$

By the considerations above $\mathrm{Av}\,\delta_{Du_j(\cdot)}$ belongs to all $H(\Omega, \overline{B(0,R)})$. By (ii) the same holds for $\mathrm{Av}\nu$. Since $\mathrm{supp}\nu(K) \subset K$ a.e. in fact $\mathrm{Av}\,\nu \in H(\Omega, K)$. If $\nu \in H(U, K)$ then $\mathrm{Av}\nu = \nu$ and hence $H(\Omega, K)$ is independent of ν.

Regarding (iv) we may suppose $F = 0$. Let $\nu_1, \nu_2 \in H_0(K)$. Let $Q_1 = (0, \lambda) \times (0,1)^{n-1}, Q_2 = (\lambda, 1) \times (0,1)^{n-1}$. By (i) there exist sequences $\{Du_{i,j}\} \subset W_0^{1,\infty}(Q_i, \mathbf{R}^m), i = 1, 2$ that generate ν_i. Hence the gradients of

$$u_j(x) = \begin{cases} u_{1,j}(x), & x \in Q_1 \\ u_{2,j}(x), & x \in Q_2 \end{cases}$$

generate

$$\nu(x) = \begin{cases} \nu_1, & x \in Q_1 \\ \nu_2, & x \in Q_2 \end{cases}.$$

By (iii) we have

$$\lambda\nu_1 + (1 - \lambda)\nu_2 = \mathrm{Av}\nu \in H_0(K). \qquad \square$$

Lemma 4.23 *(characterization of homogeneous gradient Young measures). We have*

$$H(K) = \mathcal{M}^{qc}(K).$$

Proof. Clearly $H(K) \subset \mathcal{M}^{qc}(K)$ by lower semicontinuity (see Theorem 4.4(i)). To prove the converse it suffices to consider measures with barycentre zero. Now $H_0(K)$ is weak$*$ closed and convex and the dual of $(\mathcal{M}(K), \text{weak}*)$ is $C(K)$ (see e.g. [Ru 73], Thm. 3.10). By the Hahn-Banach separation theorem it suffices to show that, for all $f \in C(K)$,

$$\langle \nu, f \rangle \geq \alpha \quad \forall \nu \in H_0(K), \tag{4.34}$$

implies

$$\langle \mu, f \rangle \geq \alpha \quad \forall \mu \in \mathcal{M}_0^{qc}(K).$$

Fix $f \in C(K)$, consider a continuous extension to $C_0(M^{m \times n})$ and let

$$f_k(F) = f(F) + k\mathrm{dist}^2(F, K).$$

We claim that

$$\lim_{k \to \infty} f_k^{qc}(0) \geq \alpha. \tag{4.35}$$

Once this is shown we are done since by definition every $\mu \in \mathcal{M}_0^{qc}(K)$ satisfies

$$\langle \mu, f \rangle = \langle \mu, f_k \rangle \geq \langle \mu, f_k^{qc} \rangle \geq f_k^{qc}(0).$$

Suppose now (4.35) was false. Then there exist $\delta > 0$ such that

$$f_k^{qc}(0) \leq \alpha - 2\delta, \quad \forall k.$$

By Theorem 4.5(iii) there exist $u_k \in W_0^{1,\infty}(Q; \mathbf{R}^m)$ such that

$$\int_Q f_k(Du_k)dy \leq \alpha - \delta. \tag{4.36}$$

In particular we may assume $u_k \rightharpoonup u$ in $W_0^{1,2}(Q, \mathbf{R}^m)$ and

$$\mathrm{dist}(Du_k, K) \to 0 \text{ in } L^1(Q).$$

By part (ii) of Zhang's lemma there exists $v_k \in W_0^{1,\infty}(Q; \mathbf{R}^m)$ such that

$$|Dv_k| \leq C, \quad |\{(Du_k \neq Dv_k\}| \to 0. \tag{4.37}$$

In particular a further subsequence of $\{Dv_k\}$ generates a gradient Young measure ν with $\mathrm{supp}\nu(x) \subset K$ and underlying deformation $u \in W_0^{1,\infty}(Q; \mathbf{R}^m)$. Thus $\mathrm{Av}\nu \in H_0(K)$ by Lemma 4.22 (iii). Since f is bounded from below we deduce from (4.37) and (4.34)

$$\liminf_{k \to \infty} \int_Q f_j(Du_k) \geq \liminf_{k \to \infty} \int_Q f_j(Dv_k)$$

$$= \int_Q \langle \nu_x, f_j \rangle dx = \langle \mathrm{Av}\nu, f_j \rangle \geq \alpha.$$

This contradicts (4.36) as $f_k \geq f_j$ if $k \geq j$, and (4.35) is proved. $\qquad \square$

Proof of Theorem 4.4. Necessity of conditions (i) - (iii) was established in Section 4.3. To prove sufficiency we first consider the case that the underlying deformation vanishes. Let

$$A = \{\nu \in L_w^\infty(\Omega, \mathcal{M}(M^{m \times n})) : \nu(x) \in \mathcal{M}_0^{qc}(K) \text{ a.e.}\}$$

denote the set of maps that satisfy (i) - (iii) with $Du = 0$. We have to show that every element of A is a gradient Young measure.

To do so we use some generalities about measurable maps to approximate the elements of A by piecewise constant maps. First note that the set of subprobability measures $\mathcal{M}_1 = \{\mu \in \mathcal{M}(M^{m \times n}) : \nu \geq 0, \|\nu\| \leq 1\}$ is weak* compact in $\mathcal{M}(M^{m \times n})$. Hence the weak* topology is metrizable on $\mathcal{M}_1$. To define a specific metric let $\{f_i\} \subset C_0(M^{m \times n})$ be a countable dense set in the unit sphere of $C_0(M^{m \times n})$ and let

$$d(\mu, \mu') = \sum_{i=1}^\infty 2^{-i} |\langle \mu - \mu', f_i \rangle|.$$

The space $(\mathcal{M}_1, d)$ is a compact metric space. Since d induces the weak* topology, a map $\nu : \Omega \to \mathcal{M}(M^{m \times n})$ that takes (a.e.) values in $\mathcal{M}_1$ is weak* measurable if and only if $\nu : \Omega \to (\mathcal{M}_1, d)$ is measurable.

The set $\{\nu \in L_w^\infty(\Omega; \mathcal{M}(M^{m \times n})) : \nu(x) \in \mathcal{M}_1 \text{ a.e.}\}$ is also weak* compact in $L_w^\infty(\Omega; \mathcal{M}(M^{m \times n}))$ (cf. the proof of Theorem 3.2). A metric $\tilde{d}$ that induces weak* convergence on that set may be defined as follows. Let $\{h_j\}$ be a countable dense set in the unit ball of $L^1(\Omega)$ and let

$$\tilde{d}(\nu, \nu') = \sum_{i,j=1}^\infty 2^{-i-j} |\langle \nu - \nu', h_j \otimes f_i \rangle|.$$

It thus follows from Proposition 4.24 below that every $\nu \in A$ can be arbitrarily well approximated in $\tilde{d}$ by a map $\tilde{\nu}$ with the following properties. There exist finitely many disjoint open sets U_i with $|\partial U_i| = 0$ such that $\tilde{\nu} = \nu_i$ on $U_i, \nu_i \in \mathcal{M}_0^{qc}(K), \tilde{\nu} = \delta_0$ on $\Omega \setminus \cup U_i$. Application of Lemma 4.22(i) to each U_i shows that $\tilde{\nu}$ is a gradient Young measure (extend the generating sequence by zero to $\Omega \setminus \cup U_i$). Hence the closure of gradient Young measures with support in $K' = K \cup \{0\}$ contains A. On the other hand the set of these Young measures is (weakly) compact (see the proof of Lemma 4.22 (ii)). Thus every $\nu \in A$ is a gradient Young measure. This finishes the proof if $Du = 0$.

The remaining case $Du \neq 0$ can now be easily treated by translation. For a measure μ define its push-forward unter translation by

$$\langle T_F\mu, f\rangle = \langle \mu, f(\cdot + F)\rangle$$

so that $T_F\delta_0 = \delta_F$. Now if ν satisfies the hypotheses of Theorem 4.4 and $\tilde{\nu}(x) = T_{-Du(x)}\nu(x)$ then $\tilde{\nu} \in A$. Hence there exists a sequence $\{Dv_j\}$ that generates $\tilde{\nu}$ and one easily verifies that $Du_j = Dv_j + Du$ generates ν (use e.g. Corollary 3.3 with $f(x, F) = g(Du(x) + F), g \in C_0(M^{m\times n})$). $\qquad\square$

Proposition 4.24 Let (X, d) be a compact metric space and $M \subset X$. Suppose that $\nu : \Omega \to X$ is measurable and $\nu(x) \in M$ a.e. Then, for every $k \in \mathbf{N}$, there exists a finite number of disjoint open sets U_i with $|\partial U_i| = 0$ and values $\nu_i \in M$ such that the map

$$\tilde{\nu} = \begin{cases} \nu_i & \text{on } U_i \\ \nu_0 & \text{on } \Omega \setminus U_i \end{cases}$$

satisfies

$$\left|\left\{x : d(\nu(x), \tilde{\nu}(x)) > \frac{1}{k}\right\}\right| < \frac{1}{k}.$$

Proof. By compactness X can be covered by a finite number of open balls B_i with radius $\frac{1}{2k}$. The sets $\tilde{E}_i = \nu^{-1}(B_i)$ are measurable. To obtain disjoint sets E_i, we define $E_1 = \tilde{E}_1$, $E_2 = \tilde{E}_2 \setminus E_1$, etc. If $|E_i| > 0$ then there exist $x_i \in E_i$ such that $\nu_i := \nu(x_i) \in M$. There exist disjoint compact sets $K_i \subset E_i$ such that

$$\sum |E_i \setminus K_i| < 1/k; \tag{4.38}$$

if $|E_i| = 0$ we take $K_i = \emptyset$. The K_i have positive distance and thus there exist disjoint open sets $U_i \supset K_i$ with $|\partial U_i| = 0$ (consider e.g. suitable sublevel sets of the distance function of K_i). Now $E_i \supset \tilde{E}_i \supset K_i$ and thus $d(\nu(x), \nu_i) < 1/k$ in K_i. The assertion follows from (4.38). $\qquad\square$

5 Exact solutions

Approximate solutions are characterized by the quasiconvex hull K^{qc} and set $\mathcal{M}^{qc}(K)$ of Young measures. The construction of exact solutions is more delicate. In view of the negative result for the two-gradient problem (see Proposition 2.1) it was widely believed that exact solutions are rather rare. Recent results suggest that many exact solutions exist but that they have to be very complicated. This is reminiscent of rigidity and flexibility results for isometric immersions and other geometric problems (see [Na 54]; [Ku 55]; [Gr 86], Section 2.4.12).

To illustrate some of the difficulties consider again the two-dimensional two-well problem (see Section 4.5)

$$Du \in K \text{ a.e. in } \Omega, \quad u = Fx \text{ on } \partial\Omega, \tag{5.1}$$

$$K = SO(2)A \cup SO(2)B, \tag{5.2}$$

$$A = \text{Id}, \quad B = \text{diag}(\lambda, \mu), \quad 0 < \lambda < 1 < \mu, \lambda\mu \geq 1. \tag{5.3}$$

If we ignore boundary conditions the simplest solutions of $Du \in K$ are simple laminates, see Figure 16. A short analysis of the rank-1 connections in K shows that such laminates are perpendicular to one of the normals n_1 or n_2, determined by the two solutions of the equation

$$QA - B = a \otimes n. \tag{5.4}$$

There is, however, no obvious way to combine the two laminates (see Fig. 17). It was thus believed that the problem (5.1) – (5.3) has no nontrivial solutions. This is false. The construction of nontrivial solutions is based on Gromov's method of convex integration.

5.1 Existence of solutions

First, one observes that the open version of the two-gradient problem admits a solution. Here and for the rest of this section we say that a map $u : \Omega \to \mathbf{R}^m$ is piecewise linear if it is Lipschitz continuous and if there exist finite or

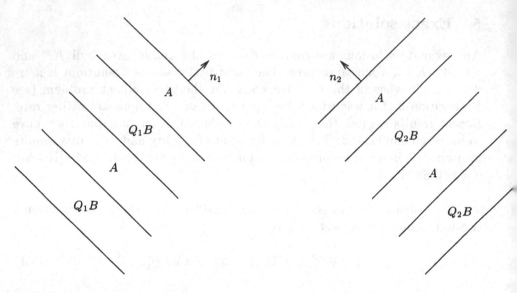

Figure 16: Two possible laminates for the two-well problem.

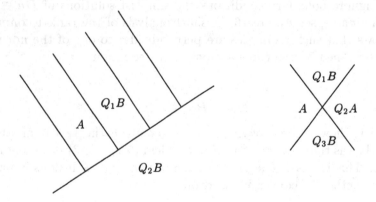

Figure 17: None of the above constructions satisfies the rank-1 condition across every interface.

countably many disjoint open sets Ω_i whose union has full measure in Ω such that $u_{|\Omega_i}$ is affine.

Lemma 5.1 *([MS 96]). Suppose that* $\mathrm{rk}(B-A) = 1, F = \lambda A + (1-\lambda)B, \lambda \in (0,1)$. *Then, for a bounded domain* Ω *and every* $\delta > 0$ *there exists a piecewise linear map* u *such that*

$$u(x) = Fx \quad \text{on} \quad \partial\Omega$$

$$\mathrm{dist}(Du, \{A, B\}) < \delta,$$

$$\sup |u(x) - Fx| < \delta.$$

Remark. It is even possible to handle certain constraints. If $n = m = 2$ and $\det A = \det B = c$ then one can achieve $\det Du = c$. How many constraints can be handled is a largely open problem.

Proof. The construction has some similarities with Fonseca's work [Fo 88], in particular her proof of Theorem 2.4. There are some differences, however, so I give the proof in [MS 96] which is slightly simpler. We will first construct a solution for a special domain U. The argument will then be finished by an application of the Vitali covering theorem.

By an affine change of variables we may assume without loss of generality that

$$A = -\lambda a \otimes e_n, \quad B = (1-\lambda)a \otimes e_n, \quad F = 0, \quad \text{and} \quad |a| = 1.$$

Let $\epsilon > 0$, let

$$V = (-1,1)^{n-1} \times ((\lambda - 1)\epsilon, \lambda\epsilon)$$

and define $v : V \to \mathbf{R}^m$ by

$$v(x) = -\epsilon\lambda(1-\lambda)a + \begin{cases} -\lambda a x_n & \text{if } x_n < 0, \\ (1-\lambda)a x_n & \text{if } x_n \geq 0. \end{cases}$$

Then $Dv \in \{A, B\}$ and $v = 0$ at $x_n = \epsilon(\lambda - 1)$ and $x_n = \epsilon\lambda$, but v does not vanish on the whole boundary ∂V. Next let

$$h(x) = \epsilon\lambda(1-\lambda)a \sum_{i=1}^{n-1} |x_i|.$$

Then h is piecewise linear and $|Dh| = \epsilon\lambda(1-\lambda)\sqrt{n-1}$. Set

$$\tilde{u} = v + h.$$

Note that $\tilde{u} \geq 0$ on ∂V and let

$$U = \{x \in V : \tilde{u}(x) < 0\}.$$

Then

$$\tilde{u}_{|U} \quad \text{is piecewise linear} \quad , \qquad \tilde{u}_{|\partial U} = 0,$$
$$\text{dist}(D\tilde{u}, \{A, B\}) \leq \epsilon\lambda(1 - \lambda)\sqrt{n - 1},$$
$$|\tilde{u}| \leq \epsilon\lambda(1 - \lambda).$$

By the Vitali covering theorem one can exhaust Ω by disjoint scaled copies of U. More precisely there exist $x_i \in \mathbf{R}^n$ and $r_i > 0$ such that the sets

$$U_i = x_i + r_i U$$

are mutually disjoint and $|\Omega \setminus \cup_i U_i| = 0$. Define u by

$$u(x) = \begin{cases} r_i \tilde{u}(\frac{x - x_i}{r_i}) & \text{if} \quad x \in U_i, \\ 0 & \text{else.} \end{cases}$$

Note that

$$Du(x) = D\tilde{u}(\frac{x - x_i}{r_i}), \quad \text{if } x \in \Omega_i.$$

It follows that u is piecewise linear, that $u_{|\partial\Omega} = 0$ and that $\text{dist}(Du, \{A, B\}) < \delta$ for a suitable $\epsilon > 0$. Moreover by choosing $r_i \leq 1$ one can also obtain the estimate for $|u - Fx|$. $\qquad\qquad\square$

Lemma 5.1 can be easily iterated, and using the notion of the lamination convex hull of a set (see Section 4.4) one obtains the following result.

Lemma 5.2 *Suppose that $U \subset M^{m \times n}$ is open. Let $v : \Omega \to \mathbf{R}^m$ be piecewise affine and Lipschitz continuous and suppose $Dv \in U^{lc}$ a.e. Then there exist $u : \Omega \to \mathbf{R}^m$ such that*

$$Du \in U \text{ a.e. in } \Omega, \quad u = v \text{ on } \partial\Omega.$$

The crucial step is the passage from open to compact sets $K \subset M^{m \times n}$. Following Gromov we say that a sequence of sets U_i is an in-approximation of K if

(i) the U_i are open and contained in a fixed ball

(ii) $U_i \subset U_{i+1}^{lc}$

(iii) $U_i \to K$ in the following sense: if $F_{i_k} \in U_{i_k}, i_k \to \infty$ and $F_{i_k} \to F$, then $F \in K$.

Theorem 5.3 *([Gr 86], p. 218; [MS 96]). Suppose that K admits an in-approximation $\{U_i\}$. Let $v \in C^1(\Omega, \mathbf{R}^m)$ with*

$$Dv \in U_1.$$

Then there exists a Lipschitz map u such that

$$Du \in K \in \Omega \text{ a.e.}, \quad u = v \text{ on } \partial\Omega.$$

Proof. The proof uses a sequence of approximations obtained by successive application of Lemma 5.2. To achieve strong convergence each approximation uses a much finer spatial scale than the previous one, similar to the construction of continuous but nowhere differentiable functions. This is one of the key ideas of convex integration.

We first construct a sequence of piecewise linear maps u_i that satisfy

$$Du_i \in U_i \quad \text{a.e},$$
$$\sup |u_{i+1} - u_i| < \delta_{i+1}, \qquad u_{i+1} = u_i \quad \text{on } \partial\Omega,$$
$$\sup |u_1 - v| < \delta/4, \qquad u_1 = v \quad \text{on } \partial\Omega.$$

To construct u_1 note that if Ω' is open and $\Omega' \subset\subset \Omega$ then $\text{dist}(Dv(x), \partial U_1) \geq c(\Omega') > 0$ for all $x \in \Omega'$. Hence it is easy to obtain $u_1|\Omega'$ by introducing a sufficiently fine triangulation. Now exhaust Ω by an increasing sequence of sets $\Omega_i \subset\subset \Omega$.

To construct u_{i+1} and δ_{i+1} from u_i and δ_i we proceed as follows. Let

$$\Omega_i = \{x \in \Omega : \text{dist}(x, \partial\Omega) > 2^{-i}\}.$$

Let ρ be a usual mollifying kernel, i.e. let ρ be smooth with support in the unit ball and $\int \rho = 1$. Let

$$\rho_\epsilon(x) = \epsilon^{-n}\rho(x/\epsilon).$$

Since the convolution $\rho_\epsilon * Du_i$ converges to u_i in $L^1(\Omega_i)$ as $\epsilon \to 0$ we can choose $\epsilon_i \in (0, 2^{-i})$ such that

$$\|\rho_{\epsilon_i} * Du_i - Du_i\|_{L^1(\Omega_i)} < 2^{-i}. \tag{5.5}$$

Let

$$\delta_{i+1} = \delta_i \epsilon_i. \tag{5.6}$$

Use Lemma 5.2 to obtain u_{i+1} such that $Du_{i+1} \in U_{i+1}$, $u_{i+1} = u_i$ on $\partial\Omega$ and

$$\sup_\Omega |u_{i+1} - u_i| < \delta_{i+1}. \tag{5.7}$$

Since $\delta_{i+1} \le \delta_i/2$ we have

$$\sum_{i=1}^\infty \delta_i \le \delta/2.$$

Thus

$$u_i \to u_\infty \quad \text{uniformly,}$$

and u_∞ is Lipschitz since the u_i are uniformly Lipschitz (by (ii) in the definition of an in-approximation). Moreover $u_\infty = v$ on $\partial\Omega$.

It only remains to show that $Du_\infty \in K$. The key point is to ensure strong convergence of Du_i. Since $||D\rho_\epsilon||_{L^1} \le C/\epsilon$ we deduce from (5.7) and (5.6)

$$
\begin{aligned}
||\rho_{\epsilon_k} * (Du_k - Du_\infty)||_{L^1(\Omega_k)} &= ||D\rho_{\epsilon_k} * (u_k - u_\infty)||_{L^1(\Omega_k)} \\
&\le \frac{C}{\epsilon_k} \sup |u_k - u_\infty| \le \frac{C}{\epsilon_k} \sum_{j=k+1}^\infty \delta_j \\
&\le 2\frac{C}{\epsilon_k}\delta_{k+1} \le C'\delta_k.
\end{aligned}
\tag{5.8}
$$

Taking into account (5.5) it follows that

$$
\begin{aligned}
||Du_k - Du_\infty||_{L^1(\Omega)} &\le C'\delta_k + 2^{-k} + ||\rho_{\epsilon_k} * Du_\infty - Du_\infty||_{L^1(\Omega_k)} \\
&\quad + ||Du_k - Du_\infty||_{L^1(\Omega\setminus\Omega_k)}.
\end{aligned}
$$

Since Du_k and Du_∞ are bounded we obtain $Du_k \to Du_\infty$ in $L^1(\Omega)$. Therefore there exists a subsequence u_{k_j} such that

$$Du_{k_j} \to Du_\infty \quad \text{a.e.}$$

It follows from the definition of an in-approximation that

$$Du_\infty \in K \quad \text{a.e.}$$

Hence $u = u_\infty$ has the desired properties. $\qquad\square$

For the two-well problem (5.1) - (5.3) one can construct an in-approximation using the explicit formula in Theorem 4.12. The details can be found in [MS 96]; for a different approach based on Baire's theorem see Dacorogna and Marcellini [DM 96a], [DM 96b], [DM 97].

Theorem 5.4 *Suppose that $\lambda\mu > 1$. Then the two-well problem* (5.1) - (5.3) *has a solution if*

$$F \in \operatorname{int} K^{lc},$$

where

$$K^{lc} = \left\{ F = (y, z) : |y| \le \frac{\lambda\mu - \det F}{\lambda\mu - 1}, |z| \le \frac{\det F - 1}{\lambda\mu - 1} \right\}.$$

Remark. A similar result holds if $\lambda\mu = 1$ provided that in the definition of an in-approximation and interior one considers relatively open sets subject to the constraint $\det F = 1$. One only needs to use the remark after Lemma 5.1 to achieve $\det Du = 1$, provided that $\det A = \det B = 1$.

A more detailed analysis shows that in the definition of an in-approximation one can replace the lamination convex hull which is based on explicit rank-1 connections by the rank-1 convex hull defined by duality with functions (see Section 4.4). This has a striking consequence for the four-gradient example

$$K = \left\{ \pm \begin{pmatrix} 1 & 0 \\ 0 & 3 \end{pmatrix}, \pm \begin{pmatrix} -3 & 0 \\ 0 & 1 \end{pmatrix} \right\}$$

discussed in Section 2.6, see in particular Figure 4. For any matrix

$$F \in K^{rc} \supset \left\{ \begin{pmatrix} F_{11} & 0 \\ 0 & F_{22} \end{pmatrix} : |F_{11}| \le 1, |F_{22}| \le 1 \right\}$$

and any open neighbourhood $U \supset K$ there exists a map $u : \Omega \to \mathbf{R}^2$ such that

$$Du \in U \quad \text{a.e. in } \Omega,$$
$$u = Fx \qquad \text{on } \partial\Omega.$$

This is true despite the fact that small neigbourhoods contain no rank-1 connections so at first glance there seems to be no way to start the construction.

This obstacle is overcome by first constructing a (piecewise linear) map that satisfies $Dv \in U^{rc}$ a.e. and $Dv \in U$ except on a set of small measure. One can then show that the exceptional set can be inductively removed.

The major outstanding problem is whether in the definition of an in-approximation one can replace the lamination convex hull (or rank-1 convex hull) by the quasiconvex hull. One key step would be to resolve the following question.

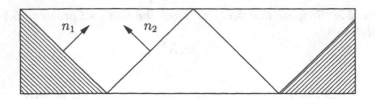

Figure 18: Structure of solutions with finite perimeter. The normals n_1, n_2 are determined by (5.4).

Conjecture 5.5 *Let K be a compact quasiconvex set, i.e. $K^{qc} = K$ and let $\nu \in \mathcal{M}^{qc}(K)$. Then for every open set $U \supset K$ there exists a sequence $u_j : (0,1)^n \to \mathbf{R}^m$ such that Du_j generates ν and $Du_j \in U$ a.e.*

The conjecture is true for compact convex sets [Mu 97a]; this refines Zhang's Lemma (see Lemma 4.21) which implies the existence of u_j such that $Du_j \in B(0, R)$ for a sufficiently large ball.

5.2 Regularity and rigidity

The construction outlined above yields very complicated solutions of the two-well problem (5.1) - (5.3). This raises the question whether the geometry of the solutions can be controlled. Consider the set

$$E = \{x \in \Omega : Du(x) \in SO(2)A\}$$

where Du takes values in one connected component of K (or one phase in the applications to crystals). The perimeter of a set $E \subset \Omega \subset \mathbf{R}^n$ is defined as

$$\mathrm{Per}\,E = \sup\left\{ \int_E \mathrm{div}\,\varphi\,dx : \varphi \in C_0^1(\Omega, \mathbf{R}^n),\ |\varphi| \leq 1 \right\}.$$

For smooth or polyhedral sets this agrees with the $(n-1)$ dimensional measure of ∂E.

Theorem 5.6 *([DM 95]). If u is a solution of (5.1) - (5.3) and if $\mathrm{Per}\,E < \infty$ then u is locally a simple laminate and ∂E consists of straight line segments that can only intersect at $\partial\Omega$.*

The proof combines geometric and measure-theoretic ideas. The geometric idea is that the Gauss curvature $K(g)$ of the pull-back metric $g = (Du)^T Du$ should vanish (in a suitable sense). Since g only takes two values this should give information on E.

One key step in the implementation of this idea is a finite perimeter version of Liouville's theorem on the rigidity of infinitesimal rotations (cf. Theorem 2.4). In this framework connected components are replaced by indecomposable components. A set A of finite perimeter is indecomposable if for every $A_1 \subset A$ with $\text{Per}\, A = \text{Per}\, A_1 + \text{Per}\, A \setminus A_1$ the set A_1 or $A \setminus A_1$ has zero measure. It can be shown that each set of finite perimeter is a union of at most countably many indecomposable components.

Theorem 5.7 *Suppose that $u : \Omega \subset \mathbf{R}^n \to \mathbf{R}$ belongs to $W^{1,\infty}(\Omega; \mathbf{R}^n)$ and that $\det Du \geq c > 0$. Suppose further that $E \subset \Omega$ has finite perimeter and*

$$Du \in SO(n) \qquad \text{a.e. in } E.$$

Then Du is constant on each indecomposable component of E.

To finish the proof of Theorem 5.6 one can decompose Du as $e^{i\Theta}g^{1/2}$ (where $g = (Du)^T Du \in \{A^T A, B^T B\}$) and analyze the jump conditions at the boundary of each indecomposable component to deduce that Θ only takes two values and solves (in the distributional sense) a wave equation with characteristic directions n_1 and n_2.

B. Kirchheim recently devised more flexible measure-theoretic arguments, and combining them with algebraic ideas he established a generalization of Theorem 5.6 to the three-well problem $K = \bigcup_{i=1}^{3} SO(3)U_i$ in three dimensions with $U_1 = \text{diag}(\lambda_1, \lambda_2, \lambda_2)$, $U_2 = \text{diag}(\lambda_1, \lambda_2, \lambda_1)$, $U_3 = \text{diag}(\lambda_2, \lambda_2, \lambda_1)$, $\lambda_i > 0$. A major additional difficulty in this case is that the gauge group $SO(3)$ is not abelian and one cannot hope to derive a linear equation for a quantity like Θ in the two-dimensional situation.

6 Length scales and surface energy

Minimization of the continuum elastic energy is a drastic simplification, in particular if a very fine mixture of phases is observed. It neglects interfacial energy as well as discreteness effects due to the atomic lattice. It is therefore not surprising that elastic energy minimization often predicts an infinitesimally fine mixture of phases (in the sense of a nontrivial Young measure), whereas in any real crystal all microstructures are of finite size.

Nonetheless elastic energy minimization does surprisingly well. It often correctly predicts the phase proportions and in combination with considerations of rank-1 compatibility the orientation of phase interfaces. It recovers in particular the predictions of the crystallographic theory of martensite. In fact one of the major achievements was to realize that the predictions of that theory can be understood as consequences of energy minimization. This allows one to bring to bear the powerful methods of the calculus of variations in the analysis of microstructures.

The problem that elastic energy minimization does not determine the length scale and fine geometry of the microstructure remains. It can be overcome by introducing a small amount of interfacial energy or higher gradient terms. One expects these contributions which penalize rapid changes to be small since otherwise a very fine structure would not arise in the first place. The most popular functionals are

$$I^\epsilon(u) = \int_\Omega W(Du)dx + \int_\Omega \epsilon^2 |D^2u|^2 dx \qquad (6.1)$$

and

$$J^\epsilon(u) = \int_\Omega W(Du)dx + \int_\Omega \epsilon |D^2u| dx. \qquad (6.2)$$

The second functional allows for jumps in the gradient and $|D^2u|$ is understood as the total variation of a Radon measure.

The small parameter $\epsilon > 0$ introduces a length scale and as $\epsilon \to 0$ both models approach (at least formally) pure elastic energy minimization. More realistic models should of course involve anisotropic terms in D^2u or more generally terms of the form $h(Du, \epsilon D^2u)$. Even the basic models (6.1) and (6.2) are, however, far from being understood for maps $u : \Omega \subset \mathbf{R}^3 \to \mathbf{R}^3$. In the following we discuss briefly two simple scalar models which already show some of the interesting effects generated by the interaction of elastic energy and surface energy.

6.1 Selection of periodic structures

As a simple one-dimensional counterpart of the two-well problem consider the problem

$$\text{Minimize } I(u) = \int\limits_0^1 (u_x^2 - 1)^2 + u^2 \, dx \tag{6.3}$$

subject to periodic boundary conditions. Clearly $I(u) > 0$ since the conditions $u = 0$ a.e. and $u_x = \pm 1$ a.e. are incompatible. On the other hand $\inf I = 0$, since a sequence of finely oscillating of sawtooth functions u_j can achieve $u_{jx} \in \{\pm 1\}, u_j \to 0$ uniformly. For any such sequence u_{jx} generates the (unique) Young measure $\nu = \frac{1}{2}\delta_{-1} + \frac{1}{2}\delta_1$ (see Section 3.2b)). Note that there are many 'different' sequences that generate this Young measure. Minimizers of the singularly perturbed functional

$$I^\epsilon(u) = \int\limits_0^1 \epsilon^2 u_{xx}^2 + (u_x^2 - 1)^2 + u^2 \, dx$$

yield a very special minimizing sequence for I.

Theorem 6.1 *If $\epsilon > 0$ is sufficiently small then every minimizer of I^ϵ (subject to periodic boundary conditions) is periodic with minimal period $P^\epsilon = 4(2\epsilon)^{1/3} + \mathcal{O}(\epsilon^{2/3})$.*

A more detailed analyis shows that the minimizers u^ϵ look approximately like a sawtooth function with slope ± 1 and involve *two* small length scales: the sawtooth has period $\sim \epsilon^{1/3}$, and its corners are rounded off on a scale $\sim \epsilon$ (see Fig. 19).

The heuristics behind the proof of Theorem 6.1 is simple and relies on two observations. First, the condition $I^\epsilon(u^\epsilon) \to 0$ enforces that u^ϵ is almost a sawtooth function with slopes ± 1. Second, a key observation of Modica and Mortola is that the first two terms of the energy combined essentially count (ϵ times) the number of changes in the slope from 1 to -1 and vice versa. Indeed the arithmetic geometric mean inequality yields for any interval $(a, b) \subset (0, 1)$

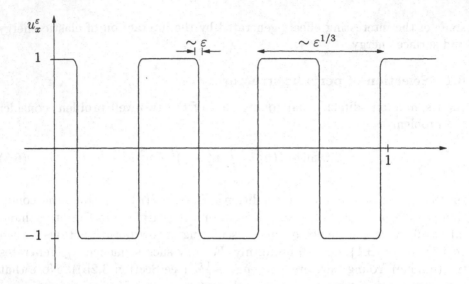

Figure 19: Sketch of u_x^ϵ for a minimizer of I^ϵ

over which u_x changes sign

$$\int\limits_a^b \epsilon^2 u_{xx}^2 + (u_x^2 - 1)^2 \, dx \geq \int\limits_a^b 2\epsilon |(u_x^2 - 1)u_{xx}| \, dx$$

$$\geq \epsilon \left| \int\limits_a^b H'(u_x) \, dx \right| \geq \epsilon \left| H(u_x(b)) - H(u_x(a)) \right|$$

$$\approx \epsilon \left| H(1) - H(-1) \right|,$$

where $H'(t) = 2|t^2 - 1|$. On the other hand the above estimates can be made sharp if one choose u as a solution of the ODE $\epsilon u_{xx} = (u_x^2 - 1)^2$, e.g. $u_x = \tanh \frac{x - x_0}{\epsilon}$.

The two observations strongly suggest that (6.3) is essentially equivalent to the following "sharp-interface problem"

$$\text{Minimize} \qquad \epsilon A_0 N + \int_0^1 u^2 \, dx \tag{6.4}$$

among periodic function with $|u_x| = 1$.

Here N denotes the number of sign changes of u_x and $A_0 = H(1) - H(-1) = 8/3$. For fixed N (6.4) is a discrete problem and a short calculation shows that in this case periodically spaced sign changes of u_x are optimal and the second term in the energy becomes $\frac{1}{12} N^{-2}$. Minimization over N yields the assertion.

The actual proof of Theorem 6.1 uses the expected analogy between (6.4) and (6.3) only as a guiding principle and proceeds by careful approximations and estimates for odes. Nonetheless it would be very useful to relate (6.4) and (6.3) in a rigorous way, also as a test case for higher dimensional problems where the fine ode methods are not available. Conventional Γ-convergence methods do not apply since the problem involves two small length scales and the passage from (6.3) to (6.4) corresponds to removing only the faster one (i.e. the smoothing of the sawtooth's corners). Recently G. Alberti and the writer developped a new approach that allows one to do that. One of the main ideas is to introduce a new variable y that corresponds to the slower scale and to view

$$v^\epsilon(x, y) := \epsilon^{-1/3} u(x + \epsilon^{1/3} y)$$

as a map V^ϵ from $(0, 1)$ into a suitable function space X via $V^\epsilon(x) = v^\epsilon(x, \cdot)$. One can endow X with a topology that makes it a compact metric space and study of the Young measure ν generated by V^ϵ. For each $x \in (0, 1)$ the measure ν_x is a probability measure on the function space X. If u^ϵ is a sequence of (almost) minimizers of I^ϵ then one can show that ν_x is supported on translates of sawtooth functions with the optimal period $4\, 2^{1/3}$.

One easily checks that the asymptotic behaviour is the same for minimizers of (6.4) and this gives a precise meaning to the assertion that (6.3) and (6.4) are asymptotically equivalent.

This approach is inspired by the idea of two-scale convergence ([Al 92], [E 92], [Ng 89]). A crucial difference is that two-scale convergence usually only applies if the period of the microstructure is fixed and possible phase shifts are controlled. This is the case if, for example, the solutions are of the form $\tilde{u}(x, \frac{x}{\epsilon^{1/3}})$ where $\tilde{u}$ is periodic in the second variable.

6.2 Surface energy and domain branching

Consider the two-dimensional scalar model problem (see [KM 92] for the relation with three-dimensional elasticity)

$$I(u) = \int_0^1 \int_0^L u_x^2 + (u_y^2 - 1)^2 dx \, dy \overset{!}{\to} \min$$

$$u = 0 \text{ on } x = 0. \tag{6.5}$$

The integrand is minimized at $Du = (u_x, u_y) = (0, \pm 1)$. The preferred gradients are incompatible with the boundary condition. The infimum of I subject to (6.5) is zero but not attained. The gradients Du_j of any minimizing sequence generate the Young measure $\frac{1}{2}\delta_{(0,-1)} + \frac{1}{2}\delta_{(0,1)}$. One possible construction of a minimizing sequence is as follows (see Fig. 20). Let s_h be a periodic sawtooth function with period h and slope ± 1 and let $u(x,y) = s_h(y)$ for $x \geq \delta$, $u(x,y) = \frac{x}{\delta} s_h(y)$ for $0 \leq x < \delta$. Then consider a limit $h \to 0, \delta \to 0$ such that h/δ remains bounded. Similar reasoning applies if we replace (6.5) by the condition that u vanishes on the whole boundary of $[0, L] \times [0, 1]$.

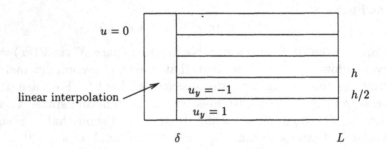

Figure 20: Construction of a minimizing sequence.

To understand the influence of regularizing terms on the length scale and the geometry of the fine scale structure we consider

$$I^\epsilon(u) = \int_0^1 \int_0^L u_x^2 + (u_y^2 - 1)^2 + \epsilon^2 u_{yy}^2 dx \, dy,$$

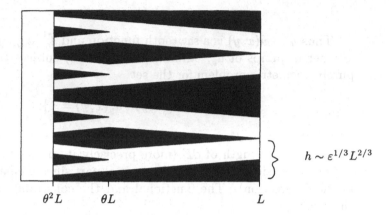

$$h \sim \epsilon^{1/3} L^{2/3}$$

$\theta^2 L \qquad \theta L \qquad\qquad\qquad L$

Figure 21: The self-similar construction with $1/4 < \Theta < 1/2$. Only two generations of refinement are shown.

subject to (6.5). Instead of the second derivatives in y one can consider other regularizing terms, e.g. $|D^2u|^2$. The derivatives in y are, however, the most important ones, since we expect that fine scale oscillations arise mainly in the y direction. It was widely believed that for small $\epsilon > 0$ the minimizers of I^ϵ look roughly like the construction $u_{h,\delta}$ depicted in Figure 20 (with the corners of the sawtooth 'rounded off' and optimal choices $\delta(\epsilon), h(\epsilon)$). This is false. Indeed a short calculation shows that $\delta(\epsilon) \sim (\epsilon L)^{1/2}$, $h(\epsilon) \sim (\epsilon L)^{1/2}$ and $I^\epsilon(u_{h,L}^\epsilon) \sim \epsilon^{1/2} L^{1/2}$. On the other hand one has

Theorem 6.2 *([Sch 94]) For $0 < \epsilon < 1$ there exists constants $c, C > 0$ such that*

$$c\epsilon^{2/3} L^{1/3} \leq \min_{u=0 \text{ at } x=0} I^\epsilon \leq C\epsilon^{2/3} L^{1/3}.$$

The upper bound is obtained by a smooth version of the self-similar construction depicted in Figure 21.

The mathematical issues become clearer if we again replace I^ϵ by a sharp interface version

$$J^\epsilon(u) = \int_0^L \int_0^1 u_x^2 + \epsilon |u_{yy}| dy dx \qquad (6.6)$$

subject to

$$|u_y| = 1 \text{ a.e.} \tag{6.7}$$

Thus $y \mapsto u(x, y)$ is a sawtooth function and $\int_0^1 |u_{yy}| dy$ denotes twice the number of jumps of u_y. Minimization of (6.6) subject to (6.7) is in fact a purely geometric problem for the set

$$E = \{(x, y) : u_y(x, y) = 1\}.$$

The first term in J^ϵ is a nonlocal energy in terms of E, while the second is essentially the length of ∂E (more precisely its projection to the x-axis; as before we consider this to be the essential part since oscillations occur mainly in the y direction). The functional and the constraint are invariant under the scaling

$$u_\lambda(x, y) = \lambda^{-1} u(\lambda^{3/2} x, \lambda y)$$

which suggests a self-similar construction with $\Theta = \left(\frac{1}{2}\right)^{3/2}$.

Theorem 6.3 *([KM 94]). For $0 < \epsilon < 1$ one has*

$$c\epsilon^{2/3} L^{1/3} \leq \min_{(6.6)(6.7)} J^\epsilon \leq C\epsilon^{2/3} L^{1/3}.$$

Moreover if $\bar{u}$ is a minimizer of J^ϵ subject to (6.6), (6.7) then

$$c\epsilon^{2/3} l^{1/3} \leq \int_0^1 \int_0^l \bar{u}_x^2 + \epsilon |\bar{u}_{yy}| dx dy \leq C\epsilon^{2/3} l^{1/3}. \tag{6.8}$$

The scaling in (6.8) is exactly the scaling predicted by the self-similar construction with $\theta = \left(\frac{1}{2}\right)^{3/2}$.

The prediction of refinement of the microstructure (domain branching) towards the boundary $x = 0$ in the simple models (6.6), (6.7) inspired new experimental investigations ([Sch 93]). In closely related models for magnetization domains in ferromagnetic materials domain branching is experimentally well established ([Li 44], [Hu 67], [Pr 76]), a rigorous mathematical analysis is just beginning to emerge ([CK 97b], [CKO 97]. Already a quick look at some of the sophisticated constructions in [Pr 76] suggests that a lot is to be discovered.

7 Outlook

There are many other interesting aspects of microstructure and I can mention only three areas: alternative descriptions of microstructure, dynamics and computation.

7.1 Alternative descriptions of microstructure

Young measures are but one way to describe microstructure and to extract 'relevant' information from a sequence of rapidly oscillating functions. They determine the asymptotic local distribution of function values but contain no information about the direction, length scale or fine geometry of the oscillations. As we saw in Section 3.3 the Young measure does not suffice to determine the limits of natural nonlocal quantities such as the magnetostatic energy or the self-correlation function.

There is an intense search for new objects that record additional information, see [Ta 95] for a survey. One such object was introduced independently by Tartar [Ta 90] and Gérard [Ge 91] under the names 'H-measure' and 'microlocal defect measure', respectively. They show that for every sequence $\{u_j\}$ that converges to zero weakly in $L^2(\Omega)$ there exists a subsequence $\{u_{j_k}\}$ and a Radon measure μ on $\bar{\Omega} \times S^{n-1}$ (the H-measure of $\{u_{j_k}\}$) such that for every pseudo-differential operator A of order zero with (sufficiently regular) symbol $a(x, \xi)$ one has

$$\langle Au_{j_k}, u_{j_k}\rangle_{L^2} \to \int_{\bar{\Omega} \times S^{n-1}} a\, d\mu.$$

For $\mathbf{R}^m$-valued sequences one similarly obtains a matrix-valued (hermitian) measure $\nu = (\nu^{ij})_{1 \le i,j \le m}$. The H-measure suffices, for example, to compute the limit of the micromagnetic energy discussed in Example 1 of Section 3.3 (the corresponding matrix valued symbol is just $a(\xi) = \frac{\xi}{|\xi|} \otimes \frac{\xi}{|\xi|}$). Other applications of the H-measure include small amplitude homogenization, compensated compactness with variable coefficients, compactness by averaging in kinetic equations and the propagation of energy concentrations in linear hyperbolic systems.

Two outstanding open problems are the relation between H-measures and Young measures (see [MT 97], [Ta 95] for partial results) and a useful generalization that allows one to compute limits of nonquadratic quantities; even the case of trilinear expressions is open.

The H-measure tracks the energy of oscillations depending on the direction, but regardless of length scales. P. Gèrard [Ge 90] introduced a variant of the H-measure, called semiclassical measure, that allows one to study the effect of oscillations on a typical length scale $h_j \to 0$ (see also [LP 93]). A completely different approach to analyze the detailed behaviour on small length scales was briefly discussed at the end of Section 6.1.

7.2 Dynamics

Three fundamental questions are:

Can realistic dynamics create microstructure?

Can one deduce a law for the evolution of microstructure from the macroscopic laws, and possibly reasonable additional assumptions?

What is a 'good' evolution law for interfaces in complex microstructures and how can one model hysteresis?

A typical setting for the first question is a dynamical system that admits a Liapunov function (such as energy or entropy) for which there exist no classical minimizer. Will the dynamics drive the Liapunov function to its infinum and hence create fine scale oscillations or will the dynamics generate compact orbits (in a suitable energy space) and thus prevent global minimization of the Liapunov function?

The papers [Ba 90a] and [BHJPS 91] give nice surveys; Friesecke and McLeod [FM 96] solved a longstanding problem by showing that one-dimensional viscoelastic dynamics with a nonconvex elastic energy does not generate microstructure.

A precise setting for the second question is as follows. Consider a sequence of rapidly oscillating initial data that generate a certain Young measure (or H-measure, semiclassical measure, ...). Is the Young measure of the solution at a later time determined by the Young measure of the initial data? In physical language this is closely related to the idea of coarse-graining. Given an evolution law for very complex pattern are there simpler laws for certain gross quantities such as the local phase average (= Young measure = one-point statistics)? If this can be achieved it can lead not only to new insights but also to huge savings in computer time and more reliable results since it is no longer needed to resolve the finest scale of the pattern.

A typical obstacle in attacking these questions is the closure problem. Often the time derivatives of certain moments of the Young measure involve higher moments. Even worse, sometimes the time derivative of the Young measure involves terms that depend on two-point or higher correlations which cannot be determined from the Young measure (see Example 2 of Section 3.3).

The first results on the evolution of Young measures and creation or non-creation of oscillations were obtained by Tartar for kinetic models and more general semilinear hyperbolic systems ([Ta 80], [Ta 81], [Ta 84], [Ta 86], [Ta 87], [MPT 85]), see [Jo 83], [JMR 95], [Mi 97] for further developments. In [FBS 94] Tartar's ideas were used to study the evolution of Young measures for a viscoelastically damped wave equation with nonmonotone stress-strain relation. Theil [Th 97] recently obtained very sharp results on this problem by a modification of the method that relies on transport theory rather than on a study of the moments of the measure. Otto derived equations for the evolution of microstructure in unstable porous media [Ot 95] and magnetic fluids [Ot 97]; further references on the evolution of microstructure include [De 96] and [HR 94].

Regarding the third question about evolution laws and hysteresis Chu and James observed that the hysteresis curves obtained in cyclic biaxial loading of a Cu-Al-Ni single crystal cannot be explained by usual kinetic laws. One alternative approach is based on metastability induced by lack of rank-1 connections [BCJ 95], another interesting route is explored in [ACJ 96]: the energy landscape in function space contains many local minima (that correspond to different microstructures) and the study of the effective evolution for such 'wiggly' potentials yields surprising conclusions already in a simple model; see also [Kin 97]. For other views of hysteresis, see the survey article of Huo and I. Müller [HM 93], the recent monograph of Brokate and Sprekels [BS 96] and the series of lectures [Br 94]; general references for hysteresis include [KP 89] and [Vi 94].

7.3 Computation

The computation of microstructure by numerical energy minimization is a very challenging task, see M. Luskin [Lu 96] for a recent survey. If microstructure is numerically observed, it often forms on the scale of the underlying mesh. Hence calculations are notoriously mesh dependent unless (expensive) regularizations are included or special care is taken.

So far most numerical schemes do not make use of analytical insights, except for scalar problems where relaxation leads not only to a drastic speed up but also to more accurate results ([CP 97]). Some other exceptions are discussed in Sections 1 and 7 of [Lu 96]. One difficulty in using analytical information in higher dimensions is that quasiconvexity while being the natural convexity notion (see Section 4.2) is still largely mysterious and no efficient algorithm for the computation of the quasiconvex hull is known. At least for rank-convex hulls there has been some progress in [MP 98] and [Do 97].

One important issue is how to represent microstructures numerically in an efficient way. Currently mostly finite element approaches are used but they require a lot of unknowns to represent simple microstructures such as an order 2 laminate (cf. Fig. 14). Ideally a good representation should both yield a high compression ratio and be well adopted to the numerical algorithm. The search for better analytical objects to describe microstructure discussed above may well be relevant here.

7.4 Some solved and unsolved problems

The following table gives an overview of the state of Problems 1 and 2 for sets K without rank-one connections. Further examples and references can be found in [Ba 90b], [BFJK 94], [Sv 95].

Acknowledgements

The writing of this article and the work reported here would not have been possible without the support and encouragement of many colleagues and friends. Particular thanks go to G. Alberti, J.M. Ball, K. Bhattacharya, G. Dolzmann, I. Fonseca, G. Friesecke, R.D. James, D. Kinderlehrer, B. Kirchheim, J. Kristensen, R.V. Kohn, V. Šverák and L. Tartar. I am also very greatful to U. Knieper and M. Krieger for their excellent typing of these notes.

I thank R. Conti, S. Hildebrandt, and M. Struwe for their invitation to prepare these lectures and for providing such a delightful and stimulating atmosphere at the C.I.M.E. summer school in Cetraro.

In the preparation of the lectures I greatly benefited from an invitation by J. Lukeš and J. Maly to present a similar set of lectures at a Spring School in Paseky in 1995.

K	Nontrivial exact soln.	Nontrivial GYM	Method	References
$\{A, B\}$	No	No	Minors or Cauchy-Riemann eqns.	[BJ 87], [Sv 91a]
$\{A, B, C\}$	No	No	Nonconvex solns. of Monge-Ampère	[Sv 91b], [Sv 92b]
finite	?	Yes	Example with four matrices	[AH 86], [Ta 93], [CT 93], [BFJK 94]
countable	Yes	Yes	Example	[Sv]
$SO(2); \mathbf{R}\,SO(2)$	No; holomorphic fns.	No	Cauchy-Riemann eqns.	[Sv 91a]
$SO(n); \mathbf{R}\,SO(n)$ $n \geq 3$	No; Möbius maps	No	Minors;degenerate elliptic eqns.	[Ki 88], [Re 68a]
$\cup_{i=1}^{k} SO(2)A_i$	No	No	Minors or elliptic regularity	[Sv 93a]
$SO(3)A \cup SO(3)B$	?	?	Special cases known	[Ma 92], [Sv 93a]
scalar conservation laws	No	No	compensated compactness; div-curl lemma	[Ta 79b]
$m \times 2$ elliptic systems	?	Yes, $m \geq 6$?, $m < 6$		[Sv 95]

Table 1: Nontrivial exact solutions and nontrivial gradient Young measures (GYM) for incompatible sets K (i.e. $\mathrm{rk}(A - B) \neq 1 \, \forall A, B \in K$).

Notes

Here I have collected some additional references to the literature without any pretention to be exhaustive or impartial.

Chapter 1 The idea to use nonlinear continuum theory for elastic crystals and solid-solid phase transformations goes back to Ericksen [Er 75], [Er 77], [Er 80], [Er 84], [Er 89] (see also [Gu 83], [Ja 81], [Pa 81], [Pi 84]) and was developed in the context of the calculus of variations by Ball and James ([BJ 87], [BJ 92]), Chipot and Kinderlehrer [CK 88], Fonseca [Fo 87] and subsequently by many others. There is a similar theory for micromagnetism ([Br 63], [DS 93], [JK 90]) and magnetostriction ([Br 66], [JK 93]).

The analytical foundations of the theory go back to the fundamental work of Morrey ([Mo 52], [Mo 66]) on lower semicontinuity (extended by Reshetnyak [Re 67], [Re 89] to problems in quasiconformal geometry and by Ball [Ba 77] to nonlinear elasticity) and to the pioneering work of Tartar on compensated compactness (partly in collaboration with Murat) and on weak convergence as a tool to pass from microscopic to macroscopic descriptions. His work in the seventies is summarized in the seminal paper [Ta 79b], some more recent developments are discussed in [Ta 90], [Ta 93] and [Ta 95], and a comprehensive treatment will appear in [Ta 98]. While the current notes focus mostly on variational problems Tartar's approach is more general. In view of applications to nonlinear partial differential equations in continuum mechanics he considers general combinations of pointwise constraints $w \in K \subset \mathbf{R}^d$ a.e. (these usually arise from constitutive equations) and differential constraints $\sum_{j,k} a_{ijk} \partial_j w^k = 0$ (or in a compact set of $W^{-1,p}$; these correspond to the balance laws). The situation considered in the current notes corresponds to the constaint curl $w = 0$.

Partially motivated by Eshelby's classic work on ellipsoidal inclusions [Es 57, Es 59, Es 61], Khachaturyan, Roitburd and Shatalov ([Kh 67], [KSh 69], [Ro 69], [Ro 78]) developed already in the sixties a theory of microstructure based on energy minimization in the context of linear elasticity; see [Kh 83] for a comprehensive treatment. Comparisons between the linear and the nonlinear theory appear in [BJ 92], [Bh 93] and [Ko 89].

For lack of space I have not been able to discuss the close relation between the variational approach to microstructures and the theory of optimal design and optimal composites. Some constructions used in optimal design closely resemble observed phase arrangements in solid-solid phase transitions. Early work in this direction includes [Ta 75], [Ta 79b], [Mu 78] and [KL 71]. A number of important papers that were previously difficult to access have recently appeared in English translation in [CK 97a]. Further references can be found both in the introduction and the individual articles of that volume as well as in the forthcoming books [Mi 98] and [Ta 98].

The approach to microstructures via energy minimization provides a new foundation for the crystallographic theory of martensite ([BM 54], [WLR 53]) and has found important applications which include: new criteria for the reversible shape-memory effect based on the possibility of self-accomodation of the transformed phase [Bh 92], bounds for the recoverable strains in polycrystals and their dependence on the symmetry of the phase transformation and material texture ([BK 96], [BK 97], [BRL 97]), a proposed design of micromachines that are based on thin films of shape-memory materials [BhJ 97] and the discovery of a new magnetostrictive material with greatly enlarged magnetostrictive constant [JW 97].

The book by Pitteri and Zanzotto [PZ 97] and the forthcoming book by Ball and James [BJ 97] as well as the collection of reviews [AMM 98] give an overview of the theory and engineering applications. More on the mathematical side, the recent book of Pedregal [Pe 97] reviews the relevance of microstructure and Young measures in various areas of application, while Roubíček's book [Ro 97] focuses more on the functional analytic aspects. Evans' notes [Ev 90] are an excellent introduction to the application of weak convergence methods to partial differential equations. Many further examples can be found in [BFJK 94] and [Sv 95].

The experimental observations described in Section 2.2 are discussed in detail in [CJ 95] and in Chu's thesis [Ch 93]; a careful comparison of theory and experiment for a variety of solid-solid phase transformations was undertaken by Hane [Ha 97].

Chapter 2 The connection with the Cauchy-Riemann equation appears in [Sv 91a]. A counterpart of Theorem 2.4 holds for quasiconformal maps, i.e. $K = \mathbf{R}^+ SO(n), n \geq 3$. In this case one is led to degenerate elliptic equations and Reshetnyak's work ([Re 68a], [Re 89]) was a breakthrough in the study of quasiconformal and quasiregular maps by pde methods. Part (i) of Theorem 2.5 was proved in [Zh] and also follows from more general results in [JO 90]; the proof given is due to Kirchheim.

Lemma 2.7 is called the 'span restriction' in [BFJK 94] because it implies (in view of Corollary 3.2) that the span of the support of a nontrivial gradient

Young measure must contain a rank-1 line. The result is essentially a special case of Theorem 3 in Tartar's work [Ta 83]. It was probably known in some form to Serre [Se 83] and is implicit in [DP 85]. The use of elliptic theory is a common idea in the theory of microstructures, see e.g. [DP 85], [Ma 92], [Sv 93b], [Sv 95].

Chapter 3 Young measures (also known as parametrized measures, relaxed controls, chattering controls or generalized curves) were invented by L.C. Young [Yo 37]; his book [Yo 69] is a delightful read (see also McShane [MS 40] for early applications of the theory and [MS 89] for a personal review). The theory was generalized to much more general domains, target spaces and integrals by Berliocchi and Lasry [BL 73], Balder [Ba 84] and many others including Kristensen [Kr 94]; recent surveys with extensive references are [Va 90] and [Va 94]. Varifolds (see [Al 66], [Al 72], [Re 68b]) are a generalization of Young measures in a geometric setting. Tartar [Ta 79b, Ta 83] introduced Young measures as a fundamental tool for the study of oscillation effects as well as compactness and existence questions in nonlinear partial differential equations. His theory of compensated compactness allows one to derive nontrivial constraints on the Young measure from the combination of pointwise and differential constraints on the generating sequence. One of the early successes of the theory were applications to conservation laws ([Ta 83], [DP 85]); for other applications see e.g. [Ev 90], [Sv 95], [Ta 98].

The presentation here follows [Ba 89]; Section 3.3 is based on [BJ 94]. Another phenomenon that Young measures cannot detect are concentration effects. Varifolds, currents [FF 60, Fe 69, GMS 89, GMS 96] or H-measures [Ge 91, Ta 90] do better in this regard; see also [FMP 97]. There are various alternative proofs of the fundamental theorem: via disintegration of measures on $\Omega \times \mathbf{R}^d$ (see e.g. [BL 73] for a much more general setting and [Ev 90] for a short proof), via L^∞ weak∗ precompactness of bounded sequences and the theory of multivalued maps (see [Sy 97]) or by consideration of countable dense sets of integrands f_j and test functions φ_k and diagonalization. Corollary 3.3 is a special case of results in [Ba 84].

Chapter 4 The fundamental connection between quasivonvexity and lower semicontinuity was discovered by Morrey (see [Mo 52], [Mo 66]). Dacorogna [Da 81], [Da 82a] discovered the relation between quasiconvexity and relaxation (see also [AF 84]); his book [Da 89] gives a comprehensive treatment of the different notions of convexity. The work of Acerbi and Fusco [AF 84] and

Marcellini [Ma 85] brought a major technical refinement with the coverage of Carathéodory integrands. Since then many further refinements and generalizations have been achieved; a selection is [ABF 96], [BFM 97], [Fo 96], [Kr 97b], [Ma 94], and many further references can be found there. For the connection between quasiconvexity, regularity and compactness see [Ev 86], [EG 87], [FH 85] and [GM 86].

Tartar has pointed out various weaknesses of quasiconvexity. First, quasiconvexity might not be necessary to obtain existence of minimizers. In view of Ekeland's variational principle [Ek 79] (which makes use of the Bishop-Phelps argument [BP 61]) one can choose minimizing sequences that satisfy in addition $\mathrm{div}\sigma_k \to 0$ in $W^{-1,q}$ where $\sigma_k = \frac{\partial f}{\partial F}(Du_k)$ is the stress and hence one does not need to verify lower semicontinuity along arbitrary sequences. To my knowledge this line of thought has not been explored in detail. Secondly it is not clear (indeed rather doubtful) whether quasiconvexity implies the stability of equilibria, i.e. whether the conditions $u_k \overset{*}{\rightharpoonup} u$ in $W^{1,\infty}$, $\sigma_k \overset{*}{\rightharpoonup} \overline{\sigma}$ in L^∞ and $\mathrm{div}\sigma_k \to \mathrm{div}\overline{\sigma}$ in $W^{-1,\infty}$ do imply $\overline{\sigma} = \frac{\partial f}{\partial F}(Du)$ (by contrast Jensen obtained a nice classification in the scalar case, see [Ta 79b], Theorem 23). Šverak has shown [Sv 95], [Sv 98] that the compactness arguments that are the cornerstone of the regularity theory for minimizers for (uniformly) quasiconvex integrals fail for solutions of the equilibrium equations. For arguments in favour of quasiconvexity, in addition to those in the text, see [BMa 84] (cf. also [Me 65], pp.128–131) and [BM 84], Theorem 5.1.

Sections 4.1 and 4.2 are partially based on [BJ 94]. In the definition of quasiconvexity often additional restrictions on the integrand are imposed. Hüsseniov [Hu 88], [Hu 95] realized that this is not necessary, see also [Fo 88]. Section 4.4 follows partially unpublished lectures by Šverák, see also [Sv 95]. Šverák's counterexample is reminicent of a counterexample by Tartar in the theory of compensated compactness (see [Ta 79b], pp.185–186). The proof of the classification result follows roughly Kinderlehrer and Pedregal's original work [KP 91] (see also [Kr 94]). Some simplifications, in particular for the nonhomogeneous case, are based on discussions with Alberti. Sychev [Sy 97] recently presented independently a similar approach for the case $1 < p < \infty$. The idea to use the Hahn-Banach theorem to characterize Young measure appears e.g. in [Ta 79b], p.152, for the case without differential constraints; in a similar vein the Krein-Milman theorem is used in [BL 73], p.148. The proof of Theorem 4.4.(i) is by now standard (see [Mo 52]), the proof of Theorem 4.5(iii) is the same as Fonseca's [Fo 88], see also [Hu 88]. Truncation arguments that are closely related to what I called Zhang's lemma were used earlier by Acerbi and Fusco [AF 84], [AF 88], based on work of Liu [Li 77].

Gradient Young measures and quasiconvexity correspond to the constraint $\operatorname{curl} v = 0$. As mentioned above, in continuum mechanics and electromagnetism one also meets more general systems of first order constraints $A(Dv) = 0$. If A satisfies a constant rank condition there is a largely parallel theory ([Da 82b], [FM 97]) (in an L^p-setting, $1 < p < \infty$) while the situation is widely open even in simple examples where this condition fails (see [Ta 93]).

Chapter 5 Most of the material is taken from [MS 96] and [Mu 97c]. The basic existence result is the theorem on p.218 of Gromov's book [Gr 86]. A detailed proof for a special case and the application to the two-well problem are described in [MS 96]. The case $K = O(3)$ is studied in [CP 95] (here some simplifications occur since $K^{lc} = \overline{\operatorname{conv}} K$); results for more general isometric maps appear in [Gr 86], Chapter 2.4.11. For variable prescribed singular values see also [CPe 97].

Chapter 6 This chapter is based on [Mu 97c]. Theorem 6.1 is taken from [Mu 93]; the $\varepsilon^{1/3}$ scaling had been predicted earlier by Tartar based on matched asymptotic expansions. The Modica-Mortola inequality [MM 77a, MM 77b] was found shortly after De Giorgi had introduced the notion of Γ-convergence [DG 75], [DGF 75], but was initially somewhat overlooked. With the growing interest in the gradient theory of phase transitions since the mid-80's (see [BF 94], [Bo 90], [FT 89], [Gu 87] [KS 89], [Mo 87] and the references therein) it later became a crucial tool. Dal Maso's book [DM 93] is a good reference on Γ-convergence with a very useful commented bibliography. The influence of surface energy on phase transformations in crystals was studied in a series of papers by Parry and others [MP 86], [Pa 87a], [Pa 87b], [Pa 89], mostly in one-dimensional situations.

References

[ACJ 96]   R. Abeyaratne, C. Chu and R.D. James, Kinetics of materials with wiggly energies: theory and application to the evolution of twinning microstructures in a Cu-Al-Ni shape-memory alloy, *Phil. Mag.* **A 73** (1996), 457–497.

[AF 84]   E. Acerbi and N. Fusco, Semicontinuity problems in the calculus of variations, *Arch. Rat. Mech. Anal.* **86** (1984), 125–145.

[AF 88]   E. Acerbi and N. Fusco, An approximation lemma for $W^{1,p}$ functions, in: *Material instabilities in continuum mechanics and related mathematical problems* (J. M. Ball, Ed.), Oxford Univ. Press, 1988, 1–5.

[AMM 98]   G. Airoldi, I. Müller, S. Miyazaki (eds.), *Shape-memory alloys: from microstructure properties*, Trans Tech Pub., in press, 1998.

[AM 97]   G. Alberti and S. Müller, in preparation.

[AD 92]   J.J. Alibert and B. Dacorogna, An example of a quasiconvex function not polyconvex in dimension two, *Arch. Rat. Mech. Anal.* **117** (1992), 155–166.

[Al 92]   G. Allaire, Homogenization and two-scale convergence, *SIAM J. Math. Anal.* **23** (1992), 1482–1518.

[Al 72]   W.K. Allard, On the first variation of a varifold, *Ann. Math.* **95** (1972), 417–491.

[Al 66]   F. J. Almgren, Plateau's problem: an invitation to varifold geometry, Benjamin, 1966.

[ABF 96]   L. Ambrosio, G. Buttazzo and I. Fonseca, Lower semicontinuity problems in Sobolev spaces with respect to a measure, *J. Math. Pures Appl.* **75** (1996), 211–224.

[AH 86]   R. Aumann and S. Hart, Bi-convexity and bi-martingales, *Israel J. Math.* **54** (1986), 159–180.

[Ba 84]   E.J. Balder, A general approach to lower semicontinuity and lower closure in optimal control theory, *SIAM J. Control and Optimization* **22** (1984), 570–598.

[Ba 77]   J.M. Ball, Convexity conditions and existence theorems in non-linear elasticity, *Arch. Rat. Mech. Anal.* **63** (1977), 337–403.

[Ba 87]   J.M. Ball, Does rank one convexity imply quasiconvexity?, in: *Metastability and incompletely posed problems*, IMA Vol. Math. Appl. **3** (S.S. Antman, J.L. Ericksen, D. Kinderlehrer and I. Müller, eds.), Springer, 1987.

[Ba 89]   J.M. Ball, A version of the fundamental theorem for Young measures, in: *PDE's and Continuum Models of Phase Transitions* (M. Rascle, D. Serre, M. Slemrod, eds.), Lecture Notes in Physics **344**, Springer, 1989, 207–215.

[Ba 90a]   J.M. Ball, Dynamics and minimizing sequences, in: *Problems involving change of type* (K. Kirchgässner, ed.), Lecture Notes in Physics **359**, Springer, 1990, 3–16.

[Ba 90b]   J.M. Ball, Sets of gradients with no rank one connections, *J. Math. Pures Appl.* **69** (1990), 241–259.

[BCJ 95]   J.M. Ball, C. Chu and R.D. James, Hysteresis in martensite phase transformations, in: Proc. ICOMAT-95, *J. de physique* IV **5**, colloque C8, 245–251.

[BHJPS 91] J.M. Ball, P.J. Holmes, R.D. James, R.L. Pego and P.J. Swart, On the dynamics of fine structure, *J. Nonlin. Sci.* **1** (1991), 17–70.

[BJ 87]   J.M. Ball and R.D. James, Fine phase mixtures as minimizers of energy, *Arch. Rat. Mech. Anal.* **100** (1987), 13–52.

[BJ 92]   J.M. Ball and R.D. James, Proposed experimental tests of a theory of fine microstructure and the two-well problem, *Phil. Trans. Roy. Soc. London* **A 338** (1992), 389–450.

[BJ 94]   J.M. Ball and R.D. James, *The mathematics of microstructure*, DMV-Seminar (unpublished lectures).

[BJ 97]   J.M. Ball and R.D. James, book in preparation.

[BMa 84]   J. M. Ball and J. E. Marsden, Quasiconvexity at the boundary, positivity of second variation and elastic stability, *Arch. Rat. Mech. Anal.* **86** (1984), 251–277.

[BM 84]   J.M. Ball and F. Murat, $W^{1,p}$-quasiconvexity and variational problems for multiple integrals, *J. Funct. Anal.* **58** (1984), 225–253.

[BKK 98]   J. M. Ball, B. Kirchheim and J. Kristensen, in preparation.

[BF 94]   A.C. Barroso and I. Fonseca, Anisotropic singular perturbations – the vectorial case, *Proc. Roy. Soc. Edinburgh* **124A** (1994), 527–571.

[BL 73]   H. Berliocchi and J.M. Lasry, Intégrandes normales et mesures paramétrées en calcul des variations, *Bull. Soc. Math. France* **101** (1973), 129–184.

[Bh 91]   K. Bhattacharya, Wedge-like microstructure in martensites, *Acta metall.* **39** (1991), 2431–2444.

[Bh 92]   K. Bhattacharya, Self-accommodation in martensite, *Arch. Rat. Mech. Anal.* **120** (1992), 201–244.

[Bh 93]   K. Bhattacharya, Comparison of geometrically nonlinear and linear theories of martensitic transformation, *Cont. Mech. Thermodyn.* **5** (1993), 205–242.

[BFJK 94]   K. Bhattacharya, N. Firoozye, R.D. James and R.V. Kohn, Restrictions on microstructure, *Proc. Roy. Soc. Edinburgh*, **124A** (1994), 843–878.

[BhJ 97]   K. Bhattacharya and R.D. James, A theory of thin films of martensite materials with applications to microactuators, to appear in *J. Mech. Phys. Solids*.

[BK 96]   K. Bhattacharya and R.V. Kohn, Symmetry, texture and the recoverable strain of shape-memory polycrystals, *Acta mater.* **44** (1996), 529–542.

[BK 97]   K. Bhattacharya and R.V. Kohn, Elastic energy minimization and the recoverable strains of polycristalline shape memory materials, *Arch. Rat. Mech. Anal.* **139** (1997), 99–180.

[BP 61]   E. Bishop and R. R. Phelps, A proof that all Banach spaces are subreflexive, *Bull. Amer. Math. Soc.* **67** (1961), 97–98.

[Bo 90]   G. Bouchitté, Singular perturbations of variational problems arising from a two phases transition model, *J. Appl. Math. Opt.* **21** (1990), 289–314.

[BFM 97]   G. Bouchitté, I. Fonseca and L. Mascarenhas, *A global method for relaxation*, preprint.

[BM 54]   J.S. Bowles and J.K. Mackanzie, The crystallography of martensite transformations, I and II, *Acta Met.* **2** (1954), 129–137, 138–147.

[Br 94]   M. Brokate et al., *Phase transitions and hysteresis*, LNM **1584**, Springer, 1994.

[BS 96]   M. Brokate and J. Sprekels, *Hysteresis and phase transitions*, Springer, 1996.

[Br 63]   W.F. Brown, *Micromagnetics*, Wiley, 1963.

[Br 66]   W.F. Brown, *Magnetoelastic interactions*, Springer Tracts in Natural Philosophy **9** (C. Truesdell, ed.), Springer, 1966.

[BRL 97]   O. Bruno, F. Reitich and P. Leo, The overall elastic energy of polycristalline martensite solids, submitted to *J. Mech. Phys. Solids*.

[CP 97]   C. Carstensen and P. Plecháč, Numerical solution of the scalar double-well problem allowing microstructure, *Math. Comp.* **66** (1997), 997–1026.

[CT 93]   E. Casadio-Tarabusi, An algebraic characterization of quasiconvex functions, *Ricerche Mat.* **42** (1993), 11–24.

[CPe 97]   P. Celada and S. Perrotta, Functions with prescribed singular values of the gradient, preprint SISSA 56/97/M, 1997.

[CP 95]   A. Cellina and S. Perrotta, On a problem of potential wells, *J. convex anal.* **2** (1995), 103–115.

[CK 97a]   A. Cherkaev, R.V. Kohn (eds.), *Topics in the mathematical modelling of composites*, Birkhäuser, 1997.

[CK 88]   M. Chipot and D. Kinderlehrer, Equilibrium configurations of crystals, *Arch. Rat. Mech. Anal.* **103** (1988), 237–277.

[CK 97b]   R. Choksi and R.V. Kohn, Bounds on the micromagnetic energy of a uniaxial ferromagnet, preprint.

[CKO 97]   R. Choksi, R.V. Kohn and F. Otto, in preparation.

[Ch 93]   C. Chu, *Hysteresis and microstructure: a study of biaxial loading on compound twins of copper-aluminium-nickel single crystals*, Ph.D. thesis, University of Minnesota, 1993.

[CJ 95]   C. Chu and R.D. James, Analysis of microstructures in Cu-14% Al-3.9% Ni by energy minimization, in: Proc. ICOMAT-95, *J. de physique* IV **5**, colloque C8, 143–149.

[Da 81]   B. Dacorogna, A relaxation theorem and its applications to the equilibrium of gases, *Arch. Rat. Mech. Anal.* **77** (1981), 359–386.

[Da 82a]   B. Dacorogna, Quasiconvexity and the relaxation of non convex variational problems, *J. Funct. Anal.* **46** (1982), 102–118.

[Da 82b]   B. Dacorogna, Weak continuity and weak lower semicontinuity of non-linear functionals, Springer LNM **922**, Springer, 1982.

[Da 89]   B. Dacorogna, *Direct methods in the calculus of variations*, Springer, 1989.

[DM 88]   B. Dacorogna and P. Marcellini, A counterexample in the vectorial calculus of variations, in: *Material instabilities in continuum mechanics and related mathematical problems* (J.M. Ball, ed.), Oxford Univ. Press, 1988, 77–83.

[DM 96a]   B. Dacorogna and P. Marcellini, Théorèmes d'existence dans les cas scalaire et vectoriel pour les équations de Hamilton-Jacobi, *C.R.A.S. Paris* **322** (1996), Serie I, 237–240.

[DM 96b]   B. Dacorogna and P. Marcellini, Sur le problème de Cauchy-Dirichlet pour les systèmes d'équations non linéaires du premier ordre, *C.R.A.S. Paris* **323** (1996), Serie I, 599–602.

[DM 97]   B. Dacorogna and P. Marcellini, General existence theorems for Hamilton-Jacobi equations in the scalar and vectorial cases, *Acta Math.* **178** (1997), 1–37.

[DM 93]   G. Dal Maso, *An introduction to Γ-convergence*, Birkhäuser, 1993.

[De 96]   S. Demoulini, Young measure solutions for a nonlinear parabolic equation of forward-backward type, *SIAM J. Math. Anal.* **27** (1996), 378–403.

[DG 75]   E. De Giorgi, Sulla convergenza di alcune successioni di integrali del tipo dell'area, *Rend. Mat.* **8** (1975), 277–294.

[DGF 75]   E. De Giorgi and T. Franzoni, Su un tipo di convergenze variationale, *Atti Accad. Naz. Lincei Rend. Cl. Sci. Fis. Mat. Natur* **58** (1975), 842–850.

[DS 93]   A. DeSimone, Energy minimizers for large ferromagnetic bodies, *Arch. Rat. Mech. Anal.* **125** (1993), 99–143.

[DP 85]   R.J. DiPerna, Compensated compactness and general systems of conversation laws, *Trans. Amer. Math. Soc.* **292** (1985), 383–420.

[DPM 87]   R.J. DiPerna and A.J. Majda, Oscillations and concentrations in weak solutions of the incompressible fluid equations, *Comm. Math. Phys.* **108** (1987), 667–689.

[Do 97]   G. Dolzmann, Numerical computations of rank-one convex envelopes, preprint.

[DM 95]   G. Dolzmann and S. Müller, Microstructures with finite surface energy: the two-well problem, *Arch. Rat. Mech. Anal.* **132** (1995), 101–141.

[E 92]   Weinan E, Homogenization of linear and nonlinear transport equations, *Comm. Pure Appl. Math.* **45** (1992), 301–326.

[Ed 65]   R.E. Edwards, *Functional analysis*, Holt, Rinehart and Winston, 1965.

[Ek 79]   I. Ekeland, Nonconvex minimization problems, *Bull. Amer. Math. Soc.* **1** (1979), 443–474.

[Er 75]   J.L. Ericksen, Equilibrium of bars, *J. Elasticity* **5** (1975), 191–201.

[Er 77]   J.L. Ericksen, Special topics in nonlinear elastostatics, in: *Advances in applied mechanics* **17** (C.-S. Yih, ed.), Academic Press, 1977, 189–244.

[Er 79]   J.R. Ericksen, On the symmetry of deformable elastic crystals, *Arch. Rat. Mech. Anal.* **72** (1979), 1–13.

[Er 80]   J.L. Ericksen, Some phase transitions in elastic crystals, *Arch. Rat. Mech. Anal.* **73** (1980), 99–124.

[Er 84]   J.L. Ericksen, The Cauchy and Born hypothesis for crystals, in: *Phase transformations and material instabilities in solids* (M.E. Gurtin, ed.), Academic Press, 1984, 61–77.

[Er 89]   J.L. Ericksen, Weak martensitic transformations in Bravais lattices, *Arch. Rat. Mech. Anal.* **107** (1989), 23–36.

[Es 57]   J.D. Eshelby, The determination of the elastic field of an ellipsoidal inclusion, and related problems, *Proc. Roy. Soc. London* **A 241** (1957), 376–396.

[Es 59]   J.D. Eshelby, The elastic field outside an ellipsoidal inclusion, *Proc. Roy. Soc. London* **A 252** (1959), 561–569.

[Es 61]   J.D. Eshelby, Elastic inclusions and inhomogenities, in: *Progress in Solid Mechanics* vol. II, North-Holland, 1961, 87–140.

[Ev 86]   L.C. Evans, Quasiconvexity and partial regularity in the calculus of variations, *Arch. Rat. Mech. Anal.* **95** (1986), 227–252.

[Ev 90]   L.C. Evans, *Weak convergence methods for nonlinear partial differential equations*, Am. Math. Soc., Providence, 1990.

[EG 87]   L.C. Evans and R. Gariepy, Blow-up, compactness and partial regularity in the calculus of variations, *Indiana Univ. Math. J.* **36** (1987), 361–371.

[EG 92]   L.C. Evans and R. Gariepy, *Measure theory and fine properties of functions*, CRC Press, 1992.

[FF 60]   H. Federer and W.H. Fleming, Normal and integral currents, Ann. Math. **72** (1960), 458–520.

[Fe 69]   H. Federer, *Geometric measure theory*, Springer, 1969.

[Fo 87]   I. Fonseca, Variational methods for elastic crystals, *Arch. Rat. Mech. Anal.* **97** (1987), 189–220.

[Fo 88]   I. Fonseca, The lower quasiconvex envelope of the stored energy function for an elastic crystal, *J. Math. Pures Appl.* **67** (1988), 175–195.

[Fo 89]   I. Fonseca, Phase transitions of elastic solid materials, *Arch. Rat. Mech. Anal.* **107** (1989), 195–223.

[Fo 96]   I. Fonseca, Variational techniques for problems in materials science, in: *Variational methods for discontinuous structures*, (R. Serapioni and F. Tomarelli, eds.), Birkhäuser, 1996, 161–175.

[FBS 94]   I. Fonseca, D. Brandon and P. Swart, Dynamics and oscillatory microstructure in a model of displacive phase transformations, in: *Progress in partial differential equations, the Metz surveys* **3** (M. Chipot, ed.), Pitman, 1994, 130–144.

[FM 97]   I. Fonseca and S. Müller, A-quasiconvexity, lower semicontinuity and Young measures, in preparation.

[FMP 97]   I. Fonseca, S. Müller and P. Pedregal, Analysis of concentration and oscillation effects generated by gradients, to appear in *SIAM J. Math. Anal.*

[FT 89]   I. Fonseca and L. Tartar, The gradient theory of phase transitions for systems with two potential wells, *Proc. Roy. Soc. Edinburgh* **111 A** (1989), 89–102.

[FM 96]   G. Friesecke and J.B. McLeod, Dynamics as a mechanism preventing the formation of finer and finer microstructure, *Arch. Rat. Mech. Anal.* **133** (1996), 199–247.

[FH 85]   N. Fusco and J. Hutchinson, Partial regularity of functions minimizing quasiconvex integrals, *manuscripta math.* **54** (1985), 121–143.

[Ge 90]   P. Gérard, Mesures semi-classique et ondes de Bloch, in: Equations aux dérivées partielles, exposé XVI, seminaire 1990–91, Ecole Polytechnique, Palaiseau.

[Ge 91]   P. Gérard, Microlocal defect measures, *Comm. PDE* **16** (1991), 1761–1794.

[GM 86]   M. Giaquinta and G. Modica, Partial regularity of minimizers of quasiconvex integrals, *Ann. IHP Analyse non linéaire* **3** (1986), 185–208.

[GMS 89]   M. Giaquinta, G. Modica and J. Souček, Cartesian currents, weak diffeomorphisms and nonlinear elasticity, *Arch. Rat. Mech. Anal.* **106** (1989), 97–159; erratum **109** (1990), 385–392.

[GMS 96]   M. Giaquinta, G. Modica and J. Souček, *Cartesian currents in the calculus of variations, part I*, preprint, 1996.

[Gr 86]   M. Gromov, *Partial differential relations*, Springer, 1986.

[Gu 83]   M.E. Gurtin, Two-phase deformations in elastic solids, *Arch. Rat. Mech. Anal.* **84** (1983), 1–29.

[Gu 87]   M.E. Gurtin, Some results and conjectures in the gradient the-
            ory of phase transitions, in: *Metastability and Incompletely
            Posed Problems* (S.S. Antman, ed.), Springer, 1987, 135–146.

[Ha 97]   K. Hane, *Microstructures in thermoelastic martensites*, Ph.D.
            thesis, University of Minnesota, 1997.

[HR 94]   K.-H. Hoffmann and T. Roubíček, Optimal control of a fine
            structure, *Appl. Math. Optim.* **30** (1994), 113–126.

[Hu 67]   A. Hubert, Zur Theorie der zweiphasigen Domänenstrukturen
            in Supraleitern und Ferromagneten, *Phys. status solidi* **24**
            (1967), 669–682.

[Hu 88]   F. Hüsseinov, Continuity of quasiconvex functions and the-
            orem on quasiconvexification, *Izv. Akad. Nauk Azerbaidzhan
            SSR Ser. Fiz.-Tekhn. Mat. Nauk* **8** (1988), 17–23.

[Hu 95]   F. Hüsseinov, Weierstrass condition for the general basic varia-
            tional problem, *Proc. Roy. Soc. Edinburgh* **125 A** (1995), 801–
            806.

[HM 93]   Y. Huo and I. Müller, Nonequilibrium thermodynamics and
            pseudoelasticity, *Cont. Mech. Thermodyn.* **5** (1993), 163–204.

[IT 69]   A. & C. Ionescu Tulcea, *Topics in the theory of liftings*,
            Springer, 1969.

[Ja 81]   R.D. James, Finite deformation by mechanical twinning, *Arch.
            Rat. Mech. Anal.* **77** (1981), 143–176.

[JK 89]   R.D. James and D. Kinderlehrer, Theory of diffusionless phase
            transitions, in: *PDEs and continuum model of phase transi-
            tions* (M. Rascle, D. Serre and M. Slemrod, eds.), Lecture Notes
            in Physics **344**, Springer, 1989.

[JK 90]   R.D. James and D. Kinderlehrer, Frustration in ferromagnetic
            materials, *Cont. Mech. Thermodyn.* **2** (1990), 215–239.

[JK 93]   R.D. James and D. Kinderlehrer, Theory of magnetostriction
            with applications to $Tb_xDy_{1-x}Fe_2$, *Phil. Mag. B* **68** (1993),
            237–274.

[JW 97]   R.D. James and M. Wuttig, Magnetostriction of martensite, to appear in *Phil. Mag. A.*

[JO 90]   M. Jodeit and P.J. Olver, On the equation grad$f = M$grad$g$, Proc. Roy. Soc. Edinburgh **116 A** (1990), 341–358.

[Jo 83]   J.L. Joly, Sur la propagation des oscillations per un système hyperbolique en dimension 1, *C.R.A.S. Paris* **296** (1983), 669–672.

[JMR 95]   J. L. Joly, G. Métivier, J. Rauch, Trilinear compensated compactness and nonlinear geometric optics, *Ann. Math.* **142** (1995), 121–169.

[Kh 67]   A. Khachaturyan, Some questions concerning the theory of phase transformations in solids, *Soviet Physics-Solid State* **8** (1967), 2163–2168.

[Kh 83]   A. Khachaturyan, *Theory of structural transformations in solids*, Wiley, 1983.

[KSh 69]   A. Khachaturyan and G. Shatalov, Theory of macroscopic periodicity for a phase transition in the solid state, *Soviet Physics JETP* **29** (1969), 557–561.

[Ki 88]   D. Kinderlehrer, Remarks about equilibrium configurations of crystals, in: *Material instabilities in continuum mechanics and related mathematical problems* (J.M. Ball, ed.), Oxford Univ. Press, 1988, 217–242.

[Kin 97]   D. Kinderlehrer, Metastability and hysteresis in active materials, submitted to: Mathematics and control in smart structures (V.K. Vardan and J. Chandra, eds.), 1997.

[KP 91]   D. Kinderlehrer and P. Pedregal, Characterization of Young measures generated by gradients, *Arch. Rat. Mech. Anal.* **115** (1991), 329–365.

[KP 94]   D. Kinderlehrer and P. Pedregal, Gradient Young measure generated by sequences in Sobolev spaces, *J. Geom. Analysis* **4** (1994), 59–90.

[Ki 97]   B. Kirchheim, in preparation.

[KL 71]   B. Klosowicz and K.A. Lurie, On the optimal nonhomogeneity of a torsional elastic bar, *Arch. Mech.* **24** (1971), 239–249.

[Ko 89]   R.V. Kohn, The relationship between linear and nonlinear variational models of coherent phase transitions, in: *Trans. 7th Army Conf. on applied mathematics and computing*, (F. Dressel, ed.), 1989.

[KM 92]   R.V. Kohn and S. Müller, Branching of twins near an austenite/twinned-martensite interface, *Phil. Mag.* **A 66** (1992), 697–715.

[KM 94]   R.V. Kohn and S. Müller, Surface energy and microstructure in coherent phase transitions, *Comm. Pure Appl. Math.* **47** (1994), 405–435.

[KS 89]   R.V Kohn and P. Sternberg, Local minimisers and singular perturbations, *Proc. Roy. Soc. Edinburgh* **111A** (1989), 69–84.

[KP 89]   M.A. Krasnosel'skiĭ and A.V. Pokrovskiĭ, *Systems with hysteresis*, Springer, 1989.

[Kr 94]   J. Kristensen, Finite functionals and Young measures generated by gradients of Sobolev functions, Ph.D. Thesis, Technical University of Denmark, Lyngby.

[Kr 97a]   J. Kristensen, On the non-locality of quasiconvexity, to appear in *Ann. IHP Anal. non linéaire.*

[Kr 97b]   J. Kristensen, Lower semicontinuity in spaces of weakly differentiable functions, preprint.

[Kr 97]   M. Kružik, On the composition of quasiconvex functions and the transposition, preprint 1997.

[Ku 55]   N.H. Kuiper, On C^1 isometric embeddings, I., *Nederl. Akad. Wetensch. Proc.* **A 58** (1955), 545–556.

[Li 44]   E. Lifshitz, On the magnetic structure of iron, *J. Phys.* **8** (1944), 337–346.

[Li 77]   F.-C. Liu, A Lusin type property of Sobolev functions, *Indiana Univ. Math. J.* **26** (1977), 645–651.

[LP 93]   P.-L. Lions and T. Paul, Sur les mesures de Wigner, *Rev. Mat. Iberoamericana* **9** (1993), 553–618.

[Lu 96]   M. Luskin, On the computation of crystalline microstructure, *Acta Num.* **5** (1996), 191–258.

[MP 86]   J. H. Maddocks and G. Parry, A model for twinning, *J. Elasticity* **16** (1986), 113–133.

[Ma 94]   J. Malý, Lower semicontinuity of quasiconvex integrals, *manuscripta math.* **85** (1994), 419–428.

[Ma 85]   P. Marcellini, Approximation of quasiconvex functions, and lower semicontinuity of multiple integrals, *manuscripta math.* **51** (1985), 1-28.

[Ma 92]   J.P. Matos, Young measures and the absence of fine microstructures in a class of phase transitions, *Europ. J. Appl. Math.* **3** (1992), 31–54.

[MP 98]   J. Matoušek and P. Plecháč, On functional separately convex hulls, *J. Discrete Comput. Geom.* **19** (1998), 105–130.

[MPT 85]   D.W. McLaughlin, G. Papanicolaou and L. Tartar, Weak limits of semilinear hyperbolic sytems with oscillating data, in: *Macroscopic modelling of turbulent flows*, Lecture Notes in physics **230**, Springer, 1985, 277–289.

[MS 40]   E.J. McShane, Generalized curves, *Duke Math. J.* **6** (1940), 513–516.

[MS 89]   E.J. McShane, The calculus of variations from the beginning through optimal control theory, *SIAM J. Control Opt.* **27** (1989), 916–939.

[Me 66]   P.-A. Meyer, *Probability and potentials*, Blaisdell, 1966.

[Me 65]   N. G. Meyers, Quasiconvexity and lower semicontinuity of multiple variational integrals of any order, *Trans. Amer. Math. Soc.* **119** (1965), 125–149.

[Mi 97]   A. Mielke, Evolution equations for Young-measure solutions of semilinear hyperbolic problems, preprint, 1997.

[Mi 98]   G. Milton, *The effective tensors of composites*, book in preparation.

[Mo 87]   L. Modica, Gradient theory of phase transitions and minimal interface criterion, *Arch. Rat. Mech. Anal.* **98** (1987), 123–142.

[MM 77a]   L. Modica and S. Mortola, Il limite nella Γ-convergenza di una famiglia di funzionali ellittici, *Boll. U.M.I* **14**-A (1977), 525–529.

[MM 77b]   L. Modica and S. Mortola, Un esempio di Γ⁻-convergenza, *Boll. U.M.I* **14**-B (1977), 285–299.

[Mo 52]   C.B. Morrey, Quasi-convexity and the lower semicontinuity of multiple integrals, *Pacific J. Math.* **2** (1952), 25–53.

[Mo 66]   C.B. Morrey, *Multiple integrals in the calculus of variations*, Springer, 1966.

[Mu 93]   S. Müller, Singular perturbations as a selection criterion for periodic minimizing sequences, *Calc. Var.* **1** (1993), 169–204.

[Mu 97a]   S. Müller, *Microstructure and the calculus of variations*, Nachdiplomvorlesung ETH Zürich, in preparation.

[Mu 97b]   S. Müller, A sharp version of Zhang's theorem on truncating sequences of gradients, submitted to *Trans. AMS*.

[Mu 97c]   S. Müller, Microstructures, phase transitions and geometry, in: *Proc. ECM2, Budapest, 1996*, Birkhäuser, to appear.

[MS 96]   S. Müller and V. Šverák, Attainment results for the two-well problem by convex integration, *Geometric analysis and the calculus of variations*, (J. Jost, ed.), International Press, 1996, 239–251.

[MS 97]   S. Müller and V. Šverák, in preparation.

[Mu 78]   F. Murat, H-convergence, mimeographed notes, 1978, based partially on the Cours Peacot by L. Tartar, 1977; translation (with authors F. Murat and L. Tartar) in [CK 97a], 21–43.

[MT 97]   F. Murat and L. Tartar, On the relation between Young measures and H-measures, in preparation.

[Na 54]   J. Nash, C^1 isometric embeddings, *Ann. Math.* **60** (1954), 383–396.

[Ng 89]   G. Nguetseng, A general convergence result for a functional related to the theory of homogenization, *SIAM J. Math. Anal.* **20** (1989), 608–623.

[Ot 95]   F. Otto, Evolution of microstructure in unstable porous media flow: a relaxational approach, preprint SFB 256, Bonn, 1995.

[Ot 97]   F. Otto, Dynamics of labyrinthine pattern formation in magnetic fluids: a mean-field theory, to appear in *Arch. Rat. Mech. Anal.*

[Pa 77]   G. Parry, On the crystallographic point groups and Cauchy symmetry, *Math. Proc. Cambridge Phil. Soc.* **82** (1977), 165–175.

[Pa 81]   G. Parry, On phase transitions involving internal strain, *Int. J. Solids Structures* **17** (1981), 361–378.

[Pa 87a]   G. Parry, On internal variable models of phase transitions, *J. Elasticity* **17** (1987), 63–70.

[Pa 87b]   G. Parry, On shear bands in unloaded crystals, *J. Mech. Phys. Solids* **35** (1987), 357–382.

[Pa 89]   G. Parry, Stable phase boundaries in unloaded crystals, *Cont. Mech. Thermodyn.* **1** (1989), 305–314.

[Pe 93]   P. Pedregal, Laminates and microstructure, *Europ. J. Appl. Math.* **4** (1993), 121–149.

[Pe 97]   P. Pedregal, *Parametrized measures and variational principles*, Birkhäuser, 1997.

[Pe 87]   R.L. Pego, Phase transitions in one-dimensional nonlinear viscoelasticity: admissibility and stability, *Arch. Rat. Mech. Anal.* **97** (1987), 353–394.

[Pi 84]   M. Pitteri, Reconciliation of local and global symmetries of crystals, *J. Elasticity* **14** (1984), 175–190.

[PZ 97]   M. Pitteri and G. Zanzotto, *Continuum models for phase transitions and twinning in crystals*, Chapman and Hall, forthcoming.

[Pr 76]   I. Privorotskii, *Thermodynamic theory of domain structures*, Wiley, 1976.

[Re 67]   Yu.G. Reshetnyak, On the stability of conformal mappings in multidimensional spaces, *Sib. Math. J.* **8** (1967), 69–85.

[Re 68a]   Yu.G. Reshetnyak, Liouville's theorem under minimal regularity assumptions, *Sib. Math. J.* **8** (1968), 631–634.

[Re 68b]   Yu.G. Reshetnyak, Weak convergence of completely additive vector functions on a set, *Sib. Math. J.* **9** (1968), 1039–1045.

[Re 89]   Yu.G. Reshetnyak, *Space mapping with bounded distorsion*, Am. Math. Soc., 1989.

[Ro 69]   A. Roitburd, The domain structure of crystals formed in the solid phase, *Soviet Physics-Solid State* **10** (1969), 2870–2876.

[Ro 78]   A. Roitburd, Martensitic transformation as a typical phase transformation in solids, *Solid State Physics* **34** (1978), 317–390.

[Ro 97]   T. Roubíček, *Relaxation in optimization theory and variational calculus*, de Gruyter, 1997.

[Ru 73]   W. Rudin, *Functional analysis*, McGraw-Hill, 1973.

[Sch 94]   C. Schreiber, Rapport de stage de D.E.A., ENS Lyon, 1994.

[Sch 93]   D. Schryvers, Microtwin sequences in thermoelastic Ni_xAl_{100-x} martensite studied by conventional and high resolution transmission electron microscopy, *Phil. Mag.* **A 68** (1993), 1017–1032.

[Se 83]   D. Serre, Formes quadratique et calcul de variations, *J. Math. Pures Appl.* **62** (1983), 177–196.

[Sv 91a]   V. Šverák, Quasiconvex functions with subquadratic growth, *Proc. Roy. Soc. London* **A 433** (1991), 723–725.

[Sv 91b]   V. Šverák, On regularity for the Monge-Ampère equations, preprint, Heriot-Watt University, 1991.

[Sv 92a]   V. Šverák, Rank-one convexity does not imply quasiconvexity, *Proc. Roy. Soc. Edinburgh* **120** (1992), 185–189.

[Sv 92b]   V. Šverák, New examples of quasiconvex functions, *Arch. Rat. Mech. Anal.* **119** (1992), 293–300.

[Sv 93a]   V. Šverák, On the problem of two wells, in: *Microstructure and phase transitions*, IMA Vol. Appl. Math. **54** (D. Kinderlehrer, R.D. James, M. Luskin and J. Ericksen, eds.), Springer, 1993, 183–189.

[Sv 93b]   V. Šverák, On Tartar's conjecture, *Ann. IHP Analyse non linéaire* **10** (1993), 405–412.

[Sv 95]   V. Šverák, Lower semicontinuity of variational integrals and compensated compactness, in: Proc. ICM 1994 (S.D. Chatterji, ed.), vol. 2, Birkhäuser, 1995, 1153–1158.

[Sv 98]   V. Šverák, In preparation.

[Sv]   V. Šverák, personal communication.

[Sy 97]   M.A. Sychev, A new approach to Young measure theory, relaxation and convergence in energy, preprint.

[Ta 75]   L. Tartar, Problèmes de contrôle des coefficients dans des équations aux derivées partielles, in: *Control theory, numerical methods and computer systems modelling* (A. Bensoussau and

J.-L. Lions, eds.), Springer, 1975, 420–426; translated (with authors F. Murat and L. Tartar) in [CK 97a], 1–8.

[Ta 79a]   L. Tartar, Compensated compactness and partial differential equations, in: *Nonlinear Analysis and Mechanics: Heriot-Watt Symposium* Vol. IV, (R. Knops, ed.), Pitman, 1979, 136–212.

[Ta 79b]   L. Tartar, Estimations de coefficients homogénéises, in: *Computing methods in applied sciences and engineering*, Lecture Notes in Mathematics **704**, Springer, 1979.

[Ta 80]   L. Tartar, Some existence theorems for semilinear hyperbolic sytems in one space variable, report # 2164, Mathematics Research Center, University of Wisconsin, 1980.

[Ta 81]   L. Tartar, Solutions oscillantes des équation de Carleman, seminar Goulaouic-Meyer-Schwartz 1980/81, exp. No. XII, Ecole Polytechnique, Palaiseau, 1981.

[Ta 83]   L. Tartar, The compensated compactness method applied to systems of conservations laws, in: *Systems of Nonlinear Partial Differential Equations*, (J.M. Ball, ed.), NATO ASI Series, Vol. C111, Reidel, 1983, 263–285.

[Ta 84]   L. Tartar, Etude des oscillations dans les équations aux derivées partielles non linéaires, in: *Trends and applications of pure mathematics to mechanics*, Lecture Notes in physics **195**, Springer, 1984, 384–412.

[Ta 86]   L. Tartar, Oscillations in nonlinear partial differential equations: compensated compactness and homogenization, in: *Nonlinear systems of partial differential equations in applied mathematics*, Amer. Math. Soc., 1986, 243–266.

[Ta 87]   L. Tartar, Oscillations and asymptotic behaviour for two semilinear hyperbolic systems, in: *Dynamics of infinite-dimensional dynamical systems*, NATO ASI Ser. F, Springer, 1987.

[Ta 90]   L. Tartar, *H*-measures, a new approach for studying homogenization, oscillations and concentration effects in partial differential equations, *Proc. Roy. Soc. Edinburgh* **A 115** (1990), 193–230.

[Ta 93]   L. Tartar, Some remarks on separately convex functions, in: *Microstructure and phase transitions*, IMA Vol. Math. Appl. **54**, (D. Kinderlehrer, R.D. James, M. Luskin and J.L. Ericksen, eds.), Springer, 1993, 191–204.

[Ta 95]   L. Tartar, Beyond Young measures, *Meccanica* **30** (1995), 505–526.

[Ta 98]   L. Tartar, *Homogenization, compensated compactness and H-measures*, CBMS-NSF conference, Santa Cruz, June 1993, lecture notes in preparation.

[Th 97]   F. Theil, Young measure solutions for a viscoelastically damped wave equation with nonmonotone stress-strain relation, to appear in *Arch. Rat. Mech. Anal.*.

[Va 90]   M. Valadier, Young measures, in: *Methods of nonconvex analysis*, LNM **1446**, Springer, 1990.

[Va 94]   M. Valadier, A course on Young measures, *Rend. Istit. Mat. Univ. Trieste* **26** (1994) suppl., 349–394.

[Vi 94]   A. Visintin, *Differential models of hysteresis*, Springer, 1994.

[WLR 53]   M. S. Wechsler, D. S. Liebermann, T. A. Read, On the theory of the formation of martensite, *Trans. AIME J. Metals* **197** (1953), 1503–1515.

[Yo 37]   L.C. Young, Generalized curves and the existence of an attained absolute minimum in the calculus of variations, *Comptes Rendues de la Société des Sciences et des Lettres de Varsovie, classe III* **30** (1937), 212–234.

[Yo 69]   L.C. Young, *Lectures on the calculus of variations and optimal control theory*, Saunders, 1969 (reprinted by Chelsea, 1980).

[Za 92]   G. Zanzotto, On the material symmetry group of elastic crystals and the Born rule, *Arch. Rat. Mech. Anal.* **121** (1992), 1–36.

[Zh 92]   K. Zhang, A construction of quasiconvex functions with linear growth at infinity, *Ann. S.N.S. Pisa* **19** (1992), 313–326.

[Zh]   K. Zhang, Rank-one connections and the three-well problem, preprint.

Parametric Surfaces of Prescribed Mean Curvature

KLAUS STEFFEN

0 Introduction

The classical Plateau problem is to find a surface of least area in Euclidean space $\mathbb{R}^3$ which is bounded by a given closed Jordan curve Γ. The Euler equation for this geometric variational problem, interpreted in differential geometric terms, expresses that the mean curvature of each solution surface vanishes. Therefore it has become customary to call a surface with zero mean curvature a minimal surface. We refer to the monographies [DHKW], [Ni1], [Ni2], [Str3] for the theory of minimal surfaces and, in particular, for a treatment of the classical Plateau problem which was first solved about 1930 by Douglas and by Radó.

In a variant of this problem E. Heinz [He1] looked for a surface with prescribed constant mean curvature spanning a given boundary curve Γ in $\mathbb{R}^3$ and he succeeded to prove a first existence theorem in this direction. More generally, one can prescribe a real valued function H on $\mathbb{R}^3$ and ask for a surface with prescribed mean curvature H, i.e. one requires that the surface at each point a of its support in $\mathbb{R}^3$ has mean curvature equal to the value $H(a)$ given there in advance. In the same way as soap films governed by surface tension and no other forces are modelled by least area (or stable) minimal surfaces, this more general Plateau problem with prescribed mean curvature can be considered as a mathematical model for physical surfaces subject to surface tension and additional external forces like pressure differences or gravity.

In these lectures we treat the Plateau problem with prescribed mean curvature. We first present, with fairly complete proofs, the classical theory for 2-dimensional parametric surfaces of prescribed mean curvature in $\mathbb{R}^3$ which originated in the early 70's. Then we describe the geometric measure theory approach to the problem for hypersurfaces in $\mathbb{R}^{n+1}$, and in the final Section we discuss the problem, for the geometric measure theory setting and also in the setting of 2-dimensional parametric surfaces, in an ambient Riemannian manifold. The results here are fairly recent (partially the work is still in progress, at the time of these lectures) and we cannot give complete proofs, but we have tried to present the main ideas and to describe clearly the approach which eventually leads to solutions. We have concentrated on the existence theory, because this is the geometric part of the problem, and we treat only results which give energy minimizing (hyper-)surfaces of prescribed mean curvature, a property analogous to the area minimality of solutions for the classical Plateau problem. We refer to Section 1 for a precise formulation of the

Plateau problem with prescribed mean curvature in the simplest case and for a broader outline of the contents of these lectures.

All our recent results which we describe here were obtained in collaboration with Frank Duzaar who, with M. Fuchs, has started looking at the Plateau problem with prescribed mean curvature from the geometric measure theory point of view. For the presentation of the 2-dimensional parametric theory we have profited a lot from our recent exposition [DS5], again jointly with Frank Duzaar. So, naturally, there is considerable overlap of the latter article and some sections of these lectures.

The author wants to thank the C.I.M.E. foundation for the invitation and for the opportunity to present these lectures in a truly delightful and stimulating session and in an unforgettable surrounding.

1 The Plateau problem

We consider (nonconstant) **parametric surfaces** $x: U \to \mathbb{R}^3$ in Euclidean space which are defined on the unit disc $U = \{(u,v) \in \mathbb{R}^2 : u^2+v^2 < 1\}$. We frequently identify (u,v) with the complex variable $w=u+iv \in \mathbb{C}$. It is convenient to assume the **conformality relations**

$$|x_u|^2 - |x_v|^2 - 2i(x_u \cdot x_v) = 0 \quad \text{on} \quad U,$$

where $x_u \cdot x_v$ denotes the Euclidean inner product of the partial derivatives of x and $|x_u|, |x_v|$ their Euclidean norm.

One advantage of this conformality condition is that we have a simple analytic expression for the mean curvature (the arithmetic mean of the principal curvatures) $h(u,v)$ of x at $(u,v) \in U$. Namely, for the Laplacean $\Delta x = x_{uu} + x_{vv}$ (applied to each component) of x we have

$$2x_u \cdot \Delta x = 2x_u \cdot x_{uu} + 2x_u \cdot x_{vv} = 2x_u \cdot x_{uu} - 2x_{vv} \cdot x_v = (|x_u|^2 - |x_v|^2)_u = 0$$

and

$$2x_v \cdot \Delta x = 2x_v \cdot x_{uu} + 2x_v \cdot x_{vv} = -2x_{uv} \cdot x_u + 2x_v \cdot x_{vv} = (|x_u|^2 - |x_v|^2)_v = 0,$$

i.e. Δx is perpendicular to the tangent plane $\mathbb{R}x_u + \mathbb{R}x_v$ at regular points of x (where Dx has maximal rank 2). By definition, the mean curvature is half the trace of the differential $-D\mathcal{N}$, where $\mathcal{N} = |x_u \wedge x_v|^{-1} x_u \wedge x_v$ is the unit normal vector field along the surface x (at regular points; $\wedge: \mathbb{R}^3 \times \mathbb{R}^3 \to \mathbb{R}^3$ denotes the usual skew-symmetric product determined by the Euclidean structure and the orientation of $\mathbb{R}^3$). Using conformality we thus have

$$2h|x_u||x_v| = -\mathcal{N}_u \cdot x_u - \mathcal{N}_v \cdot x_v = \mathcal{N} \cdot x_{uu} + \mathcal{N} \cdot x_{vv} = \mathcal{N} \cdot \Delta x,$$

and since Δx is parallel to $\mathcal{N}$ along x we obtain the following equation for the mean curvature at regular points:

$$\Delta x = 2h\, x_u \wedge x_v.$$

Now, for a given function $H: \mathbb{R}^3 \to \mathbb{R}$ (which we will always assume continuous and bounded) we say that x has **prescribed mean curvature** H if $h(u,v) = H(x(u,v))$ holds

1.1 Example. Consider a smooth H-surface x on $\overline{U}$. If we integrate the H-surface equation

$$\Delta x = 2H\, x_u \wedge x_v$$

with constant H over U and apply the Gauss-Green theorem as well as Stokes' theorem, then we obtain the identity

$$\int_{\partial U} x_r\, d\vartheta = H \int_{\partial U} x \wedge x_\vartheta\, d\vartheta\,,$$

where ϑ is the arc length on ∂U and x_r, x_ϑ denote the radial and angular derivative of x on $\overline{U}$. Since, by conformality, $|x_r| = |x_\vartheta|$ holds on ∂U, one concludes the inequality

$$\text{length}(x|_{\partial U}) \geq \left| \int_{\partial U} x_r\, d\vartheta \right| = |H| \left| \int_{\partial U} x \wedge x_\vartheta\, d\vartheta \right|.$$

Now, if x is bounded by a circle $\Gamma \subset \mathbb{R}^3$ of radius $R > 0$ in the sense of the Plateau boundary condition, then the last inequality says

$$2\pi R \geq |H| 2\pi R^2\,, \quad \text{i.e.} \quad |H| \leq \frac{1}{R}\,.$$

Therefore, for H constant with $|H| > R^{-1}$ the Plateau problem $\mathcal{P}(H, \Gamma)$ cannot have a solution which is smooth up to its boundary. □

Heinz has shown that the same conclusion holds if x is merely C^2 on U and continuous on $\overline{U}$; the subtlety then is to justify the integration by parts. Gulliver [Gu3], [Gu4] has analyzed and widely generalized the reasoning in the above example. If one uses the Laplace-Beltrami operator on the surface, then conformality is not needed, hence the argument can also be applied to prove non-existence of hypersurfaces in $\mathbb{R}^{n+1}$ which have an n-sphere as boundary and constant prescribed mean curvature exceeding the radius of the sphere. We should emphasize here that the following question is still open:

> ¿ Is every immersion or even embedding of the disc $\overline{U}$ into $\mathbb{R}^3$ with constant mean curvature $H \neq 0$ and with a circle as boundary in fact a spherical cap ?

(The answer is known to be yes under various additional assumptions on the immersion [EBMR], [Ba], [BE], [BJ], [LM]. The disc-type topological structure of the surfaces is essential; by [Kap] the answer is negative if surfaces of genus bigger than 2 are admitted.)

Since the Plateau problem $\mathcal{P}(H, \Gamma)$ is not solvable in general, one is interested in reasonable (geometric) conditions on the prescribed mean curvature and the given boundary curve which are sufficient for existence, and it is the aim of these lectures to present some old and some new existence theorems in this direction. Here is an outline:

In view of the example above, a natural conjecture is that a solution should exist if H is constant with $|H| \leq R(\Gamma)^{-1}$ or, if one is optimistic, even for variable H with $\sup |H| \leq R(\Gamma)^{-1}$, where $R(\Gamma)$ is the circumscribed radius of Γ, i.e. the radius of the smallest closed ball in $\mathbb{R}^3$ containing Γ. This conjecture has actually been proved by S. Hildebrandt in 1969 and we will describe his work in Section 2. However, there are curves Γ for which the condition $|H| \leq R(\Gamma)^{-1}$ is far too restrictive. For example, if Γ is the boundary of a long and narrow rectangle, then Γ should bound H-surfaces (resembling pieces of circular cylinders with radius $|H|^{-1}$, if H is constant) for functions H much larger than $R(\Gamma)^{-1}$. Moreover, this should remain true if the long, narrow rectangle and its boundary Γ are twisted and curled through space in a complicated way. In Sections 2 and 3 we will discuss corresponding existence theorems which have their origin in work of Gulliver & Spruck, Wente and the present author, and we point out the intimate connection between isoperimetric inequalities and the Plateau problem with prescribed mean curvature.

In Section 4 we use a different model for surfaces which comes from geometric measure theory. F. Duzaar has shown that all the existence theorems for the Plateau problem with prescribed mean curvature have analogues in this setting, where surfaces are represented by integer multiplicity rectifiable currents. Moreover, these results hold for hypersurfaces in a Euclidean space $\mathbb{R}^{n+1}$ of arbitrary dimension, and the solutions to the Plateau problem are smooth n-dimensional submanifolds with boundary in $\mathbb{R}^{n+1}$ if $n < 7$ (with a priori undetermined topology, however).

In the final sections of these lectures we treat the Plateau problem with prescribed mean curvature in a Riemannian manifold. In this situation it is no longer true that a solution exists whenever $\sup |H|$ is sufficiently small, i.e. smaller than some positive constant depending on the given boundary Γ. A great hypersphere in the sphere S^{n+1} with standard metric is a counter-example. Therefore extra geometric conditions must be imposed on Γ. It turns out that isoperimetric inequalities are crucial here again, and we discuss these for an ambient Riemannian manifold in Section 5. In Section 6 we then describe our recent results with F. Duzaar on hypersurfaces with prescribed mean curvature in a Riemannian manifold which were obtained using geometric measure theory. We close with some new existence theorems of F. Duzaar and the author for the parametric Plateau problem with prescribed mean curvature in a Riemannian 3-manifold M. In contrast with previous contributions to this problem, which assumed $M = \mathbb{R}^3$ equipped with a general Riemannian metric, we do not require here that the boundary is contained in a single coordinate patch. This work is still in progress and what we present of it in these lectures may not yet have reached its final form.

The general approach which we use here to produce (weak) H-surfaces with given boundary Γ is the direct method of the calculus of variations. To set up a variational functional with the H-surface equation as its Euler equation we observe that the Laplacian Δx is the Euler operator associated with **Dirichlet's integral**

$$\mathbf{D}(x) = \tfrac{1}{2} \int_U (|x_u|^2 + |x_v|^2) \, du dv \,,$$

and the vector field $(H \circ x) \, x_u \wedge x_v$ along the parametric surface x is the Euler operator of **a volume functional**

$$\mathbf{V}_H(x) = H\text{-volume of } x \,,$$

which will be defined precisely later. The geometric meaning of $\mathbf{V}_H(x)$ is the oriented volume enclosed by x and a fixed comparison surface with the same boundary Γ, where the volume is computed with respect to H as a weight function. A suitable variational functional thus is the **energy functional**

$$\mathbf{E}_H(x) = \mathbf{D}(x) + 2\mathbf{V}_H(x) \,,$$

and we will obtain weak solutions to the Plateau problem $\mathcal{P}(H, \Gamma)$ by minimizing this functional on suitable classes $\mathcal{S}$ of surfaces satisfying the Plateau boundary condition.

Indeed, with variations of the dependent variables one can see that minimizers are weak solutions to the H-surface equation, and with variations of the domain one can verify that minimizers also fulfill the conformality relations. In the geometric measure theory setting we use the same approach replacing $D(x)$ for parametric surfaces x by the mass $M(T)$ (total area) for integer multiplicity rectifiable currents T with codimension one.

Geometric restrictions relating H and Γ enter in the first place to secure the existence of energy minimizers in S. We need such conditions to establish the lower semicontinuity of the energy functional with respect to weak $W^{1,2}$ convergence, for example. If a minimizer x can be freely varied within the class S then it is a weak H-surface. However, in many cases obstacle conditions constraining the range of the admissible surfaces must be incorporated in the definition of S. Then the free variability of a minimizer x in S is a priori not clear as the surface x might touch the obstacle. To rule out this possibility or to deduce the H-surface equation nevertheless one needs again certain geometric conditions relating H, Γ, and the boundary mean curvature of the obstacle set, in order that an inclusion principle (a geometric maximum principle) can be applied.

We have chosen to restrict these lectures to the simplest formulation of Plateau's problem with prescribed mean curvature and, as we have already pointed out, to the existence theory based on the energy minimization method. As a consequence, we had to leave aside the discussion of many existence results which address generalizations and modifications of the problem or which use other methods. For instance, the example of a circle $\Gamma \subset \mathbb{R}^3$ with radius R which bounds (in the oriented sense) two distinct spherical caps of constant mean curvature $H \neq 0$ whenever $0 < |H| < R^{-1}$, suggests the conjecture, attributed to Rellich, that there is a fundamental non-uniqueness in the Plateau problem $\mathcal{P}(H, \Gamma)$ for constant $H \neq 0$. Namely, there should exist positive constants $h_-(\Gamma), h_+(\Gamma)$ such that for each constant $H \neq 0$ with $-h_-(\Gamma) < H < h_+(\Gamma)$ one has at least two solutions of $\mathcal{P}(H, \Gamma)$, one energy minimizing and small, the other one unstable for the energy functional and large, in a sense. This conjecture was proved with the combined work of Struwe [Str1] and the present author [Ste5] and independently, with the optimal estimate $h_\pm(\Gamma) \geq R(\Gamma)^{-1}$, by Brezis & Coron [BC]. We refer to [Str2], [Str3], [Str4], [Wa], [BR] for further progress in this non-uniqueness problem.

An intuitive idea to produce "large" H-surfaces with constant H is to minimize area in the class of surfaces $x: U \to \mathbb{R}^3$ satisfying the Plateau boundary condition and a volume constraint (i.e. the volume enclosed by x and the cone over Γ is prescribed). The solutions x, whose existence was shown by Wente [Wen2], [Wen3], are solutions to $\mathcal{P}(H, \Gamma)$ with a constant H which is, however, not prescribed but determined by the Lagrange multiplier associated with the volume constraint. Since these surfaces x are "large" if the prescribed volume is big, one can infer the existence of "large" H-surfaces with boundary Γ for all values of the constant H which do occur as Lagrange multipliers for big volumes. In [Ste2] it was shown that the set of these values accumulates at 0 from above and below, but it is not clear (and may not be true, in general) that it contains a punctured neighborhood of 0 and, hence, Rellich's conjecture could not be proved completely in this way. In the context of geometric measure theory F. Duzaar [Du3] has used the same approach to prove some results on "large" H-surfaces in general dimensions.

With regard to unstable H-surfaces we mention that there is also an existence theory for "small" unstable surfaces of prescribed mean curvature (see [He5], [Strö], [Str3], [ST]) which extends part of the extensive corresponding theory for parametric minimal surfaces.

In another direction the existence theory for the Plateau problem has been generalized replacing the unit disc U by a multiply connected domain in $\mathbb{R}^2$ or by an oriented compact surface with boundary. This is called the general Plateau problem or Plateau-Douglas problem, and the principal difficulty is that the conformal structure on the domain cannot be fixed a priori. In the energy minimization process the conformal structure therefore has to be varied and it may degenerate in the limit of a minimizing sequence. (Geometrically speaking, the surfaces in the minimizing sequence may break up into a system of surfaces of simpler topological type.) With appropriate assumptions (so-called Douglas conditions) such a behaviour can be excluded, however, and the Plateau-Douglas problem then has a solution. Another variation of the theme is to replace the Plateau boundary condition by a free boundary condition with two degrees of freedom, i.e. the surfaces are required to have (part of) their boundary on a given 2-dimensional supporting manifold. Since the additional difficulties in all these generalizations occur already in the minimal surface case $H \equiv 0$, we refer to the monography [DHKW] which also contains many discussions, hints to the literature, and bibliographical entries related to H-surfaces.

Finally, we note that many of these results of the 2-dimensional parametric theory for H-surfaces, which we have just mentioned and which will not be discussed in these lectures, have also been treated in the setting of geometric measure theory and thereby extended to H-hypersurfaces in $\mathbb{R}^{n+1}$ (see [DF1], [DF2], [Du2]-[Du4], [DS1]-[DS4]).

2   The method of bounded vector fields

One convenient way to define the volume functional mentioned in Section 1 is to represent the prescribed (continuous, bounded) mean curvature function H as the divergence $H = \operatorname{div} Z$ of a vector field Z on $\mathbb{R}^3$ and let

$$\mathbf{V}_H(x) = \int_U (Z \circ x) \cdot x_u \wedge x_v \, du dv.$$

To see the geometric meaning of $\mathbf{V}_H(x)$ we consider the 1-form ω on $\mathbb{R}^3$ which is dual to Z and note that $\operatorname{div} Z = H$ is equivalent with $d\omega = H\Omega$ where Ω is the Euclidean volume form on $\mathbb{R}^3$. (It may be preferable to work generally with ω instead of Z; we wanted to follow the historical development, however.) Then $\mathbf{V}_H(x)$ is just the integral $\int_U x^\# \omega$ of ω over the parametric surface x. If x satisfies the Plateau boundary condition for the given boundary curve Γ and if Z, ω are of class C^1, then Stokes' theorem tells us that, up to a constant depending on H and Γ only, $\mathbf{V}_H(x)$ equals the H-weighted volume enclosed by x and the cone over Γ. Since we fix the boundary Γ and different choices of Z corresponding to the same function H will change the H-volume of surfaces spanning Γ only by an irrelevant constant, we have written $\mathbf{V}_H(x)$ instead of $\mathbf{V}_Z(x)$, abusing slightly the notation.

Note that we have to assume boundedness of Z on the image of x to secure the existence of the integral above for $x \in W^{1,2}(U, \mathbb{R}^3)$. Of course, if we restrict our considerations to surfaces contained in a subset A of $\mathbb{R}^3$, then we need a bound for Z only on A. Assuming this, the above integral and hence also the energy functional $\mathbf{E}_H(x) = \mathbf{D}(x) + 2\mathbf{V}_H(x)$ (more precisely denoted $\mathbf{E}_Z(x)$) is defined for all surfaces $x \in W^{1,2}(U, A)$, i.e. for all $x \in W^{1,2}(U, \mathbb{R}^3)$ mapping (almost all of) U into A. In the sequel we will in fact need the bound $\sup_A |Z| < \frac{1}{2}$ in order to have coercivity of the energy functional on $W^{1,2}(U, A)$.

It is now routine to compute the first variation of $\mathbf{E}_H$, and one obtains, as is well-known and expected in view of the discussion in Section 1:

2.1 Proposition (first variation). *Suppose A is closed in $\mathbb{R}^3$, Z is a bounded C^1 vector field with $\operatorname{div} Z = H$ on A, and $x \in W^{1,2}(U, A)$.*

(i) If $\xi \in W_0^{1,2}(U, \mathbb{R}^3)$ is bounded with $x + t\xi \in W^{1,2}(U, A)$ for $0 < t \ll 1$, then

$$\frac{d}{dt}\bigg|_{t=0+} \mathbf{E}_H(x + t\xi) = \int_U [x_u \cdot \xi_u + x_v \cdot \xi_v + 2(H \circ x)\,\xi \cdot x_u \wedge x_v]\,du\,dv.$$

(ii) If φ_t is the flow of a C^1 vector field η on $\overline{U}$ which is tangential to ∂U along ∂U, then

$$\frac{d}{dt}\bigg|_{t=0} \mathbf{E}_H(x \circ \varphi_t) = \int_U \operatorname{Re}\left[(|x_u|^2 - |x_v|^2 - 2\mathrm{i}\,x_u \cdot x_v)\bar{\partial}\eta\right]du\,dv$$

where we have used complex notation $\bar{\partial}\eta = \frac{1}{2}(\eta_u + \mathrm{i}\eta_v)$, identifying $\mathbb{R}^2 = \mathbb{C}$.

Proof. (i) Since $\int_U (x_u \cdot \xi_u + x_v \cdot \xi_v)\,du\,dv$ is the derivative of $\mathbf{D}(x + t\xi)$ we only have to treat the derivative of the volume. For this we use the identity

$$\mathbf{V}_H(x + t\xi) - \mathbf{V}_H(x) = \int_0^t \int_U H \circ (x + s\xi)\,\xi \cdot (x_u + s\xi_u) \wedge (x_v + s\xi_v)\,du\,dv\,ds,$$

from which (i) follows by differentiation using the continuity and boundedness of H. For x, ξ smooth on $\overline{U}$ this identity is a consequence of Stokes' theorem applied to the pull back of $d\omega$ to $]0, t[\times U$ by the mapping $(s, w) \mapsto x(w) + s\xi(w)$. For general x, ξ it follows by approximation, provided Z with $\operatorname{div} Z = H$ has compact support in $\mathbb{R}^3$. The latter restriction is then removed by a further approximation (cf. [Ste3]).

(ii) We first note that the volume functional is, of course, invariant with respect to reparametrization, $\mathbf{V}_H(x \circ \varphi_t) = \mathbf{V}_H(x)$. Therefore, we only have to compute the derivative of $\mathbf{D}(x \circ \varphi_t)$. Using the transformation $w = \varphi_t(\tilde{w})$ we get

$$\int_U |D(x \circ \varphi_t)|^2\,d\tilde{w} = \int_U |Dx(D\varphi_t) \circ \varphi_t^{-1}|^2 \det(D\varphi_t^{-1})\,dw,$$

and since $D\varphi_t(\tilde{w}) = I + tD\eta(\tilde{w}) + o(t)$, we obtain by differentiation

$$\frac{d}{dt}\bigg|_{t=0} \mathbf{D}(x \circ \varphi_t) = \int_U \left[Dx \cdot (DxD\eta) - \frac{1}{2}|Dx|^2\operatorname{trace} D\eta\right]dw.$$

The claim then follows by appropriately collecting terms in the integrand. $\square$

For the proof of Proposition 2.1 it is sufficient that Z is continuous with $\operatorname{div} Z = H$ in the distributional sense on a neighborhood of A.

The integral appearing in (i) is denoted $\delta \mathbf{E}_H(x; \xi)$ and known as the **first variation** of $\mathbf{E}_H$ at x in the direction of ξ. We say that x is **critical** for the energy functional $\mathbf{E}_H$ if $\delta \mathbf{E}_H(x; \xi)$ vanishes for all ξ as in (i). This is equivalent with the H-surface equation being satisfied in the weak sense, i.e. the H-surface equation is indeed the Euler equation associated with the functional $\mathbf{E}_H$.

On the other hand, it has become customary to call x **stationary** for $\mathbf{E}_H$ with the Plateau boundary condition if the integral in (ii), referred to as the **inner first variation** of $\mathbf{E}_{II}$ at x in the direction η and denoted $\partial\mathbf{E}_H(x;\eta)$, vanishes for all η as in (ii). Here, the vanishing of $\partial\mathbf{E}_H(x;\eta)$ for all η with compact support in U means that the function $\Phi(w) = |x_u|^2 - |x_v|^2 - 2ix_u \cdot x_v$ is a (weakly, hence also classically) holomorphic function on U. (The equation $\bar{\partial}\Phi = 0$ is sometimes called the second Euler equation.) With a formal integration by parts one then sees that x is stationary iff $w^2\Phi(w)$ is real on ∂U in the weak sense. This latter condition thus is the natural boundary condition associated with the free Plateau boundary condition. It can be expressed invariantly by stating the the quadratic holomorphic differential $\Phi(w)(dw)^2$, known as the Hopf differential, is real on the boundary. But holomorphic functions on U which are real on ∂U in the weak sense are in fact constant, because they can be extended to (weakly) holomorphic functions on a neighborhood of $\overline{U}$. (This is easily verified by transforming U to the upper half plane.) Evaluating at $w = 0$ we deduce that x is stationary iff the conformality relation holds on U. (For more details of the reasoning see [DHKW], [Jo2], [Ni1], [Ni2].) We thus have the

2.2 Corollary. *x is critical and stationary for $\mathbf{E}_H$ if and only if x is a weak H-surface.*

$\square$

We next consider the minimization problem. For this, we introduce the **class of admissible surfaces** $\mathcal{S}(\Gamma)$ as the set of all $x \in W^{1,2}(U, \mathbb{R}^3)$ which satisfy the Plateau boundary condition associated with the closed, oriented, rectifiable Jordan curve Γ in $\mathbb{R}^3$. For A closed in $\mathbb{R}^3$ we further let $\mathcal{S}(\Gamma, A) = \mathcal{S}(\Gamma) \cap W^{1,2}(U, A)$. We then have the following existence theorem:

2.3 Proposition (solution of the minimum problem). *Suppose A is closed in $\mathbb{R}^3$, $\Gamma \subset A$, $\mathcal{S}(\Gamma, A) \neq \emptyset$, and Z is a continuous vector field on A with*

$$2 \sup_A |Z| = c < 1.$$

Then the variational problem

$$\mathbf{E}_Z \rightsquigarrow \min \quad on \;\; \mathcal{S}(\Gamma, A)$$

admits a solution.

Proof. From the assumed bound on $|Z|$ we obtain $2\mathbf{V}_Z(x) \le c\mathbf{D}(x)$ and the following **coercivity inequality**

$$\mathbf{E}_Z(x) \ge (1 - c)\mathbf{D}(x) \quad \text{for } x \in \mathcal{S}(\Gamma, A),$$

from which we deduce that each minimizing sequence x_k for our minimization problem is bounded in the norm of $W^{1,2}(U, \mathbb{R}^3)$ (since also the boundary values are uniformly bounded). Applying Rellich's theorem and passing to a subsequence we may assume $x_k \to x$ weakly in $W^{1,2}(U, \mathbb{R}^3)$ and almost everywhere on U, hence x maps U into A, too. The hypothesis $2 \sup_A |Z| = c < 1$ also implies that the integrand of the energy functional is convex in the derivatives, and a classical lower semicontinuity theorem of Morrey can be applied to give $\mathbf{E}_Z(x) \le \liminf_{k \to \infty} \mathbf{E}_Z(x_k)$. Since the integrand is quadratic in the derivatives, one can verify this also directly by simply writing $\mathbf{E}_Z(x_k) - \mathbf{E}_Z(x)$ as the sum of the nonnegative integral

$$\int_U \left[\tfrac{1}{2}|(x_k - x)_u|^2 + \tfrac{1}{2}|(x_k - x)_v|^2 + 2(Z \circ x_k) \cdot (x_k - x)_u \wedge (x_k - x)_v \right] du\, dv$$

and of other integrals whose integrands are products of two factors, one converging L^2 weakly to zero and the other one converging L^2 strongly as $k \to \infty$.

It remains to prove that x again satisfies the Plateau boundary condition. Actually this may not be true, because due to the action of the noncompact group of conformal automorphisms of U the minimizing sequence can degenerate in the limit. However, it is well-known from the theory of parametric minimal surfaces how to overcome this problem (see [DHKW], [Jo2], [Ni1], [Ni2], [Str3]). One fixes three points w_0, w_1, w_2 on ∂U and p_0, p_1, p_2 on Γ in the order of the orientation and imposes the three-point-condition $x_k(w_i) = p_i$, $0 \le i \le 2$. This is no restriction, because $\mathbf{E}_Z$ is invariant under reparametrization of surfaces by conformal automorphisms of U. Since Γ is a closed Jordan curve it has the following property: For every $\varepsilon > 0$ there exists $\beta > 0$ such that each subarc of Γ with end points at distance $< \beta$ and containing at most one of the points p_i in its interior has diameter less than ε. For $w \in \partial U$ and $0 < \delta < 1$ one can apply a lemma of Courant and Lebesgue to find $\delta \le \varrho \le \sqrt{\delta}$ such that the restriction of x_k to $\overline{U} \cap \partial U_\varrho(w)$ is an absolutely continuous curve of length at most $(4\pi M)^{1/2}|\log \delta|^{-1/2}$ where $M = \sup_k \mathbf{D}(x_k)$ is finite by the coercivity inequality. (This lemma is readily proved by expressing the Dirichlet integral in polar coordinates centered at w and applying Fubini's theorem and Schwarz' inequality. $U_\varrho(w)$ denotes the open disc with center w and radius ϱ.) Choosing δ so small that $U_{\sqrt{\delta}}(w) \cap \partial \overline{U}$ contains at most one of the points w_i and $(4\pi M)^{1/2}|\log \delta|^{-1/2} < \varepsilon$, we conclude from the Plateau boundary condition that x_k maps the arc $U_\varrho(w) \cap \partial U$ onto a subarc of Γ with diameter less than ε. Thus $x_k|_{\partial U}$ is an equicontinuous sequence of weakly monotonic parametrizations of Γ, and the Plateau boundary condition follows for x. $\qquad\square$

Since $\mathcal{S}(\Gamma, A)$ is invariant with respect to reparametrization of surfaces by orientation preserving C^1 diffeomorphisms of $\overline{U}$, we infer from Proposition 2.1 (ii) and the subsequent discussion that minimizers x for the variational problem in Proposition 2.3 are stationary for $\mathbf{E}_{II}$ with Plateau boundary condition and hence conformal almost everywhere on U. In the case $A = \mathbb{R}^3$ we are also free to perform in $\mathcal{S}(\Gamma)$ the variations $x + t\xi$ considered in Proposition 2.1 (i), and we deduce that x is critical and, hence, a weak H-surface in $\mathcal{S}(\Gamma)$, provided we have chosen Z to satisfy $\operatorname{div} Z = H$ on $\mathbb{R}^3$. However, this existence result is not of great value, because the bound $2 \sup |Z| < 1$ required for Z implies intolerable restrictions for the prescribed mean curvature function H. For example, it follows from the Gauss-Green theorem that Z with $\operatorname{div} Z = H$ on $\mathbb{R}^3$ must be unbounded if H is bounded away from zero (and continuous). In particular, no constant $H \not\equiv 0$ can satisfy our present assumptions!

It is therefore necessary to consider the minimization problem in $\mathcal{S}(\Gamma, A)$ with $A \ne \mathbb{R}^3$. This is then an obstacle problem, the complement $\mathbb{R}^3 \setminus A$ being the obstacle which admissible surfaces may not penetrate, and correspondingly we do not obtain a variational equation $\delta \mathbf{E}(x, \cdot) = 0$ for our minimizers but only a **variational inequality**

$$\delta \mathbf{E}_H(x; \xi) \ge 0 \quad \text{for } \xi \in W_0^{1,2} \cap L^\infty(U, \mathbb{R}^3) \text{ with } x + t\xi \in W^{1,2}(U, A), 0 < t \ll 1.$$

For instance, if A is the closure of a C^2 domain in $\mathbb{R}^3$, ν is any C^1 extension to $\mathbb{R}^3$ of the inner unit normal field along ∂A, V is a neighborhood of ∂A, and $\xi \in W_0^{1,2} \cap L^\infty(U, \mathbb{R}^3)$, then we have

$$\delta \mathbf{E}_H(x; \xi) \geq 0 \quad \text{if } \xi \cdot (\nu \circ x) \geq 0 \text{ almost everywhere on } x^{-1}(V).$$

To see this we choose $0 \leq \vartheta \in \mathcal{D}(\mathbb{R}^3, \mathbb{R})$ and observe that $(\vartheta \circ x)(\xi + \varepsilon |\xi| \nu \circ x)$ with $\varepsilon > 0$ is admissible in the variational inequality. Letting ε tend to 0 and then ϑ to the constant 1 in appropriate fashion we deduce the assertion. Note that $x^{-1}(V)$ is well defined up to a set of measure zero.

Now, to obtain a variational equation instead of an inequality for the minimizers x of $\mathbf{E}_H$ on $\mathcal{S}(\Gamma, A)$, one further ingredient of the theory is needed, namely a **geometric inclusion principle** asserting that x maps U into the interior of A. For the case of a ball A of radius R_0 in $\mathbb{R}^3$ centered at the origin such a principle can be deduced from the maximum principle for subharmonic functions. For this one introduces the function $f = |x|^2 \in W^{1,2}(U, \mathbb{R})$ and computes, for $0 \leq \eta \in W^{1,2}(U, \mathbb{R})$,

$$(\eta x)_u \cdot x_u + (\eta x)_v \cdot x_v = \tfrac{1}{2} \eta_u f_u + \tfrac{1}{2} \eta_v f_v + \eta(|x_u|^2 + |x_v|^2),$$
$$|2(H \circ x)\, \eta\, x \cdot x_u \wedge x_v| \leq \eta(|x_u|^2 + |x_v|^2)|x||H \circ x|.$$

Since $\xi = -\eta x$ is admissible in the variational inequality when $0 \leq \eta \in W_0^{1,2} \cap L^\infty(U, \mathbb{R})$ we find, for such test functions η,

$$\tfrac{1}{2} \int_U (\eta_u f_u + \eta_v f_v)\, du\, dv \leq 0,$$

provided $|x||H \circ x| \leq 1$. This means that f is a (weakly) subharmonic function, and if Γ is contained in a concentric closed ball B of radius $R < R_0$ it follows that $f \leq R^2$ on U, hence x has its image in B. Indeed, using $\eta = \max(f - R^2, 0)$ as test function one finds $D\eta = 0$ and hence $f \leq R^2$ almost everywhere on U. Actually, one needs the condition $|x||H \circ x| \leq 1$ only on $x^{-1}(A \setminus B)$.

It remains to find a vector field Z satisfying $\operatorname{div} Z = H$ and $2|Z| \leq c < 1$ on A. The simplest method is radial integration (as in the usual proof of Poincaré's Lemma), i.e.

$$Z(a) = \left(\int_0^1 H(ta)\, t^2\, dt \right) a,$$

and one has

$$\sup_A |Z| \leq \tfrac{1}{3} R \sup_A |H|$$

if A is star-shaped with respect to the origin and contained in a ball of radius R. For example, for constant H this construction gives $Z(a) = \tfrac{1}{3} Ha$, a choice that was already used by E. Heinz [He1] in the first tratment of the Plateau problem with prescribed mean curvature. Heinz obtained, for constant H and with a non-optimal bound on H in his hypotheses, a version of the following optimal result which was proved by S. Hildebrandt [Hi4] after a series of previous improvements [Wer], [Hi2], [Hi3].

2.4 Theorem (Hildebrandt). *Suppose Γ is contained in a closed ball B of radius R in $\mathbb{R}^3$ and H satisfies*

$$\sup_B |H| < \tfrac{3}{2}R^{-1} \quad \text{and} \quad \sup_{\partial B} |H| \leq R^{-1}.$$

Then there exists a weak solution x with values in B to the Plateau problem $\mathcal{P}(H, \Gamma)$.

Proof. We choose $R_0 > R$ and extend H continuously to the concentric ball A of radius R_0 such that $\sup_A |H| < \tfrac{3}{2}R_0^{-1}$ and $|a||H(a)| \leq 1$ hold for $a \in A \setminus B$. With the preceding definition of the vector field Z we then have $2\sup_A |Z| < 1$. Applying Proposition 2.3 we obtain a minimizer x for $\mathbf{E}_H$ on $\mathcal{S}(\Gamma, A)$ which is conformal and satisfies the variational inequality. (Z may only be continuous and satisfy $\operatorname{div} Z = H$ in the sense of distributions, but this suffices for the first variation formula in Proposition 2.1 (i) and for the corresponding variational inequality.) By the inclusion principle x maps U into B so that we actually have the variational equation and can then conclude the proof with Corollary 2.2. $\qquad\square$

Note that, by the reasoning preceding the theorem, x actually has its image in a smaller ball $B' \subset B$ if Γ is contained in B' and $|H|$ on $B \setminus B'$ does not exceed the reciprocal distance to the center of B'. Instead of the two inequalities for H in the hypotheses of Theorem 2.4, Hildebrandt [Hi4] used the assumption $\sup_B |H| \leq R^{-1}$, but it was pointed out in [GS3] that his proof actually gave the stronger result above. Note that, by Example 1.1, the condition $|H| \leq R^{-1}$ is optimal for constant H. Therefore we may not replace R^{-1} by a smaller constant in the second inequality $\sup_{\partial B} |H| \leq R^{-1}$ from the hypotheses of Theorem 2.4. With regard to the first inequality $\sup_B |H| < \tfrac{3}{2}R^{-1}$ we note that this is almost necessary in order to find, as in the proof above, $\mathbf{E}_H$ minimizing weak H-surfaces in $\mathcal{S}(\Gamma, B)$. Indeed, for constant H with $|H| > \tfrac{3}{2}R^{-1}$ one can produce a sequence of surfaces in $\mathcal{S}(\Gamma, B)$ with $\mathbf{E}_H$ energy approaching $-\infty$ by a construction which is based on the observation that a suitably oriented k-fold sphere of radius R has energy $k(4\pi R^2 - \tfrac{2}{3}4\pi R^3 |H|) < 0$.

While Theorem 2.4 is sharp if we wish to assert existence of solutions to $\mathcal{P}(H, \Gamma)$ for *all* Jordan curves Γ contained in the ball B, i.e. we want to express the necessary restrictions on H in terms of the outer radius R of Γ, it became soon clear after Hildebrandt's result that his method is flexible enough to cover a variety of situations where the hypothesis $\sup |H| \leq R^{-1}$ would be much too restrictive. For example, Gulliver & Spruck [GS1] (see also [Hi5]) considered the case of a cylinder $A = \{a \in \mathbb{R}^3 : |Pa| \leq R_0\}$ where $P \colon \mathbb{R}^3 \to \mathbb{R}^2 \times \{0\}$ denotes the projection. For $\mathbf{E}_H$ minimizing x in $\mathcal{S}(\Gamma, A)$ it is natural to consider $f = |P \circ x|^2$. Now, for given $0 \leq \eta \in W_0^{1,2} \cap L^\infty(U, \mathbb{R})$ the vector field $\xi = -\eta(P \circ x)$ is admissible in the variational inequality, and one computes

$$-\tfrac{1}{2}(\eta_u f_u + \eta_v f_v) = \xi_u \cdot x_u + \xi_v \cdot x_v + \eta(|Px_u|^2 + |Px_v|^2).$$

But the conformality of x, which we already know for minimizers in $\mathcal{S}(\Gamma, A)$, implies

$$|2(H \circ x)\,\xi \cdot x_u \wedge x_v| \leq 2|P \circ x||H \circ x|\eta|P(x_u \wedge x_v)|$$
$$\leq 2|P \circ x||H \circ x|\eta(|Px_u|^2 + |Px_v|^2),$$

since $1 \geq |P(a \wedge b)|^2 = 2 - |Pa|^2 - |Pb|^2$ and hence

$$|P(a \wedge b)|^2 = |P^\perp a|^2 + |P^\perp b|^2 \leq 1 \leq |Pa|^2 + |Pb|^2$$

holds for a, b orthonormal in $\mathbb{R}^3$. We deduce that f is weakly subharmonic whenever $|P \circ x||H \circ x| \leq 1$, and if Γ is contained in a concentric cylinder C of radius $R < R_0$ we can conclude that also x has its image in C, provided $\sup_{A \setminus C} |H| \leq \frac{1}{2} R_0^{-1}$ or, at least, $|P \circ x||H \circ x| \leq \frac{1}{2}$ on $x^{-1}(A \setminus C)$. Choosing the vector field

$$Z(a) = \left(\int_0^1 H(ta_1, ta_2, a_3)\, t\, dt \right) Pa$$

with div $Z = H$ and $|Z(a)| \leq \frac{1}{2} |Pa| \sup_C |H|$ we may now repeat the reasoning leading to Theorem 2.4 to obtain

2.5 Theorem (Gulliver & Spruck). *Suppose Γ is contained in a rotationally symmetric cylinder C of radius R in $\mathbb{R}^3$ and H satisfies*

$$\sup_C |H| < R^{-1} \quad and \quad \sup_{\partial C} |H| \leq \frac{1}{2} R^{-1}.$$

Then $\mathcal{P}(H, \Gamma)$ has a weak solution $x \in W^{1,2}(U, C)$. □

For constant H the assumptions reduce to $|H| \leq \frac{1}{2} R^{-1}$ and this is sharp again (see [GS1]). We note that the conformality of x was not used in the proof of the inclusion principle leading to Theorem 2.4, hence this theorem of Hildebrandt is valid in an analogous formulation also for the Dirichlet problem where one minimizes $\mathbf{E}_H$ on $W^{1,2}(U, A)$ subject to a Dirichlet boundary condition. This assertion is not true with regard to Theorem 2.5.

Clearly, Theorem 2.5 gives a better result for certain long and narrow curves Γ than Theorem 2.4, but there are also curves for which both theorems are not satisfactory. Instead of discussing the inclusion principle for further special inclusion domains, however, it is now desirable to analyze in general what conditions on the prescribed mean curvature and the domain considered in $\mathbb{R}^3$ we need for an inclusion principle. This was carried out by Hildebrandt [Hi6], [Hi7], Kaul [Kau], and Gulliver & Spruck [GS3], and it made evident that the comparison of the prescribed mean curvature H with the **boundary mean curvature** $H_{\partial A}$ of the prospective inclusion domain A is decisive for the validity of an inclusion principle. Namely, assuming that A is the closure of a C^2 domain, such a principle holds for conformal solutions to the variational inequality if (and essentially also only if, cf. [GS3]) $|H|$ does nowhere exceed on ∂A the mean curvature $H_{\partial A}$ of ∂A. (Here we define $H_{\partial A}$ with respect to the inner unit normal vector field so that $H_{\partial A} \geq 0$ for convex A.) Gulliver & Spruck used the energy minimality and continuity of x in their proof of the inclusion principle, whereas Hildebrandt assumed the variational inequality and a certain amount of smoothness (which he established for minimizers in [Hi8]). The following strong version is valid for all conformal solutions $x \in W^{1,2}(U, A)$ of the variational inequality.

2.6 Proposition (variational inequality and strong inclusion). *Suppose A is the closure of a C^2 domain in $\mathbb{R}^3$, ν is the inner unit normal on ∂A, the mean curvature $H_{\partial A}$ of ∂A with respect to ν is bounded from below, and $x \in W^{1,2}(U, A)$ is a conformal solution to the variational inequality*

$$\delta \mathbf{E}_H(x; \xi) = \int_U [x_u \cdot \xi_u + x_v \cdot \xi_v + 2(H \circ x)\, \xi \cdot x_u \wedge x_v]\, du\, dv \geq 0$$

for all $\xi \in W_0^{1,2} \cap L^\infty(U, \mathbb{R}^3)$ with $x + t\xi \in W^{1,2}(U, A)$, $0 < t \ll 1$. Then the following assertions hold:

(i) There exists a nonnegative Radon measure λ on U which is absolutely continuous with respect to Lebesgue measure $\mathcal{L}^2$ and concentrated on the coincidence set $x^{-1}(\partial A)$ such that

$$\delta \mathbf{E}_H(x; \xi) = \int_{x^{-1}(\partial A)} \xi \cdot (\nu \circ x)\, d\lambda \quad \text{for all } \xi \in W_0^{1,2} \cap L^\infty(U, \mathbb{R}^3).$$

(ii) The estimate

$$\lambda \leq \mathcal{L}^2 \, \llcorner \, \left[(|x_u|^2 + |x_v|^2)(|H| - H_{\partial A})_+ \circ x \right] \quad \text{on } x^{-1}(\partial A)$$

holds; in particular we have $\lambda = 0$ and x is a weak H-surface, if $|H| \leq H_{\partial A}$ is true pointwise on ∂A.

(iii) If $|H(a)| < H_{\partial A}(a)$ holds at some point $a \in \partial A$ and the boundary trace $x|_{\partial U}$ does not meet a neighborhood of this point then also the image of x omits a neighborhood of this point.

Here the **coincidence set** $x^{-1}(\partial A)$ is the set of $w \in U$ (defined up to a set of Lebesgue $\mathcal{L}^2$ measure zero) such that $x(w) \in \partial A$, i.e. x touches the obstacle $\mathbb{R}^3 \setminus A$ at w. Similarly, the assertion (iii) means that $x^{-1}(V)$ is a set of $\mathcal{L}^2$ measure zero for some neighborhood V of a. The proof combines arguments from [Du1] and [DS4]. We refer to [DS6] for details and here only give a brief

Sketch of Proof. The idea is to use variational vector fields $\xi = \eta(\nu \circ x)$ where the function η is nonnegative and ν is a C^1 extension of the inner unit normal fileld which coincides with the gradient ∇d of the distance d to ∂A on a neighborhood of ∂A in A. The measure λ is then obtained from the L. Schwartz theorem on positive distributions. For the estimate in (ii) one uses the conformality of x together with the definition of mean curvature $H_{\partial A}$ to obtain

$$x_u \cdot (\nu \circ x)_u + x_v \cdot (\nu \circ x)_v = (H_{\partial A} \circ x)(|x_u|^2 + |x_v|^2)$$

almost everywhere on the coincidence set. (i) and (ii) then follow by suitable partition of unity and approximation arguments.

For the strong inclusion principle (iii) one takes $\eta = \psi \max(\varepsilon - d \circ x, 0)$ with small $\varepsilon > 0$, where $d \circ x$ is the distance of x to ∂A and ψ is a cut-off function concentrated near a. A careful discussion of the terms appearing then in the variational inequality and the inequality

$$\text{trace } D\nu < -2|H|$$

near a, which is implied by the assumption $|H(a)| < H_{\partial A}(a)$ in (iii), then give the assertion $\mathcal{L}^2(x^{-1}(V)) = 0$ for some neighborhood V of a. $\square$

It can be seen from the proof that x actually has its image in an interior parallel set $A_\varepsilon = \{a \in A : \text{dist}(a, \partial A) \geq \varepsilon\}$ if the following conditions are satisfied: A has bounded principal curvatures, smooth global inner parallel surfaces to ∂A exist up to distance ε, the conformal solution $x \in W^{1,2}(U, A)$ to the variational inequality has boundary trace $x|_{\partial U}$ with values in A_ε and $|H| \leq H_A$ holds pointwise on $A \setminus A_\varepsilon$, where $H_A(a)$ denotes the mean curvature of the parallel surface to ∂A at $a \in A \setminus A_\varepsilon$. If the boundary mean curvature $H_{\partial A}$ is replaced by the minimum of the principal curvatures in all the statements of Proposition 2.6 then the conformality assumption for x may be dropped.

Combining Proposition 2.6 with Proposition 2.3, Proposition 2.1, and Corollary 2.2, we now immediately obtain the following general existence theorem of Gulliver & Spruck [GS3] (with somewhat weaker assumptions on A and ∂A here; see also Hildebrandt & Kaul [HK]). The contractibility of Γ in A ensures $\mathcal{S}(\Gamma, A) \neq \emptyset$.

2.7 Theorem (Gulliver & Spruck). *Suppose A is the closure of a C^2 domain in $\mathbb{R}^3$, the prescribed mean curvature H and the boundary mean curvature $H_{\partial A}$ of A satisfy*

$$|H| \leq H_{\partial A} \quad \text{pointwise on } \partial A,$$

and there exists a continuous vector field Z with $\operatorname{div} Z = H$ on a neighborhood of A in the distributional sense such that

$$\sup_A |Z| < \tfrac{1}{2}.$$

Then, for every Jordan curve $\Gamma \subset A$ which is contractible in A the Plateau problem $\mathcal{P}(H, \Gamma)$ has a weak solution in $W^{1,2}(U, A)$. Moreover, if $|H(a)| < H_{\partial A}(a)$ holds at some point $a \in (\partial A) \setminus \Gamma$, then each solution surface omits a neighborhood of this point. $\qquad \square$

Choosing A as a ball or a rotationally symmetric cylinder and Z correspondingly as before we recover Theorems 2.4 and 2.5 as special cases of the preceding general theorem. To give an application not covered by Theorems 2.4 and 2.5 consider A contained in a slab $[-R, R] \times \mathbb{R}^2$ of width $2R$ in $\mathbb{R}^3$. Here we can take

$$Z_1(a) = a_1 \left(\int_0^1 H(ta_1, a_2, a_3) \, dt \right), \quad Z_2(a) = Z_3(a) = 0,$$

and we have $\sup_A |Z| < \tfrac{1}{2}$, provided $\sup_A |H| < \tfrac{1}{2} R^{-1}$. If also $|H| \leq H_{\partial A}$ is valid along ∂A, then Theorem 2.7 applies. Further examples like ellipsoids or rotationally symmetric bodies bounded by Delaunay surfaces have been discussed in [GS3], [Hi6].

Another idea of Gulliver & Spruck [GS3] is to use solutions to the **nonparametric mean curvature equation**

$$\operatorname{div} \frac{\nabla f}{\sqrt{1 + |\nabla f|^2}} = 2H \quad \text{on } D \subset \mathbb{R}^3,$$

in order to obtain on $A = \overline{D}$ a C^1 vector field

$$Z = \tfrac{1}{2} \frac{\nabla f}{\sqrt{1 + |\nabla f|^2}}$$

satisfying $\operatorname{div} Z = H$ on A and $\sup_A |Z| < \tfrac{1}{2}$, provided f has bounded gradient ∇f on A. The geometric meaning of this differential equation is that the graph of the scalar function f has mean curvature $H(a)$ at the point $(a, f(a))$ for each $a \in D$. Indeed, the vertically constant unit vector field orthogonal to the graph

$$\mathcal{N}(x, y) = \frac{(-\nabla f(x), 1)}{\sqrt{1 + |\nabla f(x)|^2}}$$

satisfies, on account of $\partial \mathcal{N} / \partial y = 0$,

$$\operatorname{div} \mathcal{N}(x, y) = \operatorname{div} \frac{-\nabla f(x)}{\sqrt{1 + |\nabla f(x)|^2}},$$

and, in view of $\mathcal{N} \cdot \partial_N \mathcal{N} = 0$,

$$-\tfrac{1}{2} \operatorname{div} \mathcal{N}(x, y) = \text{mean curvature of the graph of } f \text{ at } (x, f(x)).$$

The nonparametric mean curvature equation was solved (in general dimensions) with arbitrary continuous Dirichlet boundary data by Serrin [Se] for bounded C^2 domains D and bounded C^1 functions H satisfying

$$|H| \leq H_A \quad \text{on } A.$$

Here $H_A(a)$ denotes the mean curvature of the parallel surface to ∂A through a whenever this is defined (i.e. whenever a has a unique nearest point in ∂A and the principal curvatures of ∂A there are smaller than the reciprocal distance from a to ∂A), while we set $H_A(a) = \infty$ otherwise. We call H_A the **parallel mean curvature function** of A. Serrin's result was extended to unbounded domains D with finite inner radius (the supremum of radii of balls contained in D) and with global inner parallel surfaces by Gulliver & Spruck [GS3]. Assuming a uniform bound for the C^1 norm of H on A they also established that the solution f to the nonparametric mean curvature equation with zero Dirichlet boundary conditions has a bounded C^1 gradient on $A = \overline{D}$ so that $\sup_A |Z| < \frac{1}{2}$ is valid for Z above and Theorem 2.7 is applicable. The hypotheses in the following theorem are slightly weaker than those needed in [GS3] for the reasoning just described. For a proof of this stronger version of the Gulliver & Spruck result we refer to [DS5]; we will come back to this in Section 3.

2.8 Theorem (Gulliver & Spruck). *Suppose A is the closure of a C^2 domain in $\mathbb{R}^3$ with finite inner radius and with smooth global exterior parallel surfaces. If we have*

$$|H| \leq H_A \quad \text{on } A,$$

then the conclusions of Theorem 2.7 are valid. □

We conclude this section with some comments on the **regularity theory** that has been created in order to prove that the weak solutions to the Plateau problem $\mathcal{P}(H, \Gamma)$ produced in the theorems above are in fact classical H-surfaces.

Grüter [Grü1] has shown that weak H-surfaces $x \in W^{1,2}(U, \mathbb{R}^3)$, i.e. conformal weak solutions to the H-surface equation, are continuous on U and hence, by a regularity theorem of Tomi [Tom1], also of class $C^{1,\alpha}$ on U for $0 < \alpha < 1$. It follows then from potential theory that x is of class $C^{k+2,\beta}$ on U if the prescribed mean curvature H is $C^{k,\beta}$. It is conjectured that also without the conformality assumption weak solutions to the H-surface equation (which then looses its geometric meaning) are regular, if H is bounded and continuous. With additional assumptions on H (constant, resp. Lipschitz with some decay at infinity, resp. Lipschitz with a certain decay of $|\nabla H|$) Wente [Wen1], Tomi [Tom2], and Heinz [He6], [He7] had proved this some time ago, and Bethuel [Be] could settle the conjecture for H with $|\nabla H|$ bounded in $\mathbb{R}^3$. It was also shown by Bethuel

& Ghidaglia ([BG], and a recent preprint) that the conjecture is true if H depends only on two coordinates or if $|\partial_e H(a)| \leq \text{const}(1 + |a \cdot e|)^{-1}$ for some direction e in $\mathbb{R}^3$.

With regard to boundary regularity no results are known to us in the literature which are valid for general weak solutions. However, the situation is better for parametric surfaces x which minimize an energy functional $\mathbf{E}_H$ in a class $W^{1,2}(U, A)$ subject to Dirichlet boundary conditions. Namely, if A has a certain uniform local convexity property called **quasiregularity** in [HK], [Hi8], then one can apply a classical method of Morrey to deduce that x is Hölder continuous on U and also continuous on the closed disc $\overline{U}$, provided the boundary trace $x|_{\partial U}$ is continuous. Quasiregularity of $A \subset \mathbb{R}^3$ means that A is uniformly locally biLipschitz equivalent to a convex set in the sense that there exists $\Lambda < \infty$ and for each $a \in A$ a biLipschitz map Φ from a convex set $K \subset \mathbb{R}^3$ into A such that $\text{Lip}\,\Phi \leq \Lambda$, $\text{Lip}\,\Phi^{-1} \leq \Lambda$, and the image of Φ contains the intersection of A with the ball of radius Λ^{-1} and center a.

In order to include a proof of the important fact that energy minimizing solutions to the Plateau problem are continuous up to the boundary we now briefly repeat the reasoning of Hildebrandt & Kaul [HK]. Given a disc $U' = U_r(w)$ in U such that the restriction of x to $\partial U'$ is an absolutely continuous curve of length $< 2\Lambda^{-1}$ in A we choose $a \in A$ on the trace of this curve and Φ as above, define $h \in W^{1,2}(U', \mathbb{R}^3)$ as the harmonic function with boundary trace $\Phi^{-1} \circ x|_{\partial U'}$ and use $\tilde{x} = \Phi \circ h$ on U', $\tilde{x} = x$ on $U \setminus U'$ as comparison surface. The $\mathbf{E}_H$ minimality of x and the coercivity inequality from the proof of Proposition 2.3 on U' as well as the reversed inequality (valid with $1+c$ instead of $1-c$) now imply

$$\int_{U_r(w)} |Dx|^2 \, du\, dv \leq \frac{1+c}{1-c} \int_{U_r(w)} \Lambda^2 |Dh|^2 \, du\, dv.$$

Since the Dirichlet integral of a harmonic function on the unit disc is dominated by the Dirichlet integral of its boundary trace (as can be seen from expanding into a Fourier series), one infers the inequality

$$\int_{U_r(w)} |Dx|^2 \, du\, dv \leq Lr \int_{\partial U_r(w)} |Dx|^2 \, ds$$

with $L = \frac{1+c}{1-c}\Lambda^4$. On the other hand this inequality is trivially valid with $L = \pi\Lambda^2 \mathbf{D}(x)$ in the case where $x|_{\partial U'}$ is absolutely continuous with length $\geq 2\Lambda^{-1}$. Choosing the maximum value for L we have the inequality for almost all radii $0 < r < 1-|w|$, and since it is a differential inequality for the absolutely continuous function appearing on the left, one obtains by integration

$$\int_{U_r(w)} |Dx|^2 \, du\, dv \leq \left(\frac{r}{R}\right)^{2\alpha} \int_{U_R(w)} |Dx|^2 \, du\, dv \quad \text{for } 0 < r \leq R \leq 1 - |w|$$

where $\alpha = (2L)^{-1}$. By Morrey's well known Dirichlet growth theorem this implies Hölder continuity with exponent α for x on U and, if $x|_{\partial U}$ is continuous, also continuity of x on $\overline{U}$ (see [HK] for a proof of the last assertion).

We emphasize that the arguments in the preceding paragraph are valid for minimizers to the obstacle problem and do not use the H-surface equation for x. Hildebrandt has shown in [Hi8], [Hi9] interior $C^{1,\alpha} \cap W^{2,p}$ regularity for the solutions to more general obstacle problems under appropriate smoothness assumptions on the data.

Boundary regularity of higher order for parametric H-surfaces has first been proved for constant H by Hildebrandt [Hi2] who extended his work [Hi1] on the boundary regularity of parametric minimal surfaces with Plateau boundary conditions. Then Heinz [He3], [He4], using earlier work of Heinz & Tomi [HT], showed $C^{1,\alpha}$ regularity for $0 < \alpha < 1$ and in the case $H \in C^{0,\beta}(\mathbb{R}^3)$ also $C^{2,\beta}$ regularity of x up to the boundary, if $x \in W^{1,2}(U, \mathbb{R}^3)$ is a classical (i.e. C^2 on U and continuous on $\overline{U}$) H-surface satisfying the Plateau boundary condition with a Jordan curve of class C^2 or $C^{2,\beta}$ respectively. Jäger [Jä] further improved this result requiring only class $C^{1,\alpha}$ for Γ (and finiteness of the Dirichlet energy of x). The method of Heinz is also presented in chapter 7 of [DHKW].

The question of analytic regularity being settled by the above, we now turn to the problem of **geometric regularity** which is to prove the immersed character of the solutions x to the Plateau problem $\mathcal{P}(H, \Gamma)$. (One cannot expect embeddings, in general, even if Γ is unknotted; see [GS2], however.) By the conformality relations, we must have $x_u(w_0) = 0 = x_v(w_0)$ at a point $w_0 \in \overline{U}$ where x does not have maximal rank. A device of Hartman & Winter [HW] can be applied as in [HT], [He3] to demonstrate, for H-surfaces with sufficient degree of smoothness on $\overline{U}$,

$$x_u(w) - \mathrm{i}x_v(w) = (w - w_0)^k a + o(|w - w_0|^k) \quad \text{as} \quad \overline{U} \ni w \to w_0$$

with an integer $k \geq 1$ and a vector $a \in \mathbb{C}^3$, $a \neq 0$ since x is not constant. (See also [DHKW], Chapter 8 for a thorough discussion.) In view of this expansion w_0 is called a **branch point** of the H-surface x and k its **branching order**.

Geometric regularity of x then is equivalent to the absence of branch points. Clearly the expansion above implies that branch points w_0 are isolated and that the tangent plane $\mathbb{R}x_u(w) + \mathbb{R}x_v(w)$ has a limit position as $w \to w_0$. Heinz & Hildebrandt [HH1] have given estimates on the number of branch points in terms of geometric quantities associated with the boundary curve Γ. From the expansion one also infers that x cannot be constant on an arc $\gamma \subset \partial U$. (Otherwise x would be smooth on $U \cup \gamma$, if γ is relatively open in ∂U, with $x_u = x_v = 0$ on γ contradicting the isolatedness of branch points; see [He3].) Therefore, the solutions to the Plateau problem $\mathcal{P}(H, \Gamma)$ have boundary values which are not only weakly monotonic parametrizations of Γ but really homeomorphisms from ∂U onto Γ.

Ossermann [Os] could prove the non-existence of true interior branch points for minimizing solutions to the classical Plateau problem. These are branch points $w_0 \in U$ accompanied by arcs of transversal self-intersection of x emanating from w_0. Osserman considered parametric minimal surfaces minimizing Dirichlet's integral under Plateau boundary conditions, but his arguments are valid as well for E_H minimizing H-surfaces with H Lipschitz. He did not rule out false branch points, i.e. branch points $w_0 \in U$ near which x is a branched covering of a smoothly embedded surface in $\mathbb{R}^3$. This problem was solved, using unique continuation arguments, topological devices and the minimizing property of x, by Alt [Al1], [Al2] and Gulliver [Gu2] (see also [GOR], [Gu5], [Gu6], [SW]). Consequently, energy minimizing solutions to a Plateau problem $\mathcal{P}(H, \Gamma)$ with Lipschitz mean curvature H are free of interior branch points.

The results are less complete with regard to boundary branch points $w_0 \in \partial U$. These must be of even order, as can be seen from the above expansion near branch points and the monotonicity of $x|_{\partial U}$. The only general result that excludes boundary branch points for energy minimizing solutions x to a Plateau problem $\mathcal{P}(H, \Gamma)$ is due to Gulliver & Lesley [GL] and limited to constant mean curvature H and analytic Jordan curves Γ in $\mathbb{R}^3$. The reason is that a reflection argument is used in the proof. Also, if A is the closure of a smooth domain, Γ a smooth curve on the boundary ∂A, the H-surface x has boundary Γ and values in A, and if the boundary mean curvature of A satisfies $H_{\partial A} \geq |H|$ pointwise on ∂A with strict inequality somewhere on the component of ∂A containing Γ, then one can conclude from the expansion at branch points and from the maximum principle (boundary point lemma) that boundary branch points do not exist. False boundary branch points can be excluded using the energy minimizing property of x ([Gu5], [Gu6]), for true boundary branch points this is not known, however; the subtlety of the matter is illustrated in [Gu7]. Thus the question of the geometric boundary regularity of the energy minimizing solutions has not yet found a final answer.

3   The method of isoperimetric inequalities

There are situations where all the existence theorems of Section 2 for the Plateau problem $\mathcal{P}(H, \Gamma)$ impose stronger restrictions on the prescribed mean curvature H than necessary. For example, if Γ is the boundary of a very long and narrow strip of surface which is curled and knotted quite densely in a large region of space, then Γ will not be contained in a domain with large mean boundary curvature so that the results of Section 2 are not applicable for large constant H, but one would conjecture that Γ nevertheless bounds H-surfaces for quite big values of $|H|$. A different approach to the Plateau problem, which confirmed such a conjecture and eventually lead to much more geometric insight and to several new existence theorems, was proposed by Wente [Wen1]. It is based on the isoperimetric inequality in $\mathbb{R}^3$, and it gives a solution to $\mathcal{P}(H, \Gamma)$ whenever Γ bounds a surface of small area a_Γ and $|H|$ is bounded by $\mathrm{const}\, a_\Gamma^{-1/2}$.

Wente considered constant prescribed mean curvature H and used the energy functional $\mathbf{E}_H = \mathbf{D} + 2H\mathbf{V}$, where

$$\mathbf{V}(x) = \tfrac{1}{3} \int_U x \cdot x_u \wedge x_v \, du dv$$

is the volume enclosed by the parametric surface $x \in W^{1,2}(U, \mathbb{R}^3)$ and the cone over the boundary trace of x. To ensure the existence of the integral above one needs to know that x is bounded. Wente did not want to assume this, however, because then he would eventually have had to work in a bounded domain of $\mathbb{R}^3$ and could allow only values of $|H|$ not exceeding its mean boundary curvature, in order to have the inclusion principle available. He observed that the volume $\mathbf{V}(x)$ could be well defined by continuous extension for all surfaces $x \in W^{1,2}(U, \mathbb{R}^3)$ with bounded trace $x|_{\partial U}$, although the integral representation, valid in the case $x \in L^\infty(U, \mathbb{R}^3)$, will in general not hold for unbounded x.

To define this extension one decomposes $x = y + z$ into its bounded harmonic part y and its $W_0^{1,2}$ part z. (y is the minimizer for Dirichlet's integral on $x + W_0^{1,2}(U, \mathbb{R}^3)$.) Then z is a closed surface to which we can apply the classical isoperimetric inequality, at least if z is smooth enough. Since generally the area $\mathbf{A}(z) = \int_U |z_u \wedge z_v| \, du dv$ does not exceed Dirichlet's integral $\mathbf{D}(z)$ we thus obtain the following **isoperimetric inequality**:

$$|\mathbf{V}(z)| \leq \frac{1}{\sqrt{36\pi}} \mathbf{A}(z)^{3/2} \leq \frac{1}{\sqrt{36\pi}} \mathbf{D}(z)^{3/2}.$$

From the fact that $\mathbf{V}(z)$ is a cubic form in z we now deduce that $\mathbf{V}$ can be extended continuously to $W_0^{1,2}(U, \mathbb{R}^3)$ with the same inequality. Furthermore, with an integration by parts we verify

$$\mathbf{V}(x) - \mathbf{V}(z) = \int_U y \cdot \left(\tfrac{1}{3} y_u \wedge y_v + \tfrac{1}{2} y_u \wedge z_v + \tfrac{1}{2} z_u \wedge y_v + z_u \wedge z_v\right) du\,dv$$

for smooth $x = y + z$, and we conclude that $\mathbf{V}$ can be extended to the space of $x \in W^{1,2}(U, \mathbb{R}^3)$ with $x|_{\partial U} \in L^\infty(\partial U, \mathbb{R}^3)$ as a functional that is continuous with respect to $W^{1,2}$ convergence and simultaneous L^∞ convergence of boundary traces (i.e. of the harmonic parts).

Following Wente we now attempt to minimize $\mathbf{E}_H = \mathbf{D} + 2H\mathbf{V}$ on

$$\mathcal{S}(\Gamma; \sigma) = \{x \in \mathcal{S}(\Gamma) : \mathbf{D}(x) \leq \sigma a_\Gamma\}$$

where $1 < \sigma < \infty$ and a_Γ is the **least spanning area** of Γ, i.e. the area $a_\Gamma = \mathbf{D}(y_\Gamma)$ of a parametric surface y_Γ minimizing $\mathbf{D}$ on $\mathcal{S}(\Gamma)$ (which exists by the solution of the classical Plateau problem, i.e. Proposition 2.3 with $Z = 0$ and $A = \mathbb{R}^3$). Choosing a minimizing sequence $x_k = y_k + z_k$, decomposed as above, we may assume $x_k \to x = y + z \in \mathcal{S}(\Gamma; \sigma)$ weakly in $W^{1,2}(U, \mathbb{R}^3)$ with $x_k|_{\partial U} \to x|_{\partial U}$ uniformly, hence also $y_k \to y$ uniformly on $\overline{U}$ (cf. the proof of Proposition 2.3). We let $\tilde{x}_k = y + z_k$. Expanding $\mathbf{V}(x_k) - \mathbf{V}(z_k)$ and $\mathbf{V}(\tilde{x}_k) - \mathbf{V}(z_k)$ as above one can verify $\mathbf{V}(x_k) - \mathbf{V}(\tilde{x}_k) \to 0$ and $\mathbf{V}(x_k) - \mathbf{V}(x) - \mathbf{V}(z_k - z) \to 0$ as $k \to \infty$ (see [Wen1], [Ste1] for details). Using also $\mathbf{D}(\tilde{x}_k) - \mathbf{D}(z_k - z) \to \mathbf{D}(x)$ we arrive at

$$\mathbf{E}_H(x) \leq \liminf_{k \to \infty} \mathbf{E}_H(x_k) + \limsup_{k \to \infty} \left[2|H||\mathbf{V}(z_k - z) - \mathbf{D}(z_k - z)\right],$$

and with the isoperimetric inequality we infer

$$\mathbf{E}_H(x) \leq \liminf_{k \to \infty} \mathbf{E}_H(x_k),$$

provided

$$\limsup_{k \to \infty} \frac{2|H|}{\sqrt{36\pi}} \mathbf{D}(z_k - z)^{1/2} \leq 1.$$

Taking into account $\limsup_{k \to \infty} \mathbf{D}(z_k - z) \leq \limsup_{k \to \infty} \mathbf{D}(x_k)$ and the definition of $\mathcal{S}(\Gamma; \sigma)$, we see that the last inequality is satisfied if

$$|H| \leq \frac{3\sqrt{\pi}}{\sqrt{\sigma a_\Gamma}},$$

and with this condition on H we have verified that x is a minimizer for $\mathbf{E}_H$ in $\mathcal{S}(\Gamma; \sigma)$.

To show that x is a solution to $\mathcal{P}(H, \Gamma)$ it remains to prove $\mathbf{D}(x) < \sigma a_\Gamma$ so that x can be freely varied and the variational formulas of Proposition 2.1 (which are easily checked for the present definition of volume) together with Corollary 2.2 can be applied. Following Wente again, we use the inequality $\mathbf{E}_H(x) < \mathbf{E}_H(y_\Gamma)$ for this purpose which is valid in the case $H \neq 0$, because $y_\Gamma \in \mathcal{S}(\Gamma; \sigma)$ then cannot be a minimizer for $\mathbf{E}_H$. Consequently, we have

$$\mathbf{D}(x) = \mathbf{E}_H(x) - 2H\mathbf{V}(x) < \mathbf{E}_H(y_\Gamma) - 2H\mathbf{V}(x) = a_\Gamma + 2H(\mathbf{V}(y_\Gamma) - \mathbf{V}(x)).$$

Now, x and y_Γ have the same curve Γ as oriented boundary and together they form a closed surface to which the isoperimetric inequality in $\mathbb{R}^3$ can be applied again. Therefore we have

$$|\mathbf{V}(y_\Gamma) - \mathbf{V}(x)| \le \frac{1}{\sqrt{36\pi}}(\mathbf{A}(y_\Gamma) + \mathbf{A}(x))^{3/2} \le \frac{1}{\sqrt{36\pi}}(\mathbf{D}(y_\Gamma) + \mathbf{D}(x))^{3/2},$$

and hence, recalling $\mathbf{D}(y_\Gamma) = a_\Gamma$ and $\mathbf{D}(x) \le \sigma a_\Gamma$,

$$\mathbf{D}(x) < a_\Gamma \left(1 + \frac{2|H|}{\sqrt{36\pi}}(1+\sigma)^{3/2} a_\Gamma^{1/2}\right).$$

The desired conclusion follows if the right-hand side does not exceed σa_Γ, and it is readily checked that this is true if we make the optimal choice $\sigma = 5$ and require $|H| \le c\sqrt{\pi/a_\Gamma}$ with $c = \sqrt{2/3}$. Note that this inequality for H also implies the restriction imposed earlier on H.

For constant H we have thus proved the following theorem which was originally formulated by Wente [Wen1] with constant $c \approx \frac{2}{5}$ then sharpened by the present author to $c \approx 0.52$ in [Ste1], and later further improved in [Ste4] to the above constant $c = \sqrt{2/3}$, allowing also variable H.

3.1 Theorem (Wente, Steffen). *Suppose a_Γ is the least spanning area of Γ and*

$$\sup_{\mathbb{R}^3}|H| \le \sqrt{\frac{2}{3} \cdot \frac{\pi}{a_\Gamma}}.$$

Then the Plateau problem $\mathcal{P}(H, \Gamma)$ has a weak solution. $\qquad\square$

It is clear that this result is better than the existence theorems of Section 2 for curves with a shape described at the beginning of the present section. The constant $c = \sqrt{2/3}$ is, however, probably not optimal. Considering the example of a circle Γ one is lead to the conjecture that it can be replaced by $c = 1$. This would be best possible by the non-existence theorem of Heinz [He2], cf. Example 1.1. It has been observed by Struwe [Str3, III.3] that $\sqrt{2/3}$ can be replaced by a larger constant depending on Γ, if only constant H are considered. This also follows from the proof above, because $\mathbf{E}_H$ is uniformly close to $\mathbf{E}_{\tilde{H}}$ on $\mathcal{S}(\Gamma; \sigma)$ if $|H - \tilde{H}|$ is small.

The preceding proof cannot be extended to variable prescribed mean curvature H, because then we do not have an appropriate definition of the H-volume $\mathbf{V}_H(x)$ (nor would it be a cubic form in x). Since we want to bring in the isoperimetric inequality we should look for a geometric definition of the H-volume of a surface $x \in \mathcal{S}(\Gamma)$. Intuitively, $\mathbf{V}_H(x)$ should (up to a constant) equal the integral of the function H over the subset of $\mathbb{R}^3$ which is bounded by x and a fixed reference surface y with the same boundary Γ. However, x may intersect this reference surface and also itself so that one must consider a "set with integer multiplicities" bounded by x and y, the multiplicity at $a \in \mathbb{R}^3$ being given by the degree with respect to the point a of the spherical surface which is composed from x and y. To make this intuitive idea precise we now must bring in some geometric measure theory. We will give the necessary definitions and explain their geometric meaning, but for details the reader will have to consult the standard references [Fe], [Si].

We begin by noting that each parametric surface $x \in W^{1,2}(U, \mathbb{R}^3)$ defines, by integration of 2-forms $\beta \in \mathcal{D}^2(\mathbb{R}^3)$ (i.e. smooth and with compact support) over x, an associated integer multiplicity rectifiable 2-current J_x on $\mathbb{R}^3$,

$$J_x(\beta) = \int_U x^\# \beta = \int_U \langle \beta \circ x, x_u \wedge x_v \rangle \, du dv \,,$$

which has finite mass

$$\mathbf{M}(J_x) \leq \mathbf{A}(x) \leq \mathbf{D}(x) \,.$$

Here, a 2-current T is a continuous linear functional on $\mathcal{D}^2(\mathbb{R}^3)$ and its mass is the supremum of values $T(\beta)$ on 2-forms $\beta \in \mathcal{D}^2(\mathbb{R}^3)$ with $|\beta| \leq 1$ on $\mathbb{R}^3$. It is called an integer multiplicity rectifiable 2-current if it can be represented

$$T(\beta) = \int_M \langle \beta, \vec{\eta} \rangle \, \vartheta \, d\mathcal{H}^2 \,,$$

where $\mathcal{H}^2$ is the 2-dimensional Hausdorff measure on $\mathbb{R}^3$, M is a 2-rectifiable subset of $\mathbb{R}^3$, $\vec{\eta}$ is an $\mathcal{H}^2$ measurable orientation for M, and ϑ is an $\mathcal{H}^2$ measurable multiplicity function on M with values in the positive integers and with finite $\mathcal{H}^2$ integral over M. The notion of 2-rectifiable set M is a measure theoretic generalization of embedded or immersed 2-submanifold in $\mathbb{R}^3$. It means that, up to a set of $\mathcal{H}^2$ measure zero, M is a Borel set which can be covered by countably many 2-dimensional submanifolds of class C^1. Such sets M have a 2-dimensional measure theoretic tangent space $\text{Tan}_a M$ at $\mathcal{H}^2$ almost every point $a \in M$, and if we choose an orthonormal basis $e_1(a), e_2(a)$ in $\text{Tan}_a M$ depending on a in a measurable way, then we obtain a measurable orientation $\vec{\eta} = e_1 \wedge e_2 : M \to \Lambda_2 \mathbb{R}^3$ for M. (Our terminology deviates from [Fe] in that we do not require compactness of supports, but incorporate finiteness of the mass $\mathbf{M}(T) = \int_M \vartheta \, d\mathcal{H}^2$ into the definition of an integer multiplicity rectifiable current. In the language of [Fe] this would be called a locally rectifiable 2-current of finite mass.)

Using approximation by smooth mappings in the $W^{1,2}$ norm as in [Ste3] or in the sense of a Lusin type theorem ([EG, §6.6]), one sees that the current J_x associated with $x \in W^{1,2}(U, \mathbb{R}^3)$ is indeed an integer multiplicity rectifiable 2-current. If $y \in W^{1,2}(U, \mathbb{R}^3)$ is another surface, then in the integral defining $J_x(\beta) - J_y(\beta)$ we need only integrate over the set G of $w \in U$ with $x(w) \neq y(w)$ because $Dx = Dy$ holds almost everywhere on $U \setminus G$. Hence the mass of $J_x - J_y$ can be estimated

$$\mathbf{M}(J_x - J_y) \leq \mathbf{D}_G(x) + \mathbf{D}_G(y) \quad \text{if } x = y \text{ on } U \setminus G \,,$$

where $\mathbf{D}_G(x) = \frac{1}{2} \int_G (|x_u|^2 + |x_v|^2) \, du dv$ for measurable $G \subset U$.

The boundary ∂T of a current T is generally defined by $\partial T(\alpha) = T(d\alpha)$. For J_x we find that ∂J_x is the 1-current given by integration of 1-forms over the boundary $x|_{\partial U}$, if this curve is continuous and rectifiable. This is just a version of Stokes' theorem. In particular, $J_x - J_y$ is a closed 2-current whenever the surfaces x, y satisfy the same Plateau boundary condition, i.e.

$$\partial(J_x - J_y) = 0 \quad \text{if } x, y \in \mathcal{S}(\Gamma) \,,$$

which is a precise way of expressing that x and y together form a closed surface in $\mathbb{R}^3$.

Now, every closed 2-current T on $\mathbb{R}^3$ is the boundary of a 3-current which is unique up to a constant and can be obtained from T with a cone construction as in the usual proof of Poincaré's lemma. For $J_x - J_y$ one can use the homotopy formula [Fe, 4.1.9] to obtain such a 3-current $I_{x,y}$ with boundary $J_x - J_y$ as follows:

$$I_{x,y}(\gamma) = \int_{[0,1]\times U} h^{\#}\gamma = \int_U \int_0^1 \langle \gamma \circ h, h_t \wedge h_u \wedge h_v \rangle \, dt\,du\,dv \,,$$

where $h(t, w) = tx(w) + (1 - t)y(w)$ and $\gamma \in \mathcal{D}^3(\mathbb{R}^3)$. The equation

$$\partial I_{x,y} = J_x - J_y \,, \quad \text{i.e.} \quad I_{x,y}(d\beta) = J_x(\beta) - J_y(\beta) \quad \text{for } \beta \in \mathcal{D}^2(\mathbb{R}^3) \,,$$

is just a version of the Gauss-Green theorem if x, y are smooth on $\overline{U}$, and it follows in general by approximation. We note that the above integral formula for $I_{x,y}(\gamma)$ implies the mass estimate

$$\mathbf{M}(I_{x,y}) \leq \|x - y\|_{L^\infty}(\mathbf{D}_G(x) + \mathbf{D}_G(y)) \quad \text{if } x = y \text{ on } U \setminus G.$$

Now, $I_{x,y}$ is an integer multiplicity locally rectifiable current in the top dimension and such currents have a particularly simple structure. Namely, they are representable by a locally integrable (with respect to Lebesgue measure $\mathcal{L}^3$) integer valued function $i_{x,y}$ on $\mathbb{R}^3$ in the form

$$I_{x,y}(\gamma) = \int_{\mathbb{R}^3} i_{x,y}\gamma = \int_{\mathbb{R}^3} \langle \gamma, \vec{e} \rangle \, i_{x,y} \, d\mathcal{L}^3$$

where $\vec{e} \in \bigwedge_3 \mathbb{R}^3$ is the orientation of $\mathbb{R}^3$. Geometrically speaking, $i_{x,y}$ is just a set with integer multiplicities, and $\partial I_{x,y} = J_x - J_y$ tells us that this set has the closed surface composed by x and y as its boundary. From the isoperimetric inequality in the form [Fe, 4.5.9 (31)] which is valid for such currents (and which could also be deduced from the corresponding isoperimetric inequality for sets of finite perimeter proved in [DeG] and [MM], for example), one infers

$$\mathbf{M}(I_{x,y}) = \int_{\mathbb{R}^3} |i_{x,y}| \, d\mathcal{L}^3 \leq \int_{\mathbb{R}^3} |i_{x,y}|^{3/2} \, d\mathcal{L}^3 \leq \frac{1}{\sqrt{36\pi}}\mathbf{M}(J_x - J_y)^{3/2} \,,$$

proving in particular that the mass of $I_{x,y}$ is finite. We note that $I_{x,y}$ is uniquely determined by its boundary and by the condition $\mathbf{M}(I_{x,y}) < \infty$, and we deduce

$$I_{x,y} + I_{y,z} = I_{x,z} \quad \text{for } x, y, z \in \mathcal{S}(\Gamma) \,.$$

After this excursion into geometric measure theory it is evident how we should define the **H-volume functional** associated with a prescribed mean curvature function H, namely

$$\mathbf{V}_H(x,y) = \int_{\mathbb{R}^3} H i_{x,y} \, d\mathcal{L}^3 \quad \text{for } x, y \in \mathcal{S}(\Gamma).$$

The integral exists because $i_{x,y}$ is summable on $\mathbb{R}^3$ and because we generally assume that H is bounded and continuous. (The H-volume can be defined under much weaker conditions, see [Ste3].) Approximating $H\Omega$ suitably by 3-forms $\gamma \in \mathcal{D}^3(\mathbb{R}^3)$ (recall that Ω is the volume form of $\mathbb{R}^3$) we obtain from the mass estimate for $I_{x,y}$

$$|\mathbf{V}_H(x,y)| \leq \|H\|_{L^\infty}\|x - y\|_{L^\infty}(\mathbf{D}_G(x) + \mathbf{D}_G(y)) \quad \text{if } x = y \text{ on } U \setminus G,$$

and for $\xi \in W_0^{1,2}(U,\mathbb{R}^3) \cap L^\infty(U,\mathbb{R}^3)$ we see that $\mathbf{V}_H(x+t\xi,x)$ has exactly the integral representation that we have used in the proof of Proposition 2.1 (i). Since the definition of $\mathbf{V}_H(x,y)$ and the properties of the currents $I_{x,y}$ noted above also imply

$$\mathbf{V}_H(x,y) + \mathbf{V}_H(y,z) = \mathbf{V}_H(x,z) \quad \text{for } x,\, y,\, z \in \mathcal{S}(\Gamma),$$

we deduce for the functional $x \mapsto \mathbf{V}_H(x,y)$ the same first variation formula as in Proposition 2.1 (i). Moreover, J_x, J_y and hence also $I_{x,y}$, $\mathbf{V}_H(x,y)$ are invariant with respect to orientation preserving self-diffeomorphisms of $\overline{U}$. It follows that Propositions 2.1, 2.6 and all the observations made about the variational (in-)equality in Section 2 are valid if we define the H-**energy** now by

$$\mathbf{E}_H(x) = \mathbf{D}(x) + 2\mathbf{V}_H(x,y_\Gamma) \quad \text{for } x \in \mathcal{S}(\Gamma),$$

where y_Γ is a fixed surface in the class $\mathcal{S}(\Gamma)$.

In order to minimize $\mathbf{E}_H$ on $\mathcal{S}(\Gamma)$ or on suitable subclasses we next turn to the question of lower semicontinuity of $\mathbf{E}_H$ on a $W^{1,2}$ weakly convergent sequence $x_k \to x$ with $x_k|_{\partial U} \to x|_{\partial U}$ uniformly. The volume $\mathbf{V}_H(x_k, y_\Gamma)$ is not, in general, continuous for this convergence, because a "bubble" can split off x_k in the limit carrying away a certain amount of volume that is not enclosed by the limit surface x. The idea is then that such a bubble must be parametrized over an arbitrarily small part G of U and its mass must be dominated by the jump $\limsup_{k\to\infty}(\mathbf{D}(x_k) - \mathbf{D}(x))$ of Dirichlet's integral. On the other hand, we can dominate the jump of H-volume $\limsup_{k\to\infty} 2|\mathbf{V}_H(x_k,x)|$ by the mass of the bubble, using the isoperimetric inequality and suitable hypotheses for H and the x_k. The conclusion will then be $\mathbf{E}_H(x) \le \liminf_{k\to\infty} \mathbf{E}_H(x_k)$ as desired. To persue this idea we must estimate $2\mathbf{V}_H(x_k,x)$ in terms of $\mathbf{D}_G(x_k) + \mathbf{D}_G(x)$. The following definition is useful for this (cf. [Ste3], [DS4]):

3.2 Definition. Given $0 \le c < \infty$, $0 < s \le \infty$ we say that H satisfies an **isoperimetric condition** of type c, s on A if

$$\left| \int_E 2H\, d\mathcal{L}^3 \right| \le c\,\mathbf{P}(E)$$

holds whenever $E \subset A$ is a set with finite perimeter $\mathbf{P}(E) \le s$. □

Here, by a **set with finite perimeter** $E \subset \mathbb{R}^3$, also called a Caccioppoli set, we mean a Borel set of finite $\mathcal{L}^3$-measure such that its characteristic function has distributional gradient of finite total variation. This total variation is then the perimeter $\mathbf{P}(E) < \infty$, it equals the boundary mass $\mathbf{M}(\partial[\![E]\!])$ of the 3-current given by integration of 3-forms over E, and its geometric meaning is the boundary area of E. For smooth bounded domains E one has $\mathbf{P}(E) = \mathcal{H}^2(\partial E)$, and it is sufficient to verify the inequality in the definition for such domains $E \subset A$ when A is the closure of a smooth domain. A standard decomposition theorem from geometric measure theory [Fe, 4.5.17] implies that every integer multiplicity rectifiable 3-current I on $\mathbb{R}^3$ can be decomposed into a L^1-convergent sum of (suitably oriented) sets E_k, $k \in \mathbb{Z}$, with finite perimeter such that $\sum_{k\in\mathbb{Z}} \mathbf{P}(E_k) = \mathbf{M}(\partial I)$. Applying this to our currents $I_{x,y}$ above we obtain the **isoperimetric condition for the H-volume**

$$2|\mathbf{V}_H(x,y)| \le c\,\mathbf{M}(J_x - J_y) \quad \text{for } x, y \in \mathcal{S}(\Gamma) \text{ with } \mathbf{M}(J_x - J_y) \le s,$$

whenever an isoperimetric condition of type c, s is valid for H on $\mathbb{R}^3$. If we know this condition only on $A \subset \mathbb{R}^3$ we can draw the same conclusion for x, $y \in \mathcal{S}(\Gamma, A)$, provided $I_{x,y}$ has support in A. This will be the case, for instance, if $\mathbb{R}^3 \setminus A$ has no components of finite measure, because $\mathbf{M}(I_{x,y})$ is finite and $\partial I_{x,y} = J_x - J_y$ has support in A, hence $I_{x,y}$ must be constant on each component of $\mathbb{R}^3 \setminus A$. (For a more thorough discussion of Definition 3.2 and its consequences see [Ste3], [DS4].)

We are now prepared to prove the following general existence theorem for the Plateau problem $\mathcal{P}(H, \Gamma)$, which is a strengthening of the results of [Ste3].

3.3 Theorem. *Suppose A is the closure of a C^2 domain $\mathbb{R}^3$, $\mathbb{R}^3 \setminus A$ has no components of finite measure, H satisfies an isoperimetric condition of type c, s on A, $\Gamma \subset A$, and $y_\Gamma \in \mathcal{S}(\Gamma, A)$ with $(1+\sigma)\mathbf{D}(y_\Gamma) \leq s$ for some $1 < \sigma \leq \infty$. Set $\mathcal{S}(\Gamma, A; \sigma) = \{x \in \mathcal{S}(\Gamma, A) : \mathbf{D}(x) \leq \sigma \mathbf{D}(y_\Gamma)\}$. Then the following assertions hold:*

(i) If $c < 1$ or if $c = 1$ and $\sigma < \infty$, then the variational problem

$$\mathbf{E}_H(x) = \mathbf{D}(x) + 2\mathbf{V}_H(x, y_\Gamma) \rightsquigarrow \min \quad on \quad \mathcal{S}(\Gamma, A; \sigma)$$

has a solution.

(ii) If $c \leq \dfrac{\sigma - 1}{\sigma + 1}$ or $\sigma = \infty$ then each minimizer x from (i) satisfies $\mathbf{D}(x) < \sigma \mathbf{D}(y_\Gamma)$.

(iii) If $A = \mathbb{R}^3$ or $|H| \leq H_{\partial A}$ holds along ∂A, where $H_{\partial A}$ denotes the (inward) boundary mean curvature of A, then each minimizer x from (i) with $\mathbf{D}(x) < \sigma \mathbf{D}(y_\Gamma)$ is a weak solution to the Plateau problem $\mathcal{P}(H, \Gamma)$. Moreover, if $|H(a)| < H_{\partial A}(a)$ holds at some point $a \in (\partial A) \setminus \Gamma$, then x does not meet a neighborhood of this point.

Proof. (i) We choose a minimizing sequence x_k. Then $\sup_k \mathbf{D}(x_k) < \infty$, because in the case $\sigma = \infty$ we have $c < 1$, $\mathbf{D}(x_k) + \mathbf{D}(y_\Gamma) \leq s$, and hence

$$|\mathbf{D}(x_k) - \mathbf{E}_H(x_k)| = 2|\mathbf{V}_H(x_k, y_\Gamma)| \leq c(\mathbf{D}(x_k) + \mathbf{D}(y_\Gamma)),$$

by the isoperimetric condition for the H-volume. As in the proof of Proposition 2.3 we can assume $x_k \to x \in \mathcal{S}(\Gamma, A; \sigma)$ weakly in $W^{1,2}(U, \mathbb{R}^3)$ and uniformly on ∂U.

We next want to prove lower semicontinuity of the H-energy on the sequence x_k. This would be no problem if we knew $\|x_k - x\|_{L^\infty} \to 0$ as $k \to \infty$, because then we could use the identity

$$\mathbf{V}_H(x_k, y) - \mathbf{V}_H(x, y) = \mathbf{V}_H(x_k, x)$$

and the previous estimate

$$\mathbf{M}(I_{x_k,x}) \leq \|x_k - x\|_{L^\infty}(\mathbf{D}_G(x_k) + \mathbf{D}_G(x)) \quad \text{if} \quad x_k = x \text{ on } U \setminus G.$$

But by Rellich's theorem we may assume $x_k \to x$ almost everywhere on U and, by Egoroff's theorem, uniformly on a subset of U with complement G of prescribed small measure $\mathcal{L}^3(G) < \varepsilon$. One may imagine that a "bubble" is parametrized by the x_k on G which disappears in the limit.

By a technical construction we can now produce new mappings $\tilde{x}_k \in \mathcal{S}(\Gamma, A)$ with $\tilde{x}_k(w) = x(w)$ if $w \in U \setminus G$, $\tilde{x}_k(w) = x_k(w)$ if $|x_k(w)| \geq \varepsilon^{-1}$, $\|\tilde{x}_k - x_k\|_{L^\infty} \to 0$ as $k \to \infty$, and

$$\limsup_{k \to \infty} \left(\mathbf{D}_G(\tilde{x}_k) + \mathbf{D}_G(x) \right) \leq \varepsilon + \liminf_{k \to \infty} \left(\mathbf{D}(x_k) - \mathbf{D}(x) \right).$$

Moreover, if the x_k have values in the closure A of a C^2 domain then also the $\tilde{x}_k$ may be chosen to have values in A. The idea here is that the $\tilde{x}_k$ should parametrize essentially the same bubble on G, but *coincide* with their weak limit on $U \setminus G$. We refer to [DS6] for the details of this construction and to [Ste3] for a different construction based on harmonic replacement of x_k on a (small) open set containing G.

On account of $\|\tilde{x}_k - x_k\|_{L^\infty} \to 0$ we have $\mathbf{V}_H(\tilde{x}_k, x_k) \to 0$. Choosing $2\varepsilon < \mathbf{D}(x)$ we also see

$$\mathbf{M}(J_{\tilde{x}_k} - J_x) \leq \mathbf{D}_G(\tilde{x}_k) + \mathbf{D}_G(x) \leq 2\varepsilon + \mathbf{D}(x_k) - \mathbf{D}(x) < \sigma \mathbf{D}(y_\Gamma) < s$$

for large k. Therefore, the isoperimetric condition with $c \leq 1$ implies, for k large again,

$$2|\mathbf{V}_H(\tilde{x}_k, x)| \leq \mathbf{D}_G(\tilde{x}_k) + \mathbf{D}_G(x) \leq 2\varepsilon + \mathbf{D}(x_k) - \mathbf{D}(x),$$

and we conclude

$$\mathbf{E}_H(x_k) = \mathbf{D}(x_k) + 2\mathbf{V}_H(x_k, y_\Gamma) = \mathbf{E}_H(x) + \mathbf{D}(x_k) - \mathbf{D}(x) + 2\mathbf{V}_H(x_k, x)$$
$$= \mathbf{E}_H(x) + \mathbf{D}(x_k) - \mathbf{D}(x) + 2\mathbf{V}_H(\tilde{x}_k, x) - 2\mathbf{V}_H(\tilde{x}_k, x_k) \geq \mathbf{E}_H(x) - 3\varepsilon$$

for sufficiently large k.

(ii) If y_Γ is not $\mathbf{E}_H$ minimizing in $\mathcal{S}(\Gamma, A; \sigma)$, then we have

$$\mathbf{D}(x) = \mathbf{E}_H(x) - 2\mathbf{V}_H(x, y_\Gamma) < \mathbf{E}_H(y_\Gamma) - 2\mathbf{V}_H(x, y_\Gamma)$$
$$= \mathbf{D}(y_\Gamma) - 2\mathbf{V}_H(x, y_\Gamma) \leq \mathbf{D}(y_\Gamma) + c(\mathbf{D}(x) + \mathbf{D}(y_\Gamma)) \leq \mathbf{D}(y_\Gamma)(1 + c(\sigma + 1)),$$

by the isoperimetric condition for the H-volume and the definition of $\mathcal{S}(\Gamma, A; \sigma)$. Consequently, with the assumption $1 + c(\sigma + 1) \leq \sigma$ we obtain $\mathbf{D}(x) < \sigma \mathbf{D}(y_\Gamma)$. If y_Γ is a minimizer, then the same inequality holds for $x = y_\Gamma$, because $\sigma > 1$.

(iii) Here we can apply Proposition 2.6 and repeat the proof of the corresponding assertion in Theorem 2.7, since we have already noted in connection with the present definition of H-energy that the variational inequality holds and since the additional constraint $\mathbf{D}(x) \leq \sigma \mathbf{D}(y_\Gamma)$ is not effective in view of the hypothesis in (iii). $\quad\square$

To give concrete **applications** of the preceding theorem we have to verify the isoperimetric condition. For example, we can use the isoperimetric inequality for sets of finite perimeter to obtain

$$\left| \int_E 2H \, d\mathcal{L}^3 \right| \leq 2 \sup_{\mathbf{R}^3} |H| \, \mathcal{L}^3(E) \leq 2\sqrt{\frac{s}{36\pi}} \sup_{\mathbf{R}^3} |H| \, \mathbf{P}(E),$$

if $\mathbf{P}(E) \leq s < \infty$. (Or use the definition of H-volume together with the preceding inequality $\int |i_{x,y}| \, d\mathcal{L}^3 \leq (36\pi)^{-1/2} \mathbf{M}(J_x - J_y)$.) Choosing $s = (\sigma + 1)\mathbf{D}(y_\Gamma)$, the condition on c in Theorem 3.3 (ii) reduces to

$$2(\sigma + 1)^{1/2} \sqrt{\frac{\mathbf{D}(y_\Gamma)}{36\pi}} \sup_{\mathbf{R}^3} |H| \leq \frac{\sigma - 1}{\sigma + 1},$$

and it is readily verified that $\sigma=5$ is the optimal choice and then $\sup_{\mathbb{R}^3} |H| \leq (\frac{3}{2\pi} \mathbf{D}(y_\Gamma))^{-1/2}$ is the resulting condition on H. Of course, we choose y_Γ here as a minimal surface of least area $\mathbf{D}(y_\Gamma) = a_\Gamma$ in $\mathcal{S}(\Gamma)$ and we have a **proof of Theorem 3.1** for variable mean curvature and with the additional inclusion statement from part (iii) of the preceding theorem if $y_\Gamma \in \mathcal{S}(\Gamma, A)$. We note that Dierkes [Di1], [Di2] has stated inclusion principles of a different nature which require that $\sup_{\partial A} |H_{\partial A}|$ and $\sup_A |H|$ are (quite) small compared to $a_\Gamma^{-1/2}$, but allow negative values of $H_{\partial A}$.

Another application results when we first use Hölder's inequality and then apply the isoperimetric inequality:

$$\left| \int_E 2H \, d\mathcal{L}^3 \right| \leq 2 \left[\int_E |H|^3 \, d\mathcal{L}^3 \right]^{1/3} \mathcal{L}^3(E)^{2/3} \leq 2 \left[\int_E |H|^3 \, d\mathcal{L}^3 \right]^{1/3} \frac{1}{\sqrt[3]{36\pi}} \mathbf{P}(E).$$

In fact, the same inequality holds with $\|H\|_{L^3}$ replaced by the maximum of $\|H_+\|_{L^3}$ and $\|H_-\|_{L^3}$, where $H = H_+ - H_-$ is the decomposition of H into its positive and negative part. Using Theorem 3.3 with $\sigma = s = \infty$ and $A = \mathbb{R}^3$ we obtain the following theorem proved by the author in [Ste4] (and a corresponding inclusion version in the case $A \neq \mathbb{R}^3$ is also valid):

3.4 Theorem (Steffen). *The Plateau problem* $\mathcal{P}(H, \Gamma)$ *is solvable whenever*

$$\int_{\mathbb{R}^3} |H_+|^3 \, d\mathcal{L}^3 < \frac{9\pi}{2} \quad and \quad \int_{\mathbb{R}^3} |H_-|^3 \, d\mathcal{L}^3 < \frac{9\pi}{2}. \qquad \square$$

It is interesting that no condition on the boundary curve Γ is required here. There are many other possibilities to derive an isoperimetric condition. For example we may apply Fubini's theorem to the integral $\int_E H \, d\mathcal{L}^3$ and use Schwarz' inequality and the isoperimetric inequality for parametric 2-dimensional domains to prove a cylinder version of the preceding Theorem where the conditions on H_+, H_- are replaced by

$$\sup_{t \in \mathbb{R}} \int_{\mathbb{R}^2} |H_+(r, s, t)|^2 \, dr \, ds < \pi \quad and \quad \sup_{t \in \mathbb{R}} \int_{\mathbb{R}^2} |H_-(r, s, t)|^2 \, dr \, ds < \pi.$$

The following assumption on the superlevel sets of H_+ and H_- also implies an isoperimetric condition of type $c < 1$, $s = \infty$ on $\mathbb{R}^3$:

$$\mathcal{L}^3\{a \in \mathbb{R}^3 : H_\pm(a) \geq t\} \leq \mathcal{L}^3\{a \in \mathbb{R}^3 : |a|^{-1} \geq t\} \quad \text{for } 0 < t < \infty.$$

The corresponding existence theorem for the Plateau problem should be compared with Theorem 2.8 in the case of a ball A. There, the hypothesis is $|H(a)| \leq |a|^{-1}$ for $a \in A \setminus \{0\}$, but here we must only know that the superlevel sets of the function $|H(a)|$ have no larger measure than those of the function $|a|^{-1}$. We refer to [Ste4] and [DS4] for this and for further examples where an isoperimetric condition can be verified and a corresponding existence theorem for the Plateau problem follows.

If Z is a bounded continuous vector field with $\operatorname{div} Z = H$ on a neighborhood of A, then we obtain an isoperimetric condition with $c = 2 \sup_A |Z|$ from the Gauss-Green theorem (applied to bounded smooth domains $E \subset A$, or using [Fe, 4.5.6]):

$$\left| \int_E 2H \, d\mathcal{L}^3 \right| \leq \int_{\partial E} 2|Z| \, d\mathcal{H}^2 \leq 2 \sup_A |Z| \, \mathbf{P}(E).$$

In particular, we have $c < 1$ if $\sup_A |Z| < \frac{1}{2}$, and we now recover all the existence results of Section 2 as special cases of Theorem 3.3. Moreover, for this isoperimetric condition we do not need $\operatorname{div} Z = H$ but only $\operatorname{div} Z \geq |H|$ in the distributional sense, which allows much more freedom in the construction of Z.

For example, if A is the closure of a proper C^2 domain in $\mathbb{R}^3$, $d_A \colon A \to [0, \infty[$ denotes the boundary distance and $2Z = -\nabla d_A$ its gradient, then one has $|Z| = \frac{1}{2}$ and $\operatorname{div} Z = H_A$ on $A \backslash C$ where H_A is the parallel mean curvature function of A introduced before Theorem 2.8 and $C \subset A$ is the $\mathcal{L}^3$ null set consisting of cut points and focal points of ∂A (i.e. points $a \in A$ with non-unique nearest point b in ∂A or with $|a - b|^{-1}$ equal to the maximum of the principal curvatures of ∂A at b). Moreover, it is not difficult to see that $\operatorname{div} Z \geq H_A$ holds in the sense of distributions on A (see [DS4]), and we deduce an isoperimetric condition of type c, ∞ for H on A whenever $|H| \leq c H_A$ holds on A. This immediately gives a version of Theorem 2.8.

To obtain a **proof of Theorem 2.8** as formulated in Section 2 one uses variants of the definition of Z above. For instance, assuming that the inner radius $r_A = \sup_A d_A$ is finite we define

$$2Z = -\frac{e^{\lambda(r_A - d_A)}}{\sqrt{1 + e^{2\lambda(r_A - d_A)}}} \nabla d_A \quad \text{with} \quad \lambda = 2 \sup_A |H|$$

and verify $\operatorname{div} Z \geq |H|$ when $|H| \leq H_A$ on A. An isoperimetric condition with $c = e^{\lambda r_A}(1 + e^{2\lambda r_A})^{-1/2} < 1$ follows for H. In fact we can allow $|H| \leq \sqrt{1 + \epsilon^2} H_A$ on A with $\epsilon > 0$ so small that $\lambda r_A < \sqrt{1 + \epsilon^2} \log |\epsilon|$, and we still obtain an isoperimetric condition with constant $c < 1$. Thus, we may admit in Theorem 2.8 values of $|H|$ (slightly) larger than H_A in the interior of A insisting, however, that $|H| \leq H_A$ along ∂A. To apply Theorem 3.3 we must also know that $\mathbb{R}^3 \backslash A$ has no components of finite measure. It follows from $H_{\partial A} \geq 0$ that none of these components is bounded. Unbounded components of finite measure can be excluded with suitable uniformity assumptions on A at infinity, e.g. the existence of global exterior parallel surfaces for ∂A. With these observations and Theorem 3.3 we have proved all the assertions made in Theorem 2.8 (and in fact stronger statements).

To conclude this Section we briefly indicate how to prove the **regularity of weak solutions** $x \in \mathcal{S}(\Gamma, A)$ to the Plateau problem $\mathcal{P}(\Gamma, H)$ which are obtained from Theorem 3.3. The main point is to show continuity of x on $\overline{U}$, because we then use a representation $H = \operatorname{div} Z$ with Z bounded on a contractible neighborhood W of the compact set $x(\overline{U})$ and we have, with a constant V_0 depending on Γ, Z only,

$$\mathbf{V}_H(\tilde{x}, y_\Gamma) = \int_U (Z \circ \tilde{x}) \cdot \tilde{x}_u \wedge \tilde{x}_v \, du dv - V_0 = \mathbf{V}_Z(\tilde{x}) - V_0$$

for $\tilde{x} \in \mathcal{S}(\Gamma, A)$ with image in W. (The constant is just $V_0 = \mathbf{V}_Z(y_\Gamma)$.) Moreover, for a given $w_0 \in \overline{U}$ we can make $|Z|$ as small as we like near $x(w_0)$. It follows that all the results on analytic and geometric regularity mentioned at the end of Section 2 can be applied to the present situation. (There is one exception related to the exclusion of false branch points. In this connection various authors have used the global condition $\sup |Z| < \frac{1}{2}$. An inspection of the proofs in [Al1], [Al2], [Gu2], [GOR] reveals, however, that an isoperimetric condition for H with constant $c < 1$ is actually sufficient, and this was also shown in [SW].) Now, to verify continuity of x we employ the energy minimizing property and repeat the arguments from the end of Section 2 which lead to the Dirichlet growth condition for x. The only property of the energy functional needed there was the inequality, with a constant $0 \le c < 1$,

$$\mathbf{D}_G(x) \le \frac{1+c}{1-c} \mathbf{D}_G(\tilde{x})$$

for admissible comparison surfaces $\tilde{x}$ with $x = \tilde{x}$ on $U \setminus G$. The inequality follows here from the isoperimetric condition which implies

$$\mathbf{D}_G(x) - \mathbf{D}_G(\tilde{x}) = \mathbf{E}_H(x) - \mathbf{E}_H(\tilde{x}) - 2\mathbf{V}_H(x, \tilde{x})$$
$$\le 2|\mathbf{V}_H(x, \tilde{x})| \le c(\mathbf{D}_G(x) + \mathbf{D}_G(\tilde{x})).$$

(Note that we have $c < 1$ if Theorem 3.3 is applicable to produce an energy minimizing solution of $\mathcal{P}(H, \Gamma)$. Also note that we need only consider comparison surfaces with $\mathbf{D}_G(x) + \mathbf{D}_G(\tilde{x})$ small so that the isoperimetric condition is applicable.) Therefore, if A is quasiregular we can use the device of Morrey described in Section 2 to prove Hölder continuity of x on U and continuity on $\overline{U}$.

4 Hypersurfaces with prescribed mean curvature in $\mathbb{R}^{n+1}$

The arguments from geometric measure theory used in Section 3 and the isoperimetric inequality work in all dimensions. It is therefore natural to study the Plateau problem with prescribed mean curvature in the setting of geometric measure theory with the aim to prove analogues of the existence theorems from the preceding sections for n-dimensional surfaces in $\mathbb{R}^{n+1}$ with prescribed mean curvature and boundary.

In order to give a geometric measure theory formulation of the Plateau problem for hypersurfaces in $\mathbb{R}^{n+1}$ with prescribed mean curvature we first have to say what objects we take as general **hypersurfaces**. These will be integer multiplicity rectifiable n-currents T as already explained, for $n = 2$, in Section 3. Namely, T is a continuous linear functional on $\mathcal{D}^n(\mathbb{R}^{n+1})$, the space of smooth and compactly supported n-forms, such that T has a representation

$$T(\beta) = \int_M \langle \beta, \vec{\eta} \rangle \, \vartheta \, d\mathcal{H}^n \quad \text{for } \beta \in \mathcal{D}^n(\mathbb{R}^{n+1}),$$

where $\mathcal{H}^n$ is the n-dimensional Hausdorff measure on $\mathbb{R}^{n+1}$, M is an n-rectifiable subset of $\mathbb{R}^{n+1}$, $\vec{\eta}$ is an $\mathcal{H}^n$ measurable orientation for M, and ϑ is an $\mathcal{H}^n$ summable multiplicity function on M with values in the positive integers. In particular, we require finite mass

$$\mathbf{M}(T) = \int_M \vartheta \, d\mathcal{H}^n < \infty,$$

but no compact support for T, deviating here somewhat from the standard references [Fe], [Si].

Next, we formulate the **Plateau boundary condition** for T as above. This reads $\partial T = \Gamma$ where Γ is a given closed (i.e. $\partial\Gamma = 0$) $(n-1)$-current on $\mathbb{R}^{n+1}$ and the current boundary is defined by $\partial T(\alpha) = T(d\alpha)$ for $\alpha \in \mathcal{D}^{n-1}(\mathbb{R}^{n+1})$. Of course, we must require that Γ is the boundary of some integer multiplicity rectifiable n-current, so we assume that such a current T_0 is given as a **reference hypersurface**. The Plateau boundary condition is thus

$$\partial T = \Gamma = \partial T_0 \quad \text{or} \quad \partial(T - T_0) = 0 .$$

To avoid trivialities we also assume that Γ is not the zero $(n-1)$-current.

If we work with a closed subset A of $\mathbb{R}^{n+1}$ as **inclusion domain** we require that the reference surface can be found in A, i.e. T_0 (and hence also Γ) has support in A. We also assume, for reasons that become clear below, that $\mathbb{R}^{n+1} \setminus A$ has no components of finite Lebesgue $\mathcal{L}^{n+1}$ measure. The class of **admissible surfaces** $\mathcal{T}(\Gamma, A)$ is then the set of integer multiplicity n-currents T on $\mathbb{R}^{n+1}$ with support in A and with $\partial T = \Gamma$. By our assumptions it is nonempty, $T_0 \in \mathcal{T}(\Gamma, A)$.

To proceed, we must define what it means that $T \in \mathcal{T}(\Gamma, A)$ has a given bounded and continuous function $H : \mathbb{R}^{n+1} \to \mathbb{R}$ as prescribed mean curvature. One way to characterize the mean curvature of a smooth oriented hypersurface $\mathcal{M}$ is to look at deformations $f_t(x)$, $f_0(x) = x$, of the ambient space which leave the boundary of $\mathcal{M}$ fixed. The first variation of the n-area is then, up to a factor $-n$, the surface integral of the inner product of the mean curvature vector field $\vec{h}(x)$ of $\mathcal{M}$ with the deformation vector field $g(x) = \frac{\partial}{\partial t}|_{t=0} f_t(x)$.

For non-smooth hypersurfaces like our currents T the mean curvature vector field may not exist, but one can always compute the **first variation of mass**

$$\frac{d}{dt}\Big|_{t=0} \mathbf{M}(f_{t\#} T) = \int_M (P_{\vec{\eta}} \cdot Dg) \, \vartheta \, d\mathcal{H}^n ,$$

where M, $\vec{\eta}$, ϑ are representing T as usual, $P_{\vec{\eta}}$ is the orthogonal projection of $\mathbb{R}^{n+1}$ onto the subspace $\{\xi \in \mathbb{R}^{n+1} : \xi \wedge \vec{\eta} = 0\}$ associated with $\vec{\eta}$, and $P_{\vec{\eta}}(x) \cdot Dg(x) = \text{trace}\,[P_{\vec{\eta}}(x)Dg(x)^*]$ is the inner product in $\text{Hom}(\mathbb{R}^{n+1}, \mathbb{R}^{n+1})$. For smooth oriented hypersurfaces $\mathcal{M}$ the integral above equals $-n \int_{\mathcal{M}} (\vec{h} \cdot g) \, d\mathcal{H}^n$, and we have $\vec{h} = h\nu$ where h is the mean curvature and ν the orienting unit normal vector field on $\mathcal{M}$. (Our convention here is that the mean curvature is the arithmetic mean of the principal curvatures, and it is positive for spheres oriented by their inward normal field.) Denoting by $\nu_{\vec{\eta}}(x)$ the unit vector in $\ker P_{\vec{\eta}}(x)$ with $\nu_{\vec{\eta}}(x) \wedge \vec{\eta}(x) = \vec{e}$, the orientation of $\mathbb{R}^{n+1}$, we therefore arrive at the following definition of **prescribed mean curvature** H for the integer multiplicity rectifiable n-current T represented by M, $\vec{\eta}$, ϑ:

$$\int_M (P_{\vec{\eta}} \cdot Dg) \, \vartheta \, d\mathcal{H}^n + n \int_M H (\nu_{\vec{\eta}} \cdot g) \, \vartheta \, d\mathcal{H}^n = 0$$

for every C^1 vector field g with compact support in $\mathbb{R}^{n+1} \setminus \text{spt}\, \partial T$.

This is, in the present situation, the **H-hypersurface equation** in a weak formulation introduced by W.K. Allard [All] (see also [Fe, 5.1.7], [Si, Sec.16]). For smooth oriented hypersurfaces $\mathcal{M}$ with mean curvature h it is equivalent to $h(a) = H(a)$ for all $a \in \mathcal{M}$.

The **Plateau problem** $\mathcal{P}(H, \Gamma)$ in A is to find a hypersurface with prescribed mean curvature H. A current $T \in \mathcal{T}(\Gamma, A)$ solving this problem in the sense above should be considered as a **weak solution to the Plateau problem**, and with appropriate smoothness assumptions on Γ and H one hopes to eventually prove that it corresponds to a classical solution, i.e. a smooth oriented codimension 1 submanifold $\mathcal{M}$ of $\mathbb{R}^{n+1}$ with oriented boundary representing Γ and with mean curvature $H(a)$ at every point $a \in \mathcal{M}$.

The H-hypersurface equation has a variational structure. We have already noted that the first integral appearing in the equation is the first variation of mass for a deformation $f_{t\#}T$ with initial vector field g. The second integral is the first variation of the H-weighted volume functional which we want to define next. Since $\partial(T - T_0) = 0$ and T, T_0 have finite mass there exists a unique $(n+1)$-current Q_{T,T_0} on $\mathbb{R}^{n+1}$ of finite mass with $\partial Q_{T,T_0} = T - T_0$. Like T and T_0, also Q_{T,T_0} has integer multiplicity and hence can be represented as a "set with integer multiplicities", i.e.

$$Q_{T,T_0}(\gamma) = \int_{\mathbb{R}^{n+1}} i_{T,T_0}\gamma = \int_{\mathbb{R}^{n+1}} \langle \gamma, \vec{e}\rangle \, i_{T,T_0} \, d\mathcal{L}^{n+1} \quad \text{for } \gamma \in \mathcal{D}^{n+1}(\mathbb{R}^{n+1})$$

with an integer valued $\mathcal{L}^{n+1}$ measurable function i_{T,T_0} satisfying $\int |i_{T,T_0}| \, d\mathcal{L}^{n+1} = \mathbf{M}(Q_{T,T_0})$ $< \infty$. If T and T_0 have support in A then also $\operatorname{spt} Q_{T,T_0} = \operatorname{spt} i_{T,T_0} \subset A$, because we have assumed that $\mathbb{R}^{n+1} \setminus A$ has no components of finite $\mathcal{L}^{n+1}$ measure and i_{T,T_0} must be constant on such components.

For $H : \mathbb{R}^{n+1} \supset A \to \mathbb{R}$, bounded and continuous by our general assumption, the H-**volume** enclosed by T and T_0 is then

$$\mathbf{V}_H(T, T_0) = \int_{\mathbb{R}^{n+1}} H i_{T,T_0} \, d\mathcal{L}^{n+1}.$$

We note that, as a consequence of uniqueness, the identities

$$Q_{T',T_0} = Q_{T',T} + Q_{T,T_0} \quad \text{and} \quad i_{T',T_0} = i_{T',T} + i_{T,T_0}$$

are valid for $T', T, T_0 \in \mathcal{T}(\Gamma, A)$, and hence

$$\mathbf{V}_H(T', T_0) = \mathbf{V}_H(T', T) + \mathbf{V}_H(T, T_0).$$

For smooth deformations f_t with compact support we have, by the homotopy formula,

$$\mathbf{V}_H(f_{t\#}T, T) = \int_0^t \int_M H \left\langle \Omega, \tfrac{\partial}{\partial s} f_s \wedge (\wedge_n D f_s)\vec{\eta} \right\rangle \vartheta \, d\mathcal{H}^n \, ds,$$

if T is represented by M, $\vec{\eta}$, ϑ and Ω is the volume form on $\mathbb{R}^{n+1}$. Differentiating with respect to t we find

$$\frac{d}{dt}\Big|_{t=0} \mathbf{V}_H(f_{t\#}T) = \int_M H \langle \Omega, g \wedge \vec{\eta}\rangle \, \vartheta \, d\mathcal{H}^n = \int_M H \left(\nu_{\vec{\eta}} \cdot g\right) \vartheta \, d\mathcal{H}^n,$$

which is the variational formula for the second integral in the H-hypersurface equation.

The **isoperimetric inequality** holds, by [Fe, 4.5.9 (31)] (or by [DeG], [MM, 2.2.2] when i_{T,T_0} has values in $\{0, 1\}$; but one can reduce everything to this special situation with the decomposition theorem [Fe, 4.5.17]), in the form

$$\mathbf{M}(Q_{T,T_0}) = \int_{\mathbb{R}^{n+1}} |i_{T,T_0}| \, d\mathcal{L}^{n+1} \leq \int_{\mathbb{R}^{n+1}} |i_{T,T_0}|^{1+1/n} d\mathcal{L}^{n+1} \leq \gamma_n \mathbf{M}(T - T_0)^{1+1/n},$$

the optimal constant being determined by the fact that equality holds when i_{T,T_0} is the characteristic function of a ball in $\mathbb{R}^{n+1}$ and $\mathbf{M}(T - T_0)$ the n-area of its boundary sphere. As in Definition 3.2 we will say that H satisfies an **isoperimetric condition** of type $0 \leq c < \infty$, $0 < s \leq \infty$ on A if

$$\left| n \int_E H \, d\mathcal{L}^{n+1} \right| \leq c \mathbf{P}(E)$$

holds for all sets $E \subset A$ with finite perimeter $\mathbf{P}(E) \leq s$. An immediate consequence is, by the decomposition theorem, the **isoperimetric condition for the H-volume**

$$|n\mathbf{V}_H(T, T_0)| \leq c\,\mathbf{M}\,(T - T_0),$$

whenever $T, T_0 \in \mathcal{T}(\Gamma, A)$ with $\mathbf{M}\,(T - T_0) \leq s$.

We can now repeat the scheme of the proof for Theorem 3.3 using the H-**energy**

$$\mathbf{E}_H(T) = \mathbf{M}\,(T) + n\mathbf{V}_H(T, T_0)$$

with a fixed reference current $T_0 \in \mathcal{T}(\Gamma, A)$ and attempting to minimize $\mathbf{E}_H$ on the set of hypersurfaces $T \in \mathcal{T}(\Gamma, A)$ with $\mathbf{M}\,(T) \leq \sigma\mathbf{M}\,(T_0)$. The result is the following theorem of F. Duzaar [Du2]:

4.1 Theorem (Duzaar). *Suppose A is closed in $\mathbb{R}^{n+1}$, $\mathbb{R}^{n+1} \setminus A$ has no component of finite measure, H is continuous and bounded and satisfies an isoperimetric condition of type c, s on A, Γ is a closed $(n-1)$-current on $\mathbb{R}^{n+1}$, and $T_0 \in \mathcal{T}(\Gamma, A)$ with $(1+\sigma)\mathbf{M}\,(T_0) \leq s$ for some $1 < \sigma \leq \infty$. Set $\mathcal{T}(\Gamma, A; \sigma) = \{T \in \mathcal{T}(\Gamma, A) : \mathbf{M}\,(T) \leq \sigma\mathbf{M}\,(T_0)\}$. Then the following assertions hold:*

(i) If $c < 1$ or if $c = 1$ and $\sigma < \infty$, then the H-energy attains a minimum on $\mathcal{T}(\Gamma, A; \sigma)$.

(ii) If $c \leq \dfrac{\sigma - 1}{\sigma + 1}$ or $\sigma = \infty$ then each minimizer T from (i) satisfies $\mathbf{M}\,(T) < \sigma\mathbf{M}\,(T_0)$.

(iii) If $A = \mathbb{R}^{n+1}$ or A is the closure of a C^2 domain and $|H| \leq H_{\partial A}$ holds along ∂A, where $H_{\partial A}$ denotes the (inward) boundary mean curvature of A, then each minimizer T from (i) with $\mathbf{M}\,(T) < \sigma\mathbf{M}\,(T_0)$ is a weak solution to the Plateau problem $\mathcal{P}(H, \Gamma)$. Moreover, if $|H(a)| < H_{\partial A}(a)$ holds at some point $a \in (\partial A) \setminus \text{spt}\,\Gamma$, then $a \notin \text{spt}\,T$.

Sketch of proof. Choosing a minimizing sequence $T_k \in \mathcal{T}(\Gamma, A; \sigma)$ we first deduce a bound for $\mathbf{M}\,(T_k)$ from the isoperimetric condition

$$\mathbf{E}_H(T_k) \geq \mathbf{M}\,(T_k) - n\,|\mathbf{V}_H(T_k, T_0)| \geq (1 - c)\mathbf{M}\,(T_k) - c\mathbf{M}\,(T_0),$$

hence we may assume $T_k \to T$ weakly (i.e. $T_k(\beta) \to T(\beta)$ for all $\beta \in \mathcal{D}^n(\mathbb{R}^{n+1})$), and from lower semicontinuity of mass and the closure theorem [Fe, 4.2.16] we infer $T \in \mathcal{T}(\Gamma, A; \sigma)$. The isoperimetric inequality implies a mass bound also for the Q_{T_k, T_0}, i.e. a bound for $\|i_{T_k, T_0}\|_{L^1}$. It follows then from the compactness theorem [Fe, 4.2.17] that, after passing to a subsequence, we have L^1_{loc} convergence $i_{T_k, T_0} \to i_{T, T_0}$. In fact, we need only the BV compactness theorem here [EG, 5.2], [Zi, 5.3] which is much easier than the closure and the compactness theorem for currents with positive codimension.

Now, if A is compact, then we have L^1 convergence $\|i_{T_k, T_0} - i_{T, T_0}\|_{L^1} \to 0$ as $k \to \infty$, and the H-volume is clearly continuous with respect to this convergence. By lower semicontinuity of mass we immediately deduce that T is an $\mathbf{E}_H$ minimizer in $\mathcal{T}(\Gamma, A; \sigma)$. This is in contrast with the situation in Section 3, where the H-volume $\mathbf{V}_H$ can be discontinuous for weak $W^{1,2}$ convergence $x_k \to x$ of parametric surfaces also if all the surfaces have image in a fixed compact subset of $\mathbb{R}^3$, i.e. "bubbles" can split off in bounded regions of space.

In the present context such a "bubbling phenomenon" can only occur at infinity, i.e. we might have

$$\liminf_{R \to \infty} \limsup_{k \to \infty} \left| \int_{\mathbb{R}^{n+1} \setminus B^{n+1}(0,R)} H i_{T_k,T_0} \, d\mathcal{L}^{n+1} \right| > 0.$$

($B^{n+1}(0, R)$ denotes the ball of radius R centered at the origin in $\mathbb{R}^{n+1}$.) However, we can dominate this amount of H-volume disappearing at infinity by the corresponding amount of mass disappearing at infinity, i.e. by

$$\liminf_{R \to \infty} \limsup_{k \to \infty} M\left(T_k \llcorner \left(\mathbb{R}^{n+1} \setminus B^{n+1}(0, R) \right) \right).$$

This can be proved by using standard slicing techniques of geometric measure theory to construct from the T_k n-currents supported in $\mathbb{R}^{n+1} \setminus B^{n+1}(0, R)$, which may be thought of as "bubbles" near infinity, and by then applying the isoperimetric condition with constant $c \leq 1$ to these currents. For details we refer to [Du2] and [DS4]. In any case, it is clear that, having carried out this technical point, we can again conclude lower semicontinuity of the energy $\mathbf{E}_H$ on the sequence $T_k \to T$, and (i) follows.

For (ii) the proof of corresponding statement in Theorem 3.3 can be repeated almost verbatim. Finally, (iii) is based on an inclusion principle analogous to Proposition 2.6 for solutions to the variational inequality which one obtains instead of the H-hypersurface equation if one minimizes $\mathbf{E}_H$ on $\mathcal{T}(\Gamma, A)$ with $A \neq \mathbb{R}^{n+1}$. Note that the extra constraint in the definition of $\mathcal{T}(\Gamma, A; \sigma)$ is not effective on account of the strict inequality $M(T) < \sigma M(T_0)$ derived in (ii). Since we have already omitted the details of the proof for Proposition 2.6 it would not make sense, for the purpose of these lectures, to now discuss the modifications necessary in the present context. We therefore refer to [Du2] and [DS4] for a complete proof. □

The currents T with prescribed mean curvature H produced in the preceding theorem have a variety of special properties, see [Du2], [DS4]. For example, they are indecomposable in a certain sense, and they have compact support if their boundary Γ is supported in a compact set. For applications of the theorem we note that all the arguments used in Section 3 to prove an isoperimetric condition for H are valid also in $\mathbb{R}^{n+1}$, hence we can state

4.2 Corollary. *All the existence theorems of Sections 2 and 3 for parametric H-surfaces in $A \subset \mathbb{R}^3$ with given boundary curve Γ have analogues valid for integer multiplicity rectifiable n-currents in $A \subset \mathbb{R}^{n+1}$ with prescribed mean curvature and with a given closed $(n-1)$-current as boundary.* □

For example, Hildebrandt's existence result, Theorem 2.4, in balls B of radius R is valid in $\mathbb{R}^{n+1}$ with the conditions

$$\sup_B |H| < \tfrac{n+1}{n} R^{-1} \quad \text{and} \quad \sup_{\partial B} |H| \leq R^{-1},$$

and the "$\tfrac{9\pi}{2}$ theorem", Theorem 3.4, holds in $\mathbb{R}^{n+1}$ with the assumptions

$$\int_{\mathbb{R}^{n+1}} |H_+|^{n+1} \, d\mathcal{L}^{n+1} < \left(\tfrac{n+1}{n} \right)^{n+1} \alpha(n+1) \quad \text{and} \quad \int_{\mathbb{R}^{n+1}} |H_-|^{n+1} \, d\mathcal{L}^{n+1} < \left(\tfrac{n+1}{n} \right)^{n+1} \alpha(n+1),$$

where $\alpha(n+1)$ is the $\mathcal{L}^{n+1}$ measure of the unit ball in $\mathbb{R}^{n+1}$. Further examples, e.g. the variants in $\mathbb{R}^{n+1}$ of the Wente theorem, Theorem 3.1, and the Gulliver & Spruck theorem, Theorem 2.8, have been worked out in [Du2] and [DS4].

We conclude this section with some remarks on the regularity theory that has been developed to prove that the weak H-surfaces T produced in Theorem 4.1 are in fact smooth hypersurfaces with prescribed mean curvature H. The **interior regularity** theory was started by De Giorgi (for $H \equiv 0$) and Massari (for general H) in the case of $\mathbf{E}_H$ minimizing currents which are representable as boundaries of sets with finite perimeter (the "frontiere orientate" of De Giorgi). This theory gives smoothness away from the support of the boundary with the exception of a possible singular set with small Hausdorff dimension that was estimated by Federer. All this and also the optimality of Federer's dimension estimate for the singular set is presented in the book of Massari & Miranda [MM]. For $\mathbf{E}_H$ minimizing integer multiplicity rectifiable n-currents in $\mathbb{R}^{n+1}$ one can use the decomposition theorem [Fe, 4.5.17] to reduce the regularity question to the case of $\mathbf{E}_H$ minimizing boundaries of sets with finite perimeter, provided H is locally Lipschitz continuous. This has been done by F. Duzaar [Du2], and the result is that spt T near points not contained in spt ∂T or in the small singular set is locally a smooth n-submanifold of $\mathbb{R}^{n+1}$. Without the Lipschitz condition on H the situation may be more complicated, because different smooth "leaves" of spt T could touch each other along sets with a complicated structure; see [DS2] for an interior regularity theory covering this general case.

Furthermore, we have **complete boundary regularity** of T if the boundary current Γ is defined by a smooth oriented closed submanifold of codimension 2. This was proved in the minimal surface case $H \equiv 0$ by Hardt & Simon [HS] and extended to general H in [DS3] (see also [Du4]). The energy minimizing property of T is again crucial here. The result of the complete regularity theory is then the following

4.3 Theorem (Regularity). *Suppose T is a $\mathbf{E}_H$ minimizing weak solution to the Plateau problem $\mathcal{P}(H, \Gamma)$ as in Theorem 4.1 (iii) with H locally Lipschitz, and Γ is represented by a smooth oriented $(n-1)$-submanifold C in $\mathbb{R}^{n+1}$ with multiplicity one. Then there exists a closed subset $\operatorname{sing} T$ of $\mathbb{R}^{n+1} \setminus C$ with Hausdorff dimension at most $n-7$ and discrete in the case $n=7$ (empty in the case $n<7$), such that T locally at each point $a \in \operatorname{spt} T \setminus (C \cup \operatorname{sing} T)$ is represented by a smooth oriented n-dimensional submanifold with constant integer multiplicity. Moreover, locally at each boundary point $a \in C$ either T is represented by a smooth oriented n-submanifold with boundary C and multiplicity 1, or spt T is a smooth n-submanifold without boundary locally at a and T is represented by orienting suitably and assigning integer multiplicities $m \geq 1$ and $m+1$ to the two components in which the boundary C divides spt T near a.* $\square$

Here "smooth" means class $C^{1,\alpha}$ with some $0 < \alpha < 1$ if H is merely locally Lipschitz, and $C^{k,\alpha}$ with $k \in \{2, 3, \ldots, \infty, \omega\}$ if H is of class $C^{k-2,\alpha}$ (and locally Lipschitz when $k=2$). In the case $k \geq 2$ the mean curvature of spt T is the prescribed value $H(a)$ at every point $a \in \operatorname{spt} T \setminus \operatorname{sing} T$.

We note that the case of positive multiplicities m and $m+1$ near a boundary point can really occur, i.e. the boundary of the H-hypersurface can pass through an "interior leave" of the same surface. This can be seen already in the case $H \equiv 0$ from the example where Γ is represented by two concentric circles in a plane in $\mathbb{R}^3$ with the same orientation and with multiplicity 1. The mass minimizing 2-current T for this boundary configuration is the sum of two discs with equal orientation and with multiplicity 1, and T has multiplicities 1 and 2 near each point of the inner boundary circle. However, if spt T is connected then one can conclude from the indecomposability of the $\mathbb{E}_H$ minimizing hypersurface T that it is represented by an oriented n-submanifold with boundary $\mathcal{C}$ and multiplicity 1 locally at each boundary point. Moreover, discarding components of spt $T \setminus$ spt ∂T with even multiplicity one always obtains an oriented n-submanifold of $\mathbb{R}^{n+1} \setminus \text{sing}\, T$ with prescribed mean curvature H and with boundary $\mathcal{C} = $ spt ∂T, but the boundary orientation of T is possibly not compatible with the orientation prescribed on $\mathcal{C}$ by Γ (i.e. T does not solve the Plateau problem for Γ, but for a current obtained from Γ by reversing the orientation on some components of $\mathcal{C}$).

For $n \leq 6$ the solutions to the Plateau problem $\mathcal{P}(H, \Gamma)$ obtained here are completely free of singularities, so that one obtains smooth embedded codimension 1 submanifolds in $\mathbb{R}^{n+1}$ with prescribed mean curvature and with given smooth boundary. This is interesting even for $n = 2$, because we do not know complete geometric boundary regularity for energy minimizing parametric surfaces of prescribed mean curvature with Plateau boundary conditions due to the unsolved problem of boundary branch points (see the discussion at the end of Section 2). However, in contrast with the parametric theory where we have fixed the topological type of the admitted surfaces in advance, the solutions to the Plateau problem coming from geometric measure theory have a priori undetermined topological type. They cannot be discs if the boundary curve is knotted, for instance, but there are also examples of unknotted curves in $\mathbb{R}^3$ which cannot bound embedded discs of prescribed mean curvature $H \equiv 0$, cf. [AT].

5   Isoperimetric inequalities in Riemannian manifolds

In the final Section 6 of these lectures we discuss the Plateau problem with prescribed mean curvature in a Riemannian manifold. Since the method of isoperimetric inequalities, employed in Section 3 for 2-dimensional parametric surfaces in $\mathbb{R}^3$ and in Section 4 for n-dimensional integer multiplicity rectifiable n-currents in $\mathbb{R}^{n+1}$, is of geometric nature, it will work as well in an ambient Riemannian manifold, provided we can prove isoperimetric conditions for the prescribed mean curvature under reasonable assumptions. Such conditions will depend in turn on isoperimetric inequalities in Riemannian manifolds, and we therefore review here some simple facts from the corresponding theory. We refer to [DS4, Sec.2] for a more complete treatment of the material that is needed here and to [BZ] for general information.

We assume that N is a smooth, connected, oriented and complete Riemannian manifold of dimension $n+1$, and we denote by μ the Riemannian measure on N. It is no essential restriction, by the embedding theorems of Nash (for N compact) and Gromov & Rohlin (for N complete), to assume that N is isometrically embedded as a closed subset of some Euclidean space $\mathbb{R}^{n+1+p}$, and then $\mu = \mathcal{H}^{n+1} \llcorner N$ is just the $(n+1)$-dimensional Hausdorff measure on N. We also consider a nonempty closed subset A of N (in which we will try to find our hypersurfaces of prescribed mean curvature later).

The isoperimetric inequalities we will discuss are of two types: A **linear isoperimetric inequality** is one of the form
$$\mu(E) \leq c\,\mathbf{P}(E),$$
while by a **nonlinear isoperimetric inequality** we mean
$$\mu(E) \leq \gamma\,\mathbf{P}(E)^{1+1/n}.$$

Here E denotes a set of finite perimeter $\mathbf{P}(E)$ in A, i.e. a μ measurable subset with $\mu(E) < \infty$ which has finite boundary area $\mathbf{P}(E)$ in the distributional sense (the distributional gradient field of the characteristic function χ_E of E is a vector measure of finite total variation $\mathbf{P}(E)$ on N). We usually require that these inequalities hold for a certain class of such sets E, and the smallest possible constants c or γ will then be referred to as **isoperimetric constants**. Smoothing χ_E with a standard procedure one can see that it suffices, under appropriate conditions on A (e.g. a smooth uniform neighborhood retract in N), to verify such isoperimetric inequalities for smooth subsets E of A where $\mathbf{P}(E) = \mathcal{H}^n(\partial E)$ is the classical n-area of the boundary.

We also introduce the **isoperimetric functions** $c_A(s)$ and $\gamma_A(s)$ of A, defined for $0 < s \leq \infty$,
$$c_A(s) = \inf\{c \geq 0 : \mu(E) \leq c\mathbf{P}(E) \text{ for all } E \subset A \text{ with } \mathbf{P}(E) \leq s\},$$
$$\gamma_A(s) = \inf\{\gamma \geq 0 : \mu(E) \leq \gamma\mathbf{P}(E)^{1+1/n} \text{ for all } E \subset A \text{ with } \mathbf{P}(E) \leq s\},$$

where in a formula involving the expression $\mathbf{P}(E)$ we always understand that E is a set of finite perimeter. We then have, for all sets $E \subset A$,
$$\mu(E) \leq c_A(\mathbf{P}(E))\mathbf{P}(E),$$
$$\mu(E) \leq \gamma_A(\mathbf{P}(E))\mathbf{P}(E)^{1+1/n},$$

and $c_A(s)$, $\gamma_A(s)$ are the smallest nondecreasing functions with this property. Related to the isoperimetric functions is the **isoperimetric profile** $b_A(t)$, defined for $0 \leq t < \infty$ by
$$b_A(t) = \inf\{\mathbf{P}(E) : E \subset A \text{ with } \mu(E) \geq t\}.$$

Then
$$b_A(\mu(E)) \leq \mathbf{P}(E)$$

holds for all sets $E \subset A$ with finite perimeter, and $b_A(t)$ is the largest nonincreasing function with this property. We emphasize that the value ∞ is allowed for $c_A(s)$, $\gamma_A(s)$, $b_A(t)$.

5.1 Example. Let N_κ be the simply connected $(n+1)$-manifold of constant sectional curvature κ^2, i.e. N_κ is the Euclidean $(n+1)$-sphere with radius κ^{-1} if $\kappa > 0$, $N_\kappa = \mathbb{R}^{n+1}$ if $\kappa = 0$, and N_κ is the hyperbolic $(n+1)$-space with its standard distance function scaled by a factor $|\kappa|^{-1}$ if $\kappa \neq 0$ is purely imaginary. Denote by $\alpha_\kappa(r)$ the volume of a ball of radius r in N_κ (with $|\kappa|r < \pi$ if $\kappa^2 > 0$) and by $\beta_\kappa(r)$ the n-area of its boundary sphere. Then we have
$$\alpha_\kappa(r) = \omega(n) \int_0^r \left(\frac{\sin \kappa\varrho}{\kappa}\right)^n d\varrho$$
$$\beta_\kappa(r) = \alpha_\kappa'(r) = \omega(n) \left(\frac{\sin \kappa r}{\kappa}\right)^n,$$

where $\omega(n) = \beta_0(1) = (n+1)\alpha_0(1)$ is the area of the standard n-sphere and, of course, $\kappa^{-1}\sin\kappa r = |\kappa|^{-1}\sinh|\kappa|r$ for $\kappa^2 < 0$ and $\kappa^{-1}\sin\kappa r = r$ for $\kappa = 0$. From the isoperimetric property of balls in N_κ (see [Sch], [DeG], [BZ, 10.2]) we have the following **optimal isoperimetric inequalities** in N_κ, where $c_\kappa(r)$ denotes the quotient $\alpha_\kappa(r)/\beta_\kappa(r)$:

$$\mu(E) \leq c_\kappa(r)\mathbf{P}(E) \qquad\qquad \text{if } \mu(E) \leq \alpha_\kappa(r),$$
$$\mu(E) \leq \tfrac{1}{n|\kappa|}\mathbf{P}(E) \qquad\qquad \text{if } \kappa^2 < 0,$$
$$\mu(E) \leq c_\kappa(r)\beta_\kappa(r)^{-1/n}\mathbf{P}(E)^{1+1/n} \quad \text{if } \kappa^2 > 0 \text{ and } \mu(E) \leq \alpha_\kappa(r),$$
$$\mu(E) \leq \gamma_{n+1}\mathbf{P}(E)^{1+1/n} \qquad\quad \text{if } \kappa^2 \leq 0.$$

Here $\gamma_{n+1} = \alpha_0(1)\beta_0(1)^{-1-1/n}$ is the optimal isoperimetric constant of $\mathbb{R}^{n+1}$. Note that in the last inequality we cannot have a better constant even if $\kappa^2 < 0$, because for balls in N_κ of radius r we have asymptotic equality as $r \to 0$. For $A = N_\kappa$ or A a ball in N_κ we can compute explicitly the isoperimetric functions $c_A(s)$, $\gamma_A(s)$ and the isoperimetric profile $b_A(t)$ using these optimal isoperimetric inequalities (see [DS4, 2.5]). $\qquad\qquad\square$

We now turn to linear isoperimetric inequalities in the general case. Suppose we have a vector field Z on N with (Riemannian) norm $|Z| \leq 1$ everywhere and (Riemannian) divergence $\operatorname{div} Z \geq c^{-1}$ on A with some constant $c > 0$. Then we obtain from the Gauss-Green theorem for sets E with finite perimeter in A (we need only consider smooth domains with compact closure):

$$\mu(E) \leq c\int_E \operatorname{div} Z\,d\mu = c\int_{\partial E} Z\cdot\nu_E\,d\mathcal{H}^n \leq c\,\mathbf{P}(E),$$

where ν_E is the exterior unit normal of E on the boundary ∂E (well defined $\mathcal{H}^n$ almost everywhere for sets with finite perimeter, cf. [Fe, 4.5.6]). By approximation, it suffices to have Z continuous with $\operatorname{div} Z \geq c^{-1}$ in the sense of distributions on a neighborhood of A.

If A is geodesically star-shaped with respect to some point $a \in N$, i.e. A is contained in the diffeomorphic image under the exponential map $\exp_a$ of a subset in the tangent space T_aN which is star-shaped with respect to the origin, then a natural choice for Z is the **radial unit vector field**

$$Z(\exp_a\xi) = (D_\xi\exp_a)\frac{\xi}{|\xi|} \quad \text{for } 0 \neq \xi \in T_aN \text{ with } \exp_a([0,1]\xi) \subset A.$$

Here $\operatorname{div} Z(x) = nh(x)$ is just n times the (inward) mean curvature $h(x)$ of the sphere S in N with center a and radius $\varrho(x)$, the length of the unique geodesic joining a and x in A. Indeed, Z is the exterior unit normal field on S and the radial covariant derivative of Z vanishes, hence the trace $\operatorname{div} Z(x)$ of the covariant differential of Z at x on T_xN equals the trace of its restriction to T_xS, and the latter trace is just $nh(x)$, by definition of the mean curvature. Near the origin a one easily verifies $\operatorname{div} Z \geq nh$ in the sense of distributions. Now, from the description of first variation of n-area in terms of the mean curvature vector field we infer that $nh(x)$ is the quotient of the area change at x by the volume change under geodesic homotheties of spheres with center a. Therefore, for $x = \exp_a\xi$ the mean curvature $h(x)$ can be expressed in terms of the Jacobians $J_\zeta\exp_a$ of the exponential map at points $\zeta \in T_aN$,

$$nh(x) = (J_\xi\exp_a)^{-1}\left.\frac{d}{dt}\right|_{t=0}J_{e^t\xi}\exp_a.$$

Note that these Jacobians are equal to the corresponding tangential Jacobians for spheres in $T_a N$ centered at the origin, because $D_\xi \exp_a$ preserves length on rays and orthogonality to these rays by the Gauss Lemma.

If the sectional curvature of N satisfies $\mathrm{Sec}_N \leq \kappa^2$ on A we can therefore apply the comparison theorem for Jacobians [Gü], [BZ, § 33] to infer, in the situation above,

$$h(x) \geq h_\kappa(\varrho(x)) = \kappa \cot \kappa \varrho(x),$$

where $h_\kappa(r) = \kappa \cot \kappa r$ is the constant mean curvature of a sphere of radius r in N_κ and $\varrho(x)$ is the length of the geodesic from a to x in the star-shaped region A, as before. In geometric terms the comparison theorem just says that the mean curvature of a sphere in N with $\mathrm{Sec}_N \leq \kappa^2$ is not smaller, at each point of the sphere, than the mean curvature of a sphere with the same radius in the manifold N_κ of constant sectional curvature κ^2. Denoting by $R = \sup_{x \in A} \varrho(x)$ the **maximal radius** of A we have proved:

5.2 Theorem. *Suppose A is geodesically star-shaped in N as above and A has finite maximal radius R. Then, if $\mathrm{Sec}_N \leq \kappa^2$ on A, the linear isoperimetric inequality*

$$\mu(E) \leq \tfrac{1}{n\kappa}(\tan \kappa R)\, \mathbf{P}(E)$$

holds for all sets $E \subset A$ with finite perimeter. □

From the Hadamard-Cartan theorem we deduce, in particular, the following Corollary which was apparently first stated by S.T. Yau [Ya]:

5.3 Corollary (Yau). *If N is simply connected with $\mathrm{Sec}_N \leq \kappa^2 < 0$, then*

$$\mu(E) \leq \tfrac{1}{n|\kappa|}\, \mathbf{P}(E)$$

holds for all sets $E \subset A$ with finite perimeter. □

5.4 Remark. The inequality in Theorem 5.2 is not the optimal one, because in the case of constant sectional curvature we do not have equality for balls. Using the vector field $\tilde{Z}(x) = c_\kappa(\varrho(x))Z(x)$ instead of $Z(x)$, where $c_\kappa(r) = \alpha_\kappa(r)/\beta_\kappa(r) < \frac{1}{n\kappa} \tan \kappa r$ is from Example 5.1, we find the **optimal isoperimetric inequality**

$$\mu(E) \leq c_\kappa(R)\, \mathbf{P}(E)$$

for sets with finite perimeter in the geodesically star-shaped set A with maximal radius $R < \infty$ and with $\mathrm{Sec}_N \leq \kappa^2$ on A. For $|\kappa|R$ small, the constant $c_\kappa(R)$ is by approximately a factor $\frac{n}{n+1}$ better than the constant in Theorem 5.2. □

If A is the closure of a C^2 domain in N with positive inward mean boundary curvature, then a reasonable choice of the vector field Z is the negative gradient of the distance to ∂A on A. For this vector field $\frac{1}{n}\mathrm{div}\, Z(x) = H_A(x)$ is, almost everywhere on A, the parallel mean curvature function as in Section 3, and $\mathrm{div}\, Z \geq nH$ holds in the sense of distributions on A for every locally integrable function H with $H \leq H_A$ almost everywhere (cf. [DS4, 2.10]). In this way we obtain:

5.5 Theorem. *If A is the closure of a C^2 domain in N with parallel mean curvature function $H_A \geq c^{-1}$ for some $c > 0$, then the linear isoperimetric inequality*

$$\mu(E) \leq c\, \mathbf{P}(E)$$

holds for all sets $E \subset A$ with finite perimeter. □

With regard to the behaviour of isoperimetric functions the following observation for product manifolds $N = M \times \mathbb{R}$ is of interest. Clearly $c_N(s) < \infty$ can hold only for values of s smaller than twice the n-area $\mathcal{H}^n(M)$ of M, because $E_t = M \times [0, t]$ has volume $\mu(E_t) = t\mathcal{H}^n(M) \to \infty$ as $t \to \infty$, while $\mathbf{P}(E_t) = 2\mathcal{H}^n(M)$ for all $t > 0$. Using slicing arguments one can show that the isoperimetric constant is indeed bounded for sets E with $\mathbf{P}(E) < 2\mathcal{H}^n(M)$ (see [DS4, 2.11]):

5.6 Proposition. *Suppose $N = M \times \mathbb{R}$ where M is a compact n-manifold without boundary. Then there exists $0 < c(M) < \infty$ such that*

$$c_N(s) \begin{cases} = \infty & \text{for } s \geq 2\mathcal{H}^n(M), \\ \leq c(M) & \text{for } 0 < s < 2\mathcal{H}^n(M). \end{cases}$$

□

We now turn to nonlinear isoperimetric inequalities in the Riemannian manifold N. Here we note that such an inequality always holds for sets $E \subset N$ with sufficiently small perimeter and with volume $\mu(E) \leq \frac{1}{2}\mu(N)$ (if $\mu(N) < \infty$), provided N satisfies some uniformity condition. A suitable condition is that N is **homogeneously regular** in the sense of Morrey, i.e. there exists $\Lambda \in [1, \infty[$ such that each point in N has a neighborhood which can be mapped to the unit ball in $\mathbb{R}^{n+1}$ by a biLipschitz mapping Φ with biLipschitz constant $\max\{\mathrm{Lip}\,\Phi, \mathrm{Lip}\,\Phi^{-1}\} \leq \Lambda$. (This is similar to the condition of quasiregularity explained at the end of Section 2.) Using the homogeneous regularity of N and relative isoperimetric inequalities on balls in $\mathbb{R}^{n+1}$ (cf. [Fe, 4.4.2 (2)], [Fe, 4.5.2 (1)]) one proves the following result and its corollary (see [DS4, 2.2 and 2.3]):

5.7 Proposition. *If N is homogeneously regular, then there exist $\delta > 0$ and $0 < \gamma < \infty$ such that the nonlinear isoperimetric inequality*

$$\mu(E) \leq \gamma\, \mathbf{P}(E)^{1+1/n}$$

holds for all sets $E \subset N$ with perimeter $\mathbf{P}(E) \leq \delta$ and with measure $\mu(E) \leq \frac{1}{2}\mu(N)$ (if $\mu(N) < \infty$). □

5.8 Corollary. *For A compact and $A \neq N$ or $A = N$ noncompact and homogeneously regular there exist $\delta > 0$ and $0 < \gamma < \infty$ such that the isoperimetric function of A satisfies*

$$c_A(s) \leq \gamma s^{1/n} < \infty \quad \text{for } 0 < s \leq \delta.$$

□

5.9 Remarks. Clearly γ cannot be smaller than the optimal isoperimetric constant γ_{n+1} from the Euclidean space $\mathbb{R}^{n+1}$. (Consider balls in N with radius $r \to 0$.) For simply connected $(n+1)$-dimensional manifolds N of nonpositive sectional curvature it is conjectured that the nonlinear isoperimetric inequality

$$\mu(E) \leq \gamma_{n+1} \mathbf{P}(E)^{1+1/n}$$

is valid for all sets E with finite perimeter in N. In the case $n = 1$ the conjecture is true by a classical result of A. Weil and of Beckenbach & Rado, in the case $n = 3$ it was proved by C.B. Croke who also obtained for all dimensions $n \geq 4$ the best constant so far (which is, however, larger than γ_{n+1} and presumably not the optimal one), and in the case $n = 2$ the conjecture has recently been verified by B. Kleiner (see [Cr], [Kl]). Of course, the conjecture is also true in the spaces N_κ of constant sectional curvature $\kappa^2 \leq 0$. For $n = 2$ Kleiner proved in fact that the isoperimetric profile of N, assumed simply connected with $\mathrm{Sec}_N \leq \kappa^2$, is not smaller than that of N_κ. To derive nonlinear isoperimetric inequalities in general by comparing N with spaces of constant curvature one needs $\mathrm{Sec}_N \leq \kappa^2$ for the comparison of boundary areas, but also a lower bound $\mathrm{Ric}_N \geq \lambda^2$ on the Ricci curvature for the comparison of volumes. This does not give the optimal constant. We refer to [DS4, 2.4] where further references are given. $\quad\square$

We will need the isoperimetric inequalities not only for sets E with finite perimeter in $A \subset N$ but also for integer multiplicity rectifiable $(n+1)$-currents Q on N with support in A. Such currents are represented by integer valued functions $i_Q \in L^1(N, \mu; \mathbb{R})$ of bounded variation, i.e. the distributional gradient of i_Q has finite total variation $\mathbf{M}(\partial Q)$. In fact, by the decomposition theorem [Fe, 4.5.17] (which is valid also in Riemannian manifolds) one can write i_Q as an L^1 convergent combination of characteristic functions of sets E_k, $k \in \mathbb{Z}$, with coefficients in $\{1, -1\}$, such that the perimeters $\mathbf{P}(E_k)$ add up to the boundary mass $\mathbf{M}(\partial Q)$. It is clear from this description that a linear isoperimetric inequality for sets with finite perimeter in A immediately implies a corresponding inequality for integer multiplicity rectifiable currents Q in the top dimension, i.e.

$$\mathbf{M}(Q) \leq c_A(s)\mathbf{M}(\partial Q)$$

holds if $\mathrm{spt}\, Q \subset A$ and $\mathbf{M}(\partial Q) \leq s$.

The nonlinear isoperimetric inequalities can also be extended from sets of finite perimeter to integer multiplicity rectifiable $(n+1)$-currents Q in N by applying them to level sets of the multiplicity function i_Q and using the Fleming & Rishel coarea formula ([Fe, 4.5.9 (13)], [EG, 5.5], [Zi, 5.4.4]). In this way one obtains

$$\mathbf{M}(Q) \leq \gamma_A(s)\mathbf{M}(\partial Q)^{1+1/n}$$

for integer multiplicity rectifiable $(n+1)$-currents Q on N with support in A and with boundary mass $\mathbf{M}(\partial Q) \leq s$.

The method described here to obtain inequalities for $\mathbf{M}(Q)$ from isoperimetric inequalities for sets with finite perimeter is a special case of a general procedure to derive a Sobolev type inequality from an isoperimetric inequality for sets. In fact, if

$$b(\mu(E)) \leq \mathbf{P}(E)$$

holds for all sets $E \subset A$ with finite perimeter $\mathbf{P}(E) \le s$, then for functions $f \in L^1(N, \mu; \mathbb{R})$ with bounded variation and with support in A an inequality

$$\|f\|_a \le \mathrm{Var}(\nabla f)$$

follows where $\mathrm{Var}(\nabla f) \le s$ is the (finite) total variation of the distributional gradient of f and $\| \cdot \|_a$ is the Orlicz norm associated with a function $a(s)$ that satisfies $ta^{-1}(1/t) \le b(t)$ for $t > 0$. The isoperimetric inequalities for Q above result by choosing $f = i_Q$ and the functions $b(t)$, $a(s)$ suitably. We refer to the discussion in [DS4, Sec.2] for details.

Finally we note that proposition 5.7 can be complemented by the statement that each closed integer multiplicity rectifiable n-current T on N (a concept that is readily extended from the case $N = \mathbb{R}^{n+1}$ treated in Section 4 to general Riemannian manifolds N) is in fact the boundary $T = \partial Q$ of an integer multiplicity rectifiable $(n+1)$-current Q on N of small mass, provided $\mathbf{M}(T)$ is sufficiently small ([DS4, 2.2]):

5.10 Proposition. *If N is homogeneously regular then there exist $\delta > 0$ and $0 < \gamma < \infty$ such that for each integer multiplicity rectifiable n-current T on N with $\mathbf{M}(T) \le \delta$ there exists a unique integer multiplicity rectifiable $(n+1)$-current Q with $\partial Q = T$ and with the estimate*

$$\mathbf{M}(Q) \le \gamma \mathbf{M}(T)^{1+1/n}. \qquad \Box$$

6 The Plateau problem with prescribed mean curvature in a Riemannian manifold

In this final section we want to describe how one can generalize the results of Sections 3 and 4 to a Riemannian manifold as ambient space. The Plateau problem with prescribed mean curvature for 2-dimensional parametric surfaces in a Riemannian 3-manifold was already treated by R. Gulliver [Gu1] and by S. Hildebrandt & H. Kaul [HK]. Their results are, however, restricted to boundary curves contained in·a geodesic coordinate domain on N, i.e. one works in $\mathbb{R}^3$ with a Riemannian metric instead of the Euclidean one. Here we do not want to make such a restriction as we wish to obtain, for example, a result of the type of the Wente theorem, Theorem 3.1, in a Riemannian manifold, just assuming that the given boundary curve Γ has small spanning area in N. All the results discussed in this section were obtained in recent joint work with F. Duzaar. In [DS4] we have treated the Plateau problem for hypersurfaces of prescribed mean curvature in a Riemannian manifold using geometric measure theory, and in [DS6] we prove various new existence results related to the Plateau problem for 2-dimensional parametric surfaces in a Riemannian 3-manifold which are described below (in a preliminary form).

While the general idea which we follow to solve the Plateau problem with prescribed mean curvature in a manifold N is the same as in Sections 3 and 4, there are also some fundamental differences which can be seen from the following examples. These examples are special cases of R. Gulliver's non-existence results in [Gu3] which in turn have generalized [He2].

6.1 Examples. (i) The great sphere $\Gamma = S^{n-1} \times \{(0,0)\}$ in $S^{n+1} \subset \mathbb{R}^{n+2}$ does not bound an oriented hypersurface of constant mean curvature $H \neq 0$. (The great half-sphere $S^n_+ \times \{0\}$ spanning Γ has mean curvature zero in S^{n+1}.)

(ii) If M is a compact oriented n-manifold in $\mathbb{R}^{n+1}$ without boundary, then no oriented hypersurface exists in the cylinder $N = M \times \mathbb{R}$ which is homologous to $M \times \{0\}$ and has constant mean curvature $H \neq 0$. $\qquad\qquad\square$

Note that for boundaries in $\mathbb{R}^{n+1}$ we can always solve the Plateau problem for suffi-ciently small prescribed mean curvature, by Theorem 2.4 and Corollary 4.2. In Example (i) above we have, however, a boundary in a Riemannian manifold N which does not span a hypersurface of arbitrarily small constant mean curvature $H \neq 0$ in N. Thus, when treating the Plateau problem in a Riemannian manifold we will have to find, first of all, restrictions on the given boundary Γ which ensure that it can be spanned by hypersur-faces of sufficiently small prescribed mean curvature H, and then determine bounds on $|H|$ depending on Γ which are sufficient for the existence of a solution.

We keep the general assumptions on the $(n+1)$-manifold N from Section 5, and we assume that A is a nonempty closed subset of N. It turns out that one must distinguish to cases: In the **non-closed case** we have $A \neq N$ or $A = N$ noncompact with infinite volume $\mu(N) = \infty$. The point here is that for any two n-currents T, T_0 on N which are homologous in A, i.e. $T - T_0$ is the boundary of some $(n+1)$-current on N with support in A, there is a *unique* such $(n+1)$-current Q_{T,T_0} with $\partial Q_{T,T_0} = T - T_0$ and with finite mass $\mathbf{M}\,(Q_{T,T_0}) < \infty$, because an $(n+1)$-current with zero boundary is constant on N and hence cannot have support in A, if $A \neq N$, or finite mass, if $A = N$ has infinite volume, unless it is the zero $(n+1)$-current. (This will be different in the closed case to be discussed later.) All our currents here and in the sequel are understood to be integer multiplicity rectifiable which, as in Section 4, includes the condition of finite mass but not the assumption of compact support.

Now, Q_{T,T_0} is again represented by an integer valued multiplicity function $i_{T,T_0} \in L^1(N,\mu;\mathbb{R})$ with support in A, and for continuous bounded prescribed mean curvature functions $H : A \to \mathbb{R}$ we can define the H-**volume**

$$\mathbf{V}_H(T,T_0) = \int_A H\, i_{T,T_0}\, d\mu = \langle Q_{T,T_0}, H\Omega \rangle,$$

where μ is the Riemannian measure and Ω is the volume form on N determined by the Riemannian metric and the orientation. Minimizing the H-**energy**

$$\mathbf{E}_H(T,T_0) = \mathbf{M}(T) + n\mathbf{V}_H(T,T_0)$$

on the **class of admitted hypersurfaces** $\mathcal{T}(T_0,A;\sigma)$, the set of n-currents T which are homologous in A to a fixed reference n-current T_0 with spt $T_0 \subset A$ and which satisfy the constraint $\mathbf{M}(T) \leq \sigma\mathbf{M}(T_0)$, one can now proceed as in Section 4 and prove the following general existence theorem [DS4, 3.2]. The homology condition $T - T_0 = \partial Q_{T,T_0}$ implies, in particular, $\partial T = \partial T_0$, i.e. the boundary of the reference current is the prescribed boundary for T here. For the reference current T_0 we will assume that it is homologically nontrivial in order to exclude the zero n-current from $\mathcal{T}(T_0,A;\sigma)$. This homological nontriviality condition is automatically satisfied if $\partial T_0 \neq 0$, but we also admit closed currents T_0. All the concepts which we use in the formulation of the theorem are defined for currents on an oriented Riemannian manifold exactly as in the Euclidean situation treated in Section 4. The proof of Theorem 4.1 also extends readily to the present Riemannian situation. The "linear" isoperimetric constant $c_A(s)$ was introduced in Section 5.

6.2 Theorem. *Suppose the non-closed case for $A \subset N$, T_0 is an integer multiplicity rectifiable n-current on N with support in A and not homologous to zero in A, $0<s\leq\infty$, the isoperimetric constant $c_A(s)$ is finite, $1<\sigma\leq\infty$ with $(1+\sigma)\mathbf{M}\,(T_0)\leq s$, and H is continuous and bounded with an isoperimetric condition of type c, s on A. Then,*

(i) *An H-energy minimizer exists in $\mathcal{T}(T_0, A; \sigma)$ if $c < 1$ or if $c = 1$ and $\sigma < \infty$;*

(ii) *each such minimizer T satisfies $\mathbf{M}\,(T) < \sigma\mathbf{M}\,(T_0)$ if $c \leq \dfrac{\sigma - 1}{\sigma + 1}$ or $\sigma = \infty$;*

(iii) *If $A = N$ or $|H| \leq H_{\partial A}$ holds along ∂A, where $H_{\partial A}$ denotes the (inward) boundary mean curvature of A, then each minimizer T with $\mathbf{M}\,(T) < \sigma\mathbf{M}\,(T_0)$ is an n-current of prescribed mean curvature H which is homologous to T_0 in A. Moreover, if we have $|H(a)| < H_{\partial A}(a)$ at some point $a \in (\partial A) \setminus \text{spt}\,\partial T_0$, then $a \notin \text{spt}\,T$.* □

As in Sections 3 and 4 one can consider various conditions on H which imply the required isoperimetric condition. For example, $s = \infty$ in the preceding theorem gives:

6.3 Theorem. *Suppose we are in the non-closed case with A the closure of a C^2 domain in N and $c_A(\infty) < \infty$, i.e. the linear isoperimetric inequality $\mu(E) \leq cP(E)$ holds for all sets of finite perimeter in A with a finite constant $c = c_A(\infty)$. Assume further that $\sup_A |H| < c^{-1}$ and $|H| \leq H_{\partial A}$ along ∂A (in the case $A \neq N$). Then, for any n-current T_0 on N with support in A and not homologous to zero in A there exists an n-current T on N which is homologous to T_0 in A and has prescribed mean curvature H.* □

6.4 Corollary. *If N is simply connected with $\text{Sec}_N \leq \kappa^2 < 0$ and if $\sup_N |H| < |\kappa|$, then for any closed $(n-1)$-current $\Gamma \neq 0$ on N which bounds an (integer multiplicity rectifiable) n-current there exists such a current T which has prescribed mean curvature H.* □

This follows from Corollary 5.3 and Theorem 6.3. The analogous result for 2-dimensional parametric surfaces in a 3-manifold was already proved in [Gu1] and [HK]. If we use the information collected in Example 5.1 and combine this with Theorem 5.2 and Remark 5.4, then we obtain the following second corollary which for the 2-dimensional parametric setting is contained in [HK] and, for geodesic balls A, in [Gu1]. (The assumption on H used in these papers was $\sup_A |H| \leq \kappa \cot \kappa R$, however, which is stronger than the hypotheses on H which we have here.)

6.5 Corollary. *Suppose A is geodesically star-shaped in N with maximal radius $R < \infty$, $|H| \leq H_{\partial A}$ holds along ∂A, and $\text{Sec}_N \leq \kappa^2$ on A (where $|\kappa|R < \pi$ if $\kappa^2 > 0$). If*

$$\sup_A |H| < \frac{1}{n c_\kappa(R)}$$

and Γ has support in A, then the conclusions of Corollary 6.4 are valid with $\text{spt}\,T \subset A$.

□

Another interesting corollary of Theorem 6.3 follows from the observation that the isoperimetric constant $c_A(\infty)$ is finite for compact $A \neq N$. This is a consequence of Proposition 5.7 in the non-closed case.

6.6 Corollary. *If $A \neq N$ is compact with $H_{\partial A} > 0$, then there is a constant $h(A) > 0$ such that for $\sup_A |H| \leq h(A)$ and any n-current T_0 with support in A and not homologous to zero in A there exists an n-current T on N which is homologous to T_0 in A and has prescribed mean curvature H.* $\square$

Note that we do not assume $\partial T_0 \neq 0$ here, i.e. the preceding corollary can also be applied in the case $\partial T_0 = 0$ to produce homologically nontrivial closed n-currents in A with prescribed mean curvature.

Next we assume that a linear isoperimetric inequality and simultaneously a nonlinear isoperimetric inequality are known for all sets with finite perimeter in A, i.e. we have finite isoperimetric constants $c_A(\infty)$ and $\gamma_A(\infty)$. If we use this information to estimate the isoperimetric profile of A and set up the corresponding Sobolev inequality, as indicated at the end of Section 5, then we obtain (see [DS4, 3.9]):

6.7 Theorem. *Suppose, in the non-closed case, that the isoperimetric constants $c_A(\infty)$ and $\gamma_A(\infty)$ are finite and that $|H| \leq H_{\partial A}$ holds along the boundary ∂A of the C^2 domain A in N (if $A \neq N$). If*

$$\int_{\{c_A(\infty)|H| > \frac{1}{n+1}\}} |H|^{n+1} d\mu < \frac{1}{n^{n+1}\gamma_A(\infty)^n}$$

holds, then the conclusions of Theorem 6.3 are valid. $\square$

6.8 Remarks. If N is simply connected with $\mathrm{Sec}_N \leq \kappa^2 \leq 0$ and A is contained in a ball of radius R, then in Theorem 6.7 $c_A(\infty)$ can be replaced by $c_\kappa(R) < (n|\kappa|)^{-1}$ and $\gamma_A(\infty)$ by the optimal isoperimetric constant γ_{n+1} for $\mathbb{R}^{n+1}$, provided Sec_N is constant or dim N equals 3 or 4. This follows from Example 5.1 and Remarks 5.9. Moreover, it can be seen from a discussion of the isoperimetric profile in this situation that we can improve the condition on $|H|$ in the domain of integration by a factor $\frac{n}{n+1}$ (see [DS4, 3.10]). We may thus weaken the hypothesis on H in Theorem 6.7 to

$$\int_{\{c_\kappa(R)|H| \geq \frac{1}{n}\}} |H|^{n+1} d\mu < \frac{1}{n^{n+1}\gamma_{n+1}^n},$$

where we can further replace $c_\kappa(R)$ by the larger value $(n|\kappa|)^{-1}$ to obtain a condition independent of R. This extends to the case $\mathrm{Sec}_N \leq \kappa^2 \leq 0$, and Sec_N constant or dim $N = 3$ or dim $N = 4$, simultaneously the versions of Hildebrandt's theorem and of the "$\frac{9\pi}{2}$ theorem" in $\mathbb{R}^{n+1}$ (see Corollary 4.2 and the subsequent discussion). It also improves Corollary 6.5 to the effect that instead of measure zero for $\{x \in A : c_\kappa(R)|H(x)| \geq \frac{1}{n}\}$ we only need to require a smallness integral condition for $|H|^{n+1}$ on this set. Since isoperimetric conditions are formulated as inequalities for $|\int_E H d\mu|$ it is also clear that it is sufficient to have the conditions above separately for H_+ and H_- instead of $H = H_+ + H_-$. $\square$

We next turn to the case $s < \infty$ in Theorem 6.2, i.e. we assume a linear isoperimetric inequality $\mu(E) \leq cP(E)$ with finite constant $c = c_A(s)$ only for sets $E \subset A$ with perimeter $P(E) \leq s$. Reasoning exactly as in the proof of Theorem 3.1 (as presented after Theorem 3.3) we then obtain from part (ii) of Theorem 6.2:

6.9 Theorem. *Suppose, in the non-closed case, that A is the closure of a C^2 domain in N, $|H| \leq H_{\partial A}$ holds along ∂A (if $A \neq N$), $0 < s < \infty$, the isoperimetric constant $c_A(s)$ is finite, and T_0 is an n-current on N such that $\operatorname{spt} T_0 \subset A$, $2M(T_0) < s$ and T_0 is not homologous to zero in A. Then, if*

$$\sup_A |H| < \frac{s - 2M(T_0)}{s c_A(s)},$$

there exists an n-current with prescribed mean curvature H on N which is homologous to T_0 in A. □

This is the Wente type theorem, analogous to Theorem 3.1, for Riemannian manifolds in the non-closed case. To see this we apply Corollary 5.8 and deduce here the

6.10 Corollary. *If N is homogeneously regular with infinite volume, then there exist positive constants $m(N)$, $h(N)$ such that the conclusion of the theorem is true whenever*

$$M(T_0) \leq m(N) \quad and \quad \sup_A |H| \leq \frac{h(N)}{M(T_0)^{1/n}}.$$

□

With regard to the hypothesis $2M(T_0) < s$ in Theorem 6.9 it is instructive to look at the following

6.11 Example. Suppose $N = M \times \mathbb{R}$ is the cylinder over a compact oriented n-manifold M without boundary. Then, from Proposition 5.6 we know $c_N(s) < \infty$ if and only if $s < 2\mathcal{H}^n(M)$. With such values of s Theorem 6.9 can be applied and gives the existence of H-hypersurfaces with given boundary ∂T_0 for all reference n-currents T_0 satisfying $M(T_0) < \mathcal{H}^n(M)$ and all prescribed mean curvature functions H satisfying $\sup_N |H| \leq \operatorname{const}(M)[\mathcal{H}^n(M) - M(T_0)]$ where $\operatorname{const}(M)$ is positive. On the other hand, for T_0 represented by a section $M \times \{0\}$ with multiplicity one, so that $M(T_0) = \mathcal{H}^n(M)$, we have non-existence for all constant $H \neq 0$ by Example 6.1 (i). (Here $\partial T_0 = 0$ but T_0 is not homologous to zero in N.) □

All the results described so far in this section have counterparts for **2-dimensional parametric surfaces in a Riemannian 3-manifold N**. As the reasoning is quite similar we content ourselves here with an explanation of the differences. Working still in the non-closed case we assume that either $A \subset N$ is the closure of a proper C^2 subdomain or $A = N$ is of infinite volume. In order to define the Sobolev space $W^{1,2}(U, A)$ and the H-energy for parametric surfaces of class $W^{1,2}$ in N it is convenient to assume N isometrically embedded in $\mathbb{R}^{3+p}$ as a closed subset. Then $W^{1,2}(U, A)$ is just the set of mappings $x \in W^{1,2}(U, \mathbb{R}^{3+p})$ which map (almost all of) the unit disc U into A, and the

Riemannian energy of such parametric surfaces $x : U \to N$ can be expressed as Dirichlet's integral $\mathbf{D}(x) = \frac{1}{2} \int_U (|x_u|^2 + |x_v|^2) \, du \, dv$, where x is considered as a mapping into $\mathbb{R}^{3+p}$. Given a rectifiable oriented closed Jordan curve Γ which is contractible in A, we define the Plateau boundary condition as in Section 1 and let $\mathcal{S}(\Gamma, A)$ be the set of $x \in W^{1,2}(U, A)$ which have boundary Γ in this sense. The contractibility of Γ secures $\mathcal{S}(\Gamma, A) \neq \emptyset$ so that we can find a reference surface $y \in \mathcal{S}(\Gamma, A)$. Usually y is chosen as a minimizer y_Γ for Dirichlet's integral on $\mathcal{S}(\Gamma, A)$, and then $\mathbf{D}(y) = a_\Gamma$ is the least spanning area of Γ in A.

Next we observe that each parametric surface $x \in W^{1,2}(U, A)$ defines an integer multiplicity rectifiable 2-current J_x on N exactly as in Section 3, and $\partial(J_x - J_y) = 0$ if x satisfies the same Plateau boundary condition as the reference surface y. We then want to use the integer multiplicity rectifiable 3-current $I_{x,y}$ of finite mass on N with $\partial I_{x,y} = J_x - J_y$ and $\operatorname{spt} I_{x,y} \subset A$ to define the H-volume $\mathbf{V}_H(x, y) = \langle I_{x,y}, H\Omega \rangle = \int_A i_{x,y} H \, d\mu$ where $i_{x,y}$ is the integer valued multiplicity function of $I_{x,y}$ determined by the orientation of N. Since we are in the non-closed case, $I_{x,y}$ and $i_{x,y}$ are uniquely determined by x and y. However, to secure existence we need to know that J_x and J_y are homologous in A. Of course, we can try to work in the subclass of $\mathcal{S}(\Gamma, A)$ consisting of surfaces x for which J_x is homologous to J_y. However, this subclass need not be closed in the weak $W^{1,2}$ topology, and therefore the direct method of the calculus of variations may fail. Indeed, if N contains a homologically nontrivial 2-sphere S then we can construct a sequence $x_k \in \mathcal{S}(\Gamma, A)$ such that x_k parametrizes S on the disc $|w| < 2^{-k}$ in U and converges weakly as $k \to \infty$ to some $x \in \mathcal{S}(\Gamma, A)$, with $J_{x_k} - J_x$ representing the homology class of the sphere S for all k. Intuitively speaking, a "bubble" can split off when passing to the $W^{1,2}$ weak limit and this bubble can carry away nonzero homology.

The only way out of this dilemma is to assume that all the closed 2-currents $J_x - J_y$ which we need to consider are homologically trivial in A. With this hypothesis the H-volume and the H-energy $\mathbf{E}_H(x) = \mathbf{D}(x) + 2\mathbf{V}_H(x, y)$ are well defined, and no additional difficulties arise in proving the Riemannian version of Theorem 3.3 and its consequences using the isoperimetric inequalities from Section 5. Instead of the reference current T_0 we now must consider the reference surface $y_\Gamma \in \mathcal{S}(\Gamma, A)$, of course, and the hypothesis $(1+\sigma)\mathbf{M}(T_0) \leq s$ has to be replaced by $(1+\sigma)\mathbf{D}(y_\Gamma) \leq s$. Since in the proof of Theorem 3.3 only closed 2-currents of mass $\leq (1+\sigma)\mathbf{D}(y_\Gamma)$ appear, we can state:

6.12 Theorem. *All the previous results of this section are true in an analogous formulation for the 2-dimensional parametric Plateau problem with prescribed mean curvature in a Riemannian 3-manifold N, if we assume the non-closed case and add the hypothesis that in the inclusion domain $A \subset N$ every closed integer multiplicity rectifiable 2-current T of mass $\mathbf{M}(T) \leq s$ is homologous to zero.* □

We refer to [DS5], [DS6] for more precise statements. A further analysis reveals that the closed 2-currents one needs to deal with and also the "bubbles" that can split off in the minimization process are of spherical type, i.e. representable by a map in $W^{1,2}(S^2, N)$. It is therefore sufficient to know the homological triviality condition only for such spherical 2-currents T of mass not exceeding s and the isoperimetric condition for H only on the integer multiplicity rectifiable 3-currents Q with $\partial Q = T$, $\mathbf{M}(Q) < \infty$, and $\operatorname{spt} Q \subset A$. The example $N = S^1 \times \mathbb{R}^2$, where every spherical 2-current can be lifted to the universal covering $\mathbb{R}^3$, shows that such a "spherical isoperimetric condition" is sometimes less restrictive than a general isoperimetric condition required for all sets $E \subset A$ with finite perimeter.

Of course, the homological triviality assumption in Theorem 6.12 is satisfied if A is geodesically star-shaped as in Corollaries 6.4 and 6.5. Furthermore, it holds if we choose s small enough, by Proposition 5.10. This means that we can state a result of Wente type like Corollary 6.10 also for the 2-dimensional parametric Plateau problem in a 3-manifold without any extra assumptions on the homology of N.

It remains to discuss the closed case, i.e. $A = N$ is a compact $(n+1)$-manifold without boundary. Then the obvious problem with the definition of the H-volume is that integer multiplicity rectifiable $(n+1)$-currents Q on N are not uniquely determined by their boundary and the condition of finite mass. Indeed, we may add integer multiples of $[N]$, the current defined by integration of $(n+1)$-forms on N, without affecting the boundary or the finiteness of the mass of Q. Therefore, the previous definitions of the H-volume are well defined only up to integer multiples of the integral $\int_N H \, d\mu$. The problem disappears if we assume that H has mean value zero on N, but this excludes constant $H \neq 0$, which is undesirable. Another way out is to admit only integer multiplicity rectifiable $(n+1)$-currents Q with mass smaller than half the volume of N, i.e. $\mathbf{M}(Q) < \frac{1}{2}\mu(N)$. Such Q are then uniquely determined by their boundary ∂Q, because adding a nonzero integer multiple of $[N]$ would produce a current of mass larger than $\frac{1}{2}\mu(N)$.

Fixing an integer multiplicity rectifiable n-current T_0 on N which is not homologous to zero we therefore now work in the class $\mathcal{T}(T_0; \sigma)$ of n-currents T on N with mass $\mathbf{M}(T) \leq \sigma \mathbf{M}(T_0)$ such that $T - T_0 = \partial Q$ holds for a (necessarily unique) integer multiplicity rectifiable $(n+1)$-current Q on N with $\mathbf{M}(Q) < \frac{1}{2}\mu(N)$. For $T \in \mathcal{T}(T_0; \sigma)$ the H-volume $\mathbf{V}_H(T, T_0) = \langle Q, H\Omega \rangle$ is then well defined, but when minimizing the H-energy on $\mathcal{T}(T_0; \sigma)$ we face another problem: It might happen that for a minimizing sequence T_k and the corresponding Q_k with $\partial Q_k = T_k - T_0$ we have $\mathbf{M}(Q_k) \to \frac{1}{2}\mu(N)$, and the weak limit T of (a subsequence of) the T_k does not belong to $\mathcal{T}(T_0; \sigma)$. This can be excluded, however, if we have the isoperimetric inequality

$$\min\{\mu(E), \mu(N \setminus E)\} \leq \tilde{c}_N(s)\mathbf{P}(E)$$

for all sets $E \subset N$ with finite perimeter $\mathbf{P}(E) \leq s$, where the constant satisfies $s\tilde{c}_N(s) < \frac{1}{2}\mu(N)$ and for the mass of T_0 we assume $(1+\sigma)\mathbf{M}(T_0) \leq s$. We can then proceed as in the proof of Theorem 6.9 and obtain (see [DS4, 4.3]):

6.13 Theorem. *Suppose N is a compact $(n+1)$-manifold without boundary, $0 < s < \infty$, an isoperimetric inequality $\mu(E) \leq \tilde{c}_N(s)\mathbf{P}(E)$ holds for all $E \subset N$ with $\mathbf{P}(E) \leq s$ and $\mu(E) < \frac{1}{2}\mu(N)$, and $s\tilde{c}_N(s) < \frac{1}{2}\mu(N)$. Then, if T_0 is a homologically nontrivial n-current on N of mass $\mathbf{M}(T_0) < \frac{1}{2}s$ and if H satisfies*

$$\sup_N |H| < \frac{s - 2\mathbf{M}(T_0)}{s\tilde{c}_N(s)},$$

there exists an n-current T on N which is homologous to T_0 and has prescribed mean curvature H. □

6.14 Corollary. *For closed manifolds N the assertions of Corollary 6.10 are also true.* □

We have thus established a result of Wente type in general Riemannian manifolds, compact or noncompact and homogeneously regular. We note that the condition on H in the hypotheses of Theorem 6.13 is just one of several reasonable assumptions which imply an isoperimetric condition of type $c < 1$, s for H. One could replace this, e.g., by an integral condition as in Theorem 6.7. Also, as explained after Theorem 6.12, it is sufficient to know the isoperimetric inequality in N and the isoperimetric condition on H only for integer multiplicity rectifiable 3-currents of mass $< \frac{1}{2}\mu(N)$ which are bounded by spherical 2-currents.

With regard to the 2-dimensional parametric Plateau problem with prescribed mean curvature in a closed 3-manifold N we can proceed as above assuming that closed integer multiplicity rectifiable 2-currents T on N of mass $\mathbf{M}(T) \leq s$ are boundaries $T = \partial Q$ with the integer multiplicity rectifiable 3-current Q on N satisfying $\mathbf{M}(Q) < \frac{1}{2}\mu(N)$ (so that Q is unique). For $x \in \mathcal{S}(\Gamma, N)$ with $\mathbf{D}(x) \leq \sigma\mathbf{D}(y)$ and $(1+\sigma)\mathbf{D}(y) \leq s$ the H-volume $\mathbf{V}_H(x, y) = \langle I_{x,y}, H\Omega \rangle$ is then well defined choosing $I_{x,y}$ as the integer multiplicity rectifiable 3-current on N with $\partial I_{x,y} = J_x - J_y$ and $\mathbf{M}(I_{x,y}) < \frac{1}{2}\mu(N)$. In the minimizing process for the H-energy we now have similar problems as above. When $x_k \to x$ $W^{1,2}$ weakly in $\mathcal{S}(\Gamma, A)$, then it could happen that $I_{x_k,y}$ does not converge to $I_{x,y}$ as $k \to \infty$. Using the technical construction already employed in the proof of Theorem 3.3, one can show, however, that this phenomenon does not occur if $s\tilde{c}_N(s) < \frac{1}{2}\mu(N)$. The result is (see [DS5], [DS6]):

6.15 Theorem. *The analogue of Theorem 6.13 is valid for the 2-dimensional parametric Plateau problem with prescribed mean curvature in a closed 3-manifold N, with the added hypothesis that each closed integer multiplicity rectifiable 2-current on N of mass not exceeding s is the boundary of an integer multiplicity rectifiable 3-current with mass smaller than half the volume of N.* □

As a consequence, we have the following **Riemannian version of Wente's theorem**:

6.16 Theorem. *Suppose the 3-manifold N is compact or noncompact and homogeneously regular. Then there exist positive constants $a(N)$, $h(N)$ such that for each oriented closed rectifiable Jordan curve Γ in N with least spanning area $a_\Gamma \leq a(N)$ and each continuous function H on N with $\sup_N |H| \leq h(N)\, a_\Gamma^{-1/2}$ the Plateau problem with boundary Γ and prescribed mean curvature H has a weak solution $x \in W^{1,2}(U, N)$.* □

With regard to the hypotheses of Theorem 6.13 the following example is instructive:

6.17 Example. For the standard sphere $N = S^{n+1}$ we have $s\tilde{c}_N(s) < \frac{1}{2}\mathcal{H}^{n+1}(S^{n+1})$ if and only if $s < \mathcal{H}^n(S^n)$. The condition $\mathbf{M}(T_0) < \frac{1}{2}s$ is satisfied with some $s < \mathcal{H}^n(S^n)$ if and only if $\mathbf{M}(T_0)$ is smaller than the mass of a great half sphere in S^{n+1}. For a given closed integer multiplicity rectifiable $(n-1)$-current Γ on S^{n+1} of mass $\mathbf{M}(\Gamma) < \mathcal{H}^{n-1}(S^{n-1})$ such T_0 can be found with $\partial T_0 = \Gamma$ if spt Γ is contained in a great n-sphere (and maybe also without this restriction, cf. [Alm]). For such Γ Theorem 6.13 yields the existence of an integer multiplicity rectifiable n-current T on S^{n+1} with boundary Γ and prescribed mean curvature H, if $\sup |H|$ is sufficiently small (smaller than a positive constant depending on

Γ). On the other hand, for Γ corresponding to a sphere $S^{n-1} \times \{(0,0)\}$ and for constant $H \neq 0$ no such current T exists by Example 6.1 (i). □

Thus, Theorem 6.13 is best possible with regard to the conditions on Γ that are necessary to secure the existence of solutions to the Plateau problem for some constant prescribed mean curvature $H \neq 0$. (It is most likely not optimal, however, with respect to the conditions on the prescribed mean curvature functions H that are admitted.)

An analysis of the proofs for Theorems 6.9 and 6.13 has lead to the following theorem, the last one we state here, on the **perturbation of minimal hypersurfaces to hypersurfaces of prescribed mean curvature** [DS4, 4.6, 4.7].

6.18 Theorem. *Suppose the $(n+1)$-manifold N is compact or noncompact and homogeneously regular. If T_0 is an integer multiplicity rectifiable n-current on N which is uniquely mass minimizing in its homology class, then for some constant $h(N, T_0) > 0$ and each continuous function H on N with $\sup |H| \leq h(N, T_0)$ there exists an integer multiplicity rectifiable n-current T on N which is homologous to T_0 and has prescribed mean curvature H.* □

Instead of the unique minimizing property of T_0 it suffices, in fact, that for no other mass minimizer T_1 the current $T_1 - T_0$ bounds an integer multiplicity rectifiable $(n+1)$-current of mass t where $t = \frac{1}{2}\mu(N)$ if N is compact and $t > 0$ is prescribed arbitrarily otherwise. From Example 6.1 one can see that some conditions of this kind are really necessary in order to perturb the minimal hypersurface T_0 to a homologous hypersurface with small constant mean curvature $H \neq 0$. The theorem applies, in particular, to nonzero closed n-currents T_0 which are uniquely minimizing in their homology class. We refer to the work of M. Toda [To1], [To2] on closed 2-dimensional parametric surfaces with constant prescribed mean curvature in a compact 3-manifold.

We conclude with some comments on the **regularity** of the weak solutions to the Plateau problem with prescribed mean curvature in a Riemannian manifold which we have produced in the results described in this section. For integer multiplicity rectifiable n-currents on an $(n+1)$-manifold N the weak form of the H-hypersurface equation looks exactly as in Section 4 except that inner products, orientations, and projections now have to be taken in the tangent spaces of N and the deformation vector fields must be tangent to N. For our H-energy minimizing currents this variational equation (or the corresponding variational inequality) is valid, and the regularity theory described in Theorem 4.3 applies. (Although in [HS] and [DS3] only the Euclidean situation is treated, it is clear from the methods used in the proofs that an underlying Riemannian metric, which can be assumed arbitrarily close to the Euclidean metric for the purpose of regularity theory, does not affect the results.) Under appropriate smoothness assumptions on the data we can thus assert that spt $T \setminus$ spt ∂T is, up to a possible singular set of codimension at least 7, a smooth submanifold of N with prescribed mean curvature H, and we have complete boundary regularity in the sense of Theorem 4.3.

For 2-dimensional $\mathbf{E}_H$ minimizing parametric surfaces x in a 3-manifold N the variational equation is

$$\Delta^N x = 2(H \circ x)\, x_u \wedge_N x_v \quad \text{on } U,$$

where Δ^N is the Euler operator associated with the Riemannian energy functional for surfaces $x : U \to N$ (Dirichlet's integral) and $\wedge_N : T_a N \times T_a N \to T_a N$ denotes the skewsymmetric product on the tangent spaces to N which is determined by the Riemannian metric and the orientation on N. The in general nonlinear operator Δ^N is well known from the theory of harmonic mappings (cf. [EL], [Jo1], [Jo2]) and called the **tension field operator** for mappings from the Euclidean disc U to the Riemannian manifold N. Note that $\Delta^N x(w)$ is just the projection of $\Delta x(w)$ from $\mathbb{R}^{3+n}$ onto $T_{x(w)}N$ if N is isometrically embedded in $\mathbb{R}^{3+p}$ and x is smooth. For $x \in W^{1,2}(U, N)$ the H-surface equation above has to be interpreted in a weak sense, i.e.

$$\int_U \left[x_u \cdot \nabla_u^N \xi + x_v \cdot \nabla_v^N \xi + 2(H \circ x)\, \xi \cdot x_u \wedge_N x_v \right] du\, dv = 0$$

for all $\xi \in W_0^{1,2}(U, \mathbb{R}^{3+p}) \cap L^\infty(U, \mathbb{R}^{3+p})$ tangential to N along x,

the latter condition meaning that $\xi(w) \in T_{x(w)}N$ for (almost all) $w \in U$. The covariant derivatives $\nabla_u^N \xi(w)$ appearing here are defined by projecting $\xi_u(w)$ onto $T_{x(w)}N$. Since x_u, x_v are tangential to N along x we could, of course, simply write ξ_u, ξ_v instead of $\nabla_u^N \xi$, $\nabla_v^N \xi$ above. This would obscure, however, the intrinsic meaning of the weak H-surface equation. As in Proposition 2.1 (ii) we can also calculate the inner first variation and derive from it the conformality relations for our H-energy minimizers x which now read

$$|x_u|_N^2 - |x_v|_N^2 = (x_u, x_v)_N = 0 \quad \text{on } U,$$

where the subscripts indicate that norms and inner products are to be taken with respect to the Riemannian metric on N.

In order to proceed with the regularity theory for the weak H-surface equation above we need to rewrite it as a nonlinear elliptic system by introducing coordinates on the 3-manifold N. However, x will in general not have values in a single coordinate domain. Indeed, we have stressed that we do not want to assume such a condition for the boundary curve Γ in the Plateau problem. Therefore we have to prove continuity of x on $\overline{U}$ first. Fortunately we can do this using the H-energy minimality on $S(\Gamma, A)$ and Morrey's method exactly as in Section 2, provided A is quasiregular (see the discussion at the end of Section 2). With this, the regularity problem can be localized in the 3-manifold N, and introducing coordinates there we obtain the weak H-surface equation and the conformality relations in the form (see [Gu1], [HK])

$$x_{uu}^l + x_{vv}^l + (\Gamma_{ij}^l \circ x)(x_u^i x_u^j + x_v^i x_v^j) = 2(H \circ x)\sqrt{\gamma \circ x}\, \varepsilon_{ijk} (g^{lk} \circ x)(x_u^i x_v^j - x_v^i x_u^j),$$
$$(g_{ij} \circ x)(x_u^i x_u^j - x_v^i x_v^j) = (g_{ij} \circ x) x_u^i x_v^j = 0,$$

where x^l, g_{ij}, and Γ_{ij}^l are the coordinate components of x, of the Riemannian metric g, and of the Levi-Civita connection (Christoffel symbols of g) on N, we have set $(g^{lk}) = (g_{ij})^{-1}$, $\gamma = \det(g_{ij})$, $\varepsilon_{ijk} = \frac{1}{2}(i-j)(j-k)(k-i)$, and summation over repeated indices from 1 to 3 is understood.

For x continuous and of class $W^{1,2}$ this Riemannian H-surface equation still has to be interpreted in the usual weak sense. Interior $C^{k+2,\beta}$ regularity of conformal weak solutions x was proved in [HK], [Hi8], assuming H of class $C^{k,\beta}$. The simplest method is [Grü2]. Boundary regularity was established by E. Heinz & S. Hildebrandt [HH2]. They showed in the case $H \equiv 0$ how to deal with the conformality relations in the Riemannian situation using the earlier regularity theorems of Heinz mentioned at the end of Section 2. Their reasoning can also be applied to H-surfaces in a 3-manifold, because the extra terms appearing when $H \not\equiv 0$ are of the same quadratic nature in x_u, x_v as the terms which come from the Levi-Civita connection and are already present in the case $H \equiv 0$ (see also [DHKW, Chap.7]).

Furthermore, in [HH2] an asymptotic expansion was derived near a branch point of x in $\overline{U}$ (a point where x_u, x_v vanish simultaneously), from which it follows that branch points are isolated and that $x|_{\partial U}$ is a homeomorphism of ∂U onto Γ. Interior branch points were ruled out for energy minimizing H-surfaces in Riemannian manifolds by R. Gulliver [Gu2]. (He used the energy functional $\mathbf{E}_Z$ associated with a vector field Z on N of size $\sup_M |Z| < \frac{1}{2}$, but an isoperimetric condition with $c < 1$ for H is in fact sufficient; cf. [SW].) Except for the case of minimal surfaces in an analytic Riemannian 3-manifold [GL] it is not known, however, whether there can exist a (true) boundary branch point in an energy minimizing solution to the Plateau problem with prescribed mean curvature in a Riemannian 3-manifold.

References

[All] Allard, W.K.: On the first variation of a varifold. Ann. Math. **95** (1972), 417–491.

[Alm] Almgren, F.J.: Optimal isoperimetric inequalities. Indiana Univ. Math. J. **35** (1986), 451–547.

[AT] Almgren, F.J., Thurston, W.P.: Examples of unknotted curves which bound only surfaces of high genus within their convex hull. Ann. Math. **105** (1977), 527–538.

[Al1] Alt, H.W.: Verzweigungspunkte von H–Flächen, I. Math. Z. **127** (1972), 333–362.

[Al2] Alt, H.W.: Verzweigungspunkte von H–Flächen, II. Math. Ann. **201** (1973), 33–55.

[Ba] Barbosa, J.L.: Constant mean curvature surfaces bounded by a planar curve. Matematica Contemporanea **1** (1991), 3–15.

[BJ] Barbosa, J.L., Jorge, L.P.: Stable H-surfaces whose boundary is $S^1(1)$. An. Acad. Bras. Ci. **66** (1994), 259–263.

[Be] Bethuel, F.: Un résultat de régularité pour les solutions de l'équation des surfaces à courbure moyenne prescrite. C.R. Acad. Sci. Paris **314** (1992), 1003–1007.

[BG] Bethuel, F., Ghidaglia, J.M.: Improved regularity of solutions to elliptic equations involving Jacobians and applications. J. Math. Pures et Appliquées **72** (1993), 441–474.

[BR] Bethuel, F., Rey, O.: Multiple solutions to the Plateau problem for nonconstant mean curvature. Duke Math. J. **73** (1994), 593–646.

[BC] Brézis, H.R., Coron, M.: Multiple solutions of H–systems and Rellich's conjecture. Commun. Pure Appl. Math. **37** (1984), 149–187.

[BE] Brito, F., Earp, R.: Geometric configurations of constant mean curvature surfaces with planar boundary. An. Acad. Bras. Ci. **63** (1991), 5–19.

[BZ] Burago, Y.D., Zalgaller, V.A.: Geometric inequalities. Springer–Verlag, New York Heidelberg Berlin, 1988.

[Cr]   *Croke C.B.*: A sharp four dimensional isoperimetric inequality. Comment. Math. Helvetici **59** (1984), 187–192.

[DeG]   *De Giorgi, E.*: Sulla proprietà isoperimetrica dell' ipersfera, nelle classe degli insiemi avanti frontiera orientata di misura finita. Atti. Accad. Naz. Lincei, ser 1, **5** (1958), 33–44.

[Di1]   *Dierkes, U.*: Plateau's problem for surfaces of prescribed mean curvature in given regions. Manuscr. Math. **56** (1986), 313–331.

[Di2]   *Dierkes, U.*: A geometric maximum principle for surfaces of prescribed mean curvature in Riemannian manifolds. Z. Anal. Anwend. **8** (2) (1989), 97–102.

[DHKW] *Dierkes, U., Hildebrandt, S., Küster, A., Wohlrab, O.*: Minimal surfaces vol.1, vol.2. Grundlehren math. Wiss. 295, 296. Springer–Verlag, Berlin Heidelberg New York, 1992.

[Du1]   *Duzaar, F.*: Variational inequalities and harmonic mappings. J. Reine Angew. Math. **374** (1987), 39–60.

[Du2]   *Duzaar, F.*: On the existence of surfaces with prescribed mean curvature and boundary in higher dimensions. Ann. Inst. Henri Poincaré (Anal. Non Lineaire) **10** (1993), 191–214.

[Du3]   *Duzaar, F.*: Hypersurfaces with constant mean curvature and prescribed area. Manuscr. Math. **91** (1996), 303–315.

[Du4]   *Duzaar, F.*: Boundary regularity for area minimizing currents with prescribed volume. To appear in J. Geometric Analysis (1998?).

[DF1]   *Duzaar, F., Fuchs, M.*: On the existence of integral currents with prescribed mean curvature vector. Manuscr. Math. **67** (1990), 41–67.

[DF2]   *Duzaar, F., Fuchs, M.*: A general existence theorem for integral currents with prescribed mean curvature form. Bolletino U.M.I. (7) **6**-B (1992), 901–912.

[DS1]   *Duzaar, F., Steffen, K.*: Area minimizing hypersurfaces with prescribed volume and boundary. Math. Z. **209** (1992), 581–618.

[DS2]   *Duzaar, F., Steffen, K.*: λ minimizing currents. Manuscr. Math. **80** (1993), 403–447.

[DS3]   *Duzaar, F., Steffen, K.*: Boundary regularity for minimizing currents with prescribed mean curvature. Calc. Var. **1** (1993), 355–406.

[DS4]   *Duzaar, F., Steffen, K.*: Existence of hypersurfaces with prescribed mean curvature in Riemannian mannifolds. Indiana Univ. Math. J. **45** (1996), 1045–1093.

[DS5]   *Duzaar, F., Steffen, K.*: The Plateau problem for parametric surfaces with prescribed mean curvature. Geometric analysis and the calculus of variations (dedicated to S. Hildebrandt, ed. J. Jost), 13–70, International Press, Cambridge MA, 1996.

[DS6]   *Duzaar, F., Steffen, K.*: Parametric surfaces of least H-energy in a Riemannian manifold. Preprint No. 284, SFB 288 Differential Geometry and Quantum Physics, TU Berlin, 1997.

[EBMR] *Earp, R., Brito, F., Meeks III, W.H., Rosenberg, H.*: Structure theorems for constant mean curvature surfaces bounded by a planar curve. Indiana Univ. Math. J. **40** (1991), 333–343.

[EL]   *Eells, J., Lemaire, L.*: A report on harmonic maps. Bull. London Math. Soc. **10** (1978), 1–68. Another report on harmonic maps. Bull. London Math. Soc. **20** (1988), 385–542.

[EG]   *Evans, L.C., Gariepy, L.F.*: Measure theory and fine properties of functions. CRC Press, Boca Raton Ann Arbor London, 1992.

[Fe]   *Federer, H.*: Geometric measure theory. Springer–Verlag, Berlin Heidelberg New York, 1969.

[Grü1] *Grüter, M.*: Regularity of weak H–surfaces. J. Reine Angew. Math. **329** (1981), 1–15.

[Grü2] *Grüter, M.*: Eine Bemerkung zur Regularität stationärer Punkte von konform invarianten Variationsintegralen. Manuscr. Math. **55** (1986), 451–453.

[Gü]   *Günther, P.*: Einige Vergleichssätze über das Volumenelement eines Riemannschen Raumes. Publ. Math. Debrecen **7** (1960), 258–287.

[Gu1]   *Gulliver, R.*: The Plateau problem for surfaces of prescribed mean curvature in a Riemannian manifold. J. Differ. Geom. **8** (1973), 317–330.

[Gu2]   *Gulliver, R.*: Regularity of minimizing surfaces of prescribed mean curvature. Ann. Math. **97** (1973), 275–305.

[Gu3]   *Gulliver, R.*: On the non–existence of a hypersurface of prescribed mean curvature with a given boundary. Manuscr. Math. **11** (1974), 15–39.

[Gu4]   *Gulliver, R.*: Necessary conditions for submanifolds and currents with prescribed mean curvature vector. Seminar on minimal submanifolds, ed. E. Bombieri, Princeton, 1983.

[Gu5]   *Gulliver, R.*: Branched immersions of surfaces and reduction of topological type. I. Math. Z. **145** (1975), 267–288.

[Gu6]   *Gulliver, R.*: Branched immersions of surfaces and reduction of topological type. II. Math. Ann. **230** (1977), 25–48.

[Gu7]   *Gulliver, R.*: A minimal surface with an atypical boundary branch point. Differential Geometry, 211–228, Pitman Monographs Surveys Pure Appl. Math. 52, Longman Sci. Tech., Harlow, 1991.

[GL]   *Gulliver, R., Lesley, F.D.*: On boundary branch points of minimizing surfaces. Arch. Ration. Mech. Anal. **52** (1973), 20–25.

[GOR]   *Gulliver, R., Osserman, R., Royden, H.L.*: A theory of branched immersions of surfaces. Am. J. Math. **95** (1973), 750–812.

[GS1]   *Gulliver, R., Spruck, J.*: The Plateau problem for surfaces of prescribed mean curvature in a cylinder. Invent. Math. **13** (1971), 169–178.

[GS2]   *Gulliver, R., Spruck, J.*: Surfaces of constant mean curvature which have a simple projection. Math. Z. **129** (1972), 95–107.

[GS3]   *Gulliver, R., Spruck, J.*: Existence theorems for parametric surfaces of prescribed mean curvature. Indiana Univ. Math. J. **22** (1972), 445–472.

[HS]   *Hardt, R., Simon, L.*: Boundary regularity and embedded solutions for the oriented Plateau problem. Ann. Math. **110** (1979), 439–486.

[HW]   *Hartmann, P., Winter, A.*: On the local behaviour of solutions of nonparabolic partial differential equations. Amer. J. Math. **75** (1953), 449-476.

[He1]   *Heinz, E.*: Über die Existenz einer Fläche konstanter mittlerer Krümmung mit gegebener Berandung. Math. Ann. **127** (1954), 258–287.

[He2]   *Heinz, E.*: On the non–existence of a surface of constant mean curvature with finite area and prescribed rectifiable boundary. Arch. Rat. Mech. Anal. **35** (1969), 249–252.

[He3]   *Heinz, E.*: Ein Regularitätssatz für Flächen beschränkter mittlerer Krümmung. Nachr. Akad. Wiss. Gött., II. Math.-Phys. Kl. (1969), 107–118.

[He4]   *Heinz, E.*: Über das Randverhalten quasilinearer elliptischer Systeme mit isothermen Parametern. Math. Z. **113** (1970), 99–105.

[He5]   *Heinz, E.*: Unstable surfaces of constant mean curvature. Arch. Ration. Mech. Anal. **38** (1970), 257–267.

[He6]   *Heinz, E.*: Ein Regularitätssatz für schwache Lösungen nichtlinearer elliptischer Systeme. Nachr. Akad. Wiss. Gött., II. Math.-Phys. Kl. (1975), 1–13.

[He7]   *Heinz., E.*: Über die Regularität schwacher Lösungen nichtlinarer elliptischer Systeme. Nachr. Akad. Wiss. Gött., II. Math.-Phys. Kl. (1985), 1–15.

[HH1]   *Heinz, E., Hildebrandt, S.*: Some remarks on minimal surfaces in Riemannian manifolds. Commun. Pure Appl. Math. **23** (1970), 371–377.

[HH2]   *Heinz, E., Hildebrandt, S.*: On the number of branch points of surfaces of bounded mean curvature. J. Differ. Geom. **4** (1970), 227–235.

[HT]   *Heinz, E., Tomi, F.*: Zu einem Satz von S. Hildebrandt über das Randverhalten von Minimalflächen. Math. Z. **111** (1969), 372–386.

[Hi1]   *Hildebrandt, S.*: Boundary behavior of minimal surfaces. Arch. Ration. Mech. Anal. **35** (1969), 47–82.

[Hi2]   *Hildebrandt, S.*: Über Flächen konstanter mittlerer Krümmung. Math. Z. **112** (1969), 107–144.

[Hi3]   *Hildebrandt, S.*: On the Plateau problem for surfaces of prescribed mean curvature. Commun. Pure Appl. Math. **23** (1970), 97–114.

[Hi4]   *Hildebrandt, S.*: Randwertprobleme für Flächen mit vorgeschriebener mittlerer Krümmung und Anwendungen auf die Kapillaritätstheorie I. Math. Z. **112** (1969), 205–213.

[Hi5]   *Hildebrandt, S.*: Über einen neuen Existenzsatz für Flächen vorgeschriebener mittlerer Krümmung. Math. Z. **119** (1971), 267–272.

[Hi6]   *Hildebrandt, S.*: Einige Bemerkungen über Flächen beschränkter mittlerer Krümmung. Math. Z. **115** (1970), 169–178.

[Hi7]   *Hildebrandt, S.*: Maximum principles for minimal surfaces and for surfaces of continuous mean curvature. Math. Z. **128** (1972), 253–269.

[Hi8]   *Hildebrandt, S.*: On the regularity of solutions of two–dimensional variational problems with obstructions. Commun. Pure Appl. Math. **25** (1972), 479–496.

[Hi9]   *Hildebrandt, S.*: Interior $C^{1+\alpha}$-regularity of solutions of two–dimensional variational problems with obstacles. Math. Z. **131** (1973), 233–240.

[HK]   *Hildebrandt S., Kaul, H.*: Two–dimensional variational problems with obstructions, and Plateau's problem for H-surfaces in a Riemannian manifold. Commun. Pure Appl. Math. **25** (1972), 187–223.

[Jä]   *Jäger, W.*: Das Randverhalten von Flächen beschränkter mittlerer Krümmung bei $C^{1,\alpha}$– Rändern. Nachr. Akad. Wiss. Gött., II. Math.-Phys. Kl. (1977), 45–54.

[Jo1] *Jost, J.*: Lectures on harmonic maps (with applications to conformal mappings and minimal surfaces). Lect. Notes Math. **1161**, Springer–Verlag, Berlin Heidelberg New York (1985), 118–192.

[Jo2] *Jost, J.*: Two–dimensional geometric variational problems. Wiley–Interscience, Chichester New York, 1991.

[Kap] *Kapouleas, N.*: Compact constant mean curvature surfaces in Euclidean three-space. J. Differ. Geom. **33** (1991), 683–715.

[Kau] *Kaul, H.*: Ein Einschließungssatz für H–Flächen in Riemannschen Mannigfaltigkeiten. Manuscr. Math. **5** (1971), 103–112.

[Kl]   *Kleiner, B.*: An isoperimetric comparison theorem. Invent. Math. **108** (1992), 37–47.

[LM]   *López, S., Montiel, S.*: Constant mean curvature discs with bounded area. Proc. Amer. Math. Soc. **123** (1995), 1555–1558.

[MM]   *Massari, U., Miranda, M.*: Minimal surfaces of codimension one. North-Holland Mathematical Studies 91, Amsterdam New York Oxford, 1984.

[Ni1] *Nitsche, J.C.C.*: Vorlesungen über Minimalflächen. Grundlehren math. Wiss., vol. 199. Springer–Verlag, Berlin Heidelberg New York, 1975.

[Ni2] *Nitsche, J.C.C.*: Lectures on minimal surfaces, vol 1: Introduction, fundamentals, geometry and basic boundary problems. Cambridge Univ. Press, 1989.

[Os]   *Osserman, R.*: A proof of the regularity everywhere of the classical solution to Plateau's problem. Ann. Math. **91** (1970), 550–569.

[Sch] *Schmidt, E.*: Beweis der isoperimetrischen Eigenschaft der Kugel im hyperbolischen und sphärischen Raum jeder Dimensionszahl. Math. Z. **49** (1943/44), 1–109.

[ST]   *Schüffler, K., Tomi, F.*: Ein Indexsatz für Flächen konstanter mittlerer Krümmung. Math. Z. **182** (1983), 245–258.

[Se]   *Serrin, J.*: The problem of Dirichlet for quasilinear elliptic differential equations in many independent variables. Phil. Trans. Royal Soc. London **264** (1969), 413–419.

[Si]   *Simon, L.*: Lectures on geometric measure theory. Proc. CMA, Vol. 3, ANU Canberra, 1983.

[Ste1] *Steffen, K.*: Flächen konstanter mittlerer Krümmung mit vorgegebenem Volumen oder Flächeninhalt. Arch. Ration. Mech. Anal. **49** (1972), 99–128.

[Ste2] *Steffen, K.*: Ein verbesserter Existenzsatz für Flächen konstanter mittlerer Krümmung. Manuscr. Math. **6** (1972), 105–139.

[Ste3] *Steffen, K.*: Isoperimetric inequalities and the problem of Plateau. Math. Ann. **222** (1976), 97–144.

[Ste4] *Steffen, K.*: On the existence of surfaces with prescribed mean curvature and boundary. Math. Z. **146** (1976), 113–135.

[Ste5] *Steffen, K.*: On the nonuniqueness of surfaces with prescribed constant mean curvature spanning a given contour. Arch. Ration. Mech. Anal. **94** (1986), 101–122.

[SW]   Steffen, K., Wente, H.: The non-existence of branch points in solutions to certain classes of Plateau type variational problems. Math. Z. **163** (1978), 211–238.

[Strö]   Ströhmer, G.: Instabile Flächen vorgeschriebener mittlerer Krümmung. Math. Z. **174** (1980), 119–133.

[Str1]   Struwe, M.: Nonuniqueness in the Plateau problem for surfaces of constant mean curvature. Arch. Ration. Mech. Anal. **93** (1986), 135–157.

[Str2]   Struwe, M.: Large H–surfaces via the mountain-pass-lemma. Math. Ann. **270** (1985), 441–459.

[Str3]   Struwe, M.: Plateau's problem and the calculus of variations. Mathematical Notes 35, Princeton University Press, Princeton, New Jersey, 1988.

[Str4]   Struwe, M.: Multiple solutions to the Dirichlet problem for the equation of prescribed mean curvature. Moser-Festschrift, Academic Press, 1990.

[To1]   Toda, M.: On the existence of H-surfaces into Riemannian manifolds. Calc. Var. **5** (1997), 55–83.

[To2]   Toda, M.: Existence and non-existence results of H-surfaces into 3-dimensional Riemannian manifolds. Comm. in Analysis and Geometry **4** (1996), 161–178.

[Tom1]   Tomi, F.: Ein einfacher Beweis eines Regularitätssatzes für schwache Lösungen gewisser elliptischer Systeme. Math. Z. **112** (1969), 214–218.

[Tom2]   Tomi, F.: Bemerkungen zum Regularitätsproblem der Gleichung vorgeschriebener mittlerer Krümmung. Math. Z. **132** (1973), 323–326.

[Wa]   Wang, G.: The Dirichlet problem for the equation of prescribed mean curvature. Ann. Inst. Henri Poincaré (Anal. Non Linéaire) **9** (1992), 643–655.

[Wen1]   Wente, H.: An existence theorem for surfaces of constant mean curvature. J. Math. Anal. Appl. **26** (1969), 318–344.

[Wen2]   Wente, H.: A general existence theorem for surfaces of constant mean curvature. Math. Z. **120** (1971), 277–288.

[Wen3]   Wente, H.: An existence theorem for surfaces in equilibrium satisfying a volume constraint. Arch. Ration. Mech. Anal. **50** (1973), 139–158.

[Wer]   Werner, H.: Das Problem von Douglas für Flächen konstanter mittlerer Krümmung. Math. Ann. **133** (1957), 303–319.

[Ya]   Yau, S.T.: Isoperimetric constants and the first eigenvalue of a compact Riemannian manifold. Ann. Sci. Éc. Norm. Sup. **83** (1975), 487–507.

[Zi]   Ziemer, W.P.: Weakly differentiable functions. Springer–Verlag, New York Berlin Heidelberg, 1989.

Mathematisches Institut der Heinrich–Heine–Universität Düsseldorf
Universitätstraße 1, D-40225 Düsseldorf, Germany
steffen@cs.uni-duesseldorf.de

G. ALBERTI, Dipartimento di Matematica Applicata, Via Bonanno 25-B, 56126 Pisa, Italia
R: ALICANDRO, SISSA, Via Beirut 2-4, 34014 Trieste, Italia
L. ALMEIDA, CMLA ENS Cachan, 61 Av. du President Wilson, 94235 Cachan Cedex, France
S. BALDO, Dipartimento di Matematica, Via Sommarive 14, 38050 Povo, Trento, Italia
G. BELLETTINI, Dipartimentodi Matematica Applicata, Via Bonanno 25-B, 56126 Pisa, Italia
F. BERNATZKI, Fakultat f. Mathematik, Ruhr-Universität Bochüm, 44780 Bochüm, Germany
K. BLACHOWSKI, Inst. of Mathematics, University of Lodz, ul Banacha 22, 90-238 Lodz, Poland
A. BRAIDES, SISSA, Via Beirut, 2-4, 34014 Trieste, Italia
E. CABIB, Dipartimento di Ingegneria Civile, Via delle Scienze 208, 33100 Udine, Italia
P. CALDIROLI, SISSA, Via Beirut 2-4, 34014 Trieste, Italia
A.M. CANDELA, Dipartimento di Matematica, Via Orabona 4, 70125 Bari, Italia
A. CANINO, Dipartimento di Matematica, Univ. della Calabria, 87036 Arcavacata di Rende (CS), Italia
G. CERAMI, Dipartimento di Matematica, Via Archirafi 34, 90123 Palermo, Italia
A. CHAMBOLLE, SISSA, Via Beirut 2-4, 34014 Trieste, Italia
R. CHIAPPINELLI, Dipartimento di Matematica, Via del Capitano 15, 53100 Siena, Italia
G. CORTESANI, SISSA, Via Beirut 2-4, 34014 Trieste, Italia
G. CRASTA, Dipartimento di Matematica pura ed appl., Via Campi 213/B, 41100 Modena, Italia
M. CZARNECKI,Inst. of Mathematics, Univesity of Lodz,ul. Banacha 22, 90-238 Lodz, Poland
P. DALL'AGLIO, SISSA, Via Beirut 2-4, 34014 Trieste, Italia
G. DAL MASO, SISSA, Via Beirut 2-4, 34014 Trieste, Italia
V. DE CICCO, Dipartimento di Metodi e Modelli Matematici, Via A. Scarpa 16, 00161 Roma, Italia
S. DELLADIO, Dipartimento di Matematica, Via Sommarive 14, 38050 Povo, Trento, Italia
U. DIERKES, Mathematisches Institut, Univ. Bonn, Beringstrasse 6, 53115 Bonn, Germany
F. DOVERI, SISSA, Via Beirut 2-4, 34014 Trieste, Italia
H.I. ELIASSON, University of Iceland, Hjardarhaga 2-6, 107 Reykjavik, Iceland
R. FILIPPUCCI, Dipartimento di Matematica, Via Vanvitelli, 1, 06123 Perugia, Italia
I. FRAGALA', Dipartimento di Matematica, Via Buonarroti 2, 56127 Pisa, Italia
A. GARRONI, Dipartimento di Matematica, Univ. La Sapienza, P.le A. Moro 2, 00185 Roma, Italia
S. GELLI, SISSA, Via Beirut 2-4, 34014 Trieste, Italia
U. GIANAZZA, Dipartimento di Matematica, Via Abbiategrasso 215, 27100 Pavia, Italia
M. GIAQUINTA, Dipartimento di Matematica applicata, Via S. Marta 3, 50139 Firenze, Italia
G. GREGORI, Dept. of Mathematics, ETH-Zentrum, Ramistrasse 101, 8092 Zürich, Switzerland
M. GRUTER Fachbereich Mathematik, Univ. des Saarlandes, Im Stadtwald, 66123 Saarbrucken, Germany
P. GUASONI, Scuola Normale Superiore, Piazza dei Cavalieri 7, 56126 Pisa, Italia
S. GUSTAFSON, Dept. of Math., Univ. of Toronto, 100 St. George St., Toronto, ON Canada M5S 1A1
M. HOLDER, Math: Inst. d. Universität, AB Analysis, Auf der Morgenstelle 10, 72076 Tubingen, Germany
M. KRUZIK, Institute of Inform. Theory and Automation, Acad. of Sciences, Prague, Czech Republic
E. KUWERT, Mathematisches Institut, Univ. Bonn, Wegelerstrasse 10, 53115 Bonn, Germany
T. ILMANEN, Mathematics Dept., Northwestern University, Evanston, IL 60201, USA
S. LANCELOTTI, Dip.to di Mat., Univ. Cattolica del Sacro Cuore, Via Trieste 17, 25121 Brescia, Italia
G.P. LEONARDI, Dipartimento di Matematica, Università di Trento, 38050 Povo, Trento, Italia
E. LITSYN, Dept. of Theoretical Math., Weizmann Institute of Science, 76100 Rehovot, Israel
M. LONGOBARDI, Dip.to di Ing. dell'Inform., Via Ponte Don Melillo, 84084 Fisciano, Salerno, Italia
R. LOPEZ, Depto de Geometria y Topologia, Fac. de Ciencias,Univ. de Granada, 18071 Granada, Spain
A. MALUSA, Ist. Mat., Fac. Arch., Univ. di Napoli Federico II,Via Monteoliveto 3, 80134 Napoli, Italia
E. MANGINO, Dipartimento di Matematica, Via Orabona 4, 70125 Bari, Italia
C. MANISCALCO, Dipartimento di Matematica, Via Archirafi 34, 90123 Palermo, Italia
C. MANTEGAZZA, Scuola Normale Superiore, Piazza dei Cavalieri 7, 56126 Pisa, Italia

M. MARZOCCHI, Via Burago 28, 25087 Salò, Brescia, Italia

G. MODICA, Dipartimento di Matematica Applicata, Via S. Marta 3, 50139 Firenze, Italia

V. MOROZ, Mechanic-Mathematics Fac., Belorussian State Univ., Skoryny av. 4, Minsk 220080, Belarus

D. MUCCI, Dipartimento di Matematica, Viale Morgagni 67/a, 50134 Firenze, Italia

C. MUSCIANO, Via Monte Velino 15, 67100 L'Aquila, Italia

M. NOVAGA, Scuola Normale Superiore, Piazza dei Cavalieri, 7, 56126 Pisa, Italia

E. PAOLINI, Scuola Normale Superiore, Piazza dei Cavalieri 7, 56126 Pisa, Italia

U. PERRI, Dipartimento di Matematica, Univ. della Calabria, 87036 Arcavacata di Rende (CS), Italia

A. PIACENTINI, Dipartimento di Matematica, Univ., Via Saldini 50, 20133 Milano, Italia

S. RENTZMANN, Texas A&M Univ., Dept. of Math., Milner Hall, College Station 77843, USA

J. REULING, Univ. des Saarlandes, Fach. 9 Math., Gebaude 27, 66041 Saarbrucken, Germany

R. RIGGER, Math. Inst. d. Univ., Auf der Morgenstelle 10, 72076 Tübingen, Germany

A.K. RHODE, Argelanderstr. 58, 53115 Bonn, Germany

S. SALUR, Department of Mathematics , Michigan State University, East Lansing, MI 48824, USA

M.C. SALVATORI, Dipartimento di Matematica, Via Vanvitelli 1, 06100 Perugia, Italia

G. SAVARE', Istituto di Analisi Numerica del CNR, Via Abbiategrasso 209, 27100Pavia, Italia

M. SIMON, Mathematisches Institut d. Univ., Auf der Morgenstelle 10, 72076 Tübingen, Germany

E. STEPANOV, Scuola Normale Superiore, Piazza dei Cavalieri 7, 56126 Pisa, Italia

K.-T. STURM, Mathematisches Institut, Bismarckstrasse 1a, 91054 Erlangen, Germany

A.-M. TELEMAN, Institut für Mathematik d. Univ., Winterthurerstr. 190, Zurich, Switzerland

R. TOADER, SISSA, Via Beirut 2-4, 34014 Trieste, Italia

M. TODA, Mathematische Institut d. Univ., Beringstr. 4, 53115 Bonn, Germany

G. TUROWSKI, Mathematische Institut d. Univ., Wegelerstr. 10, 53115 Bonn, Germany

J. URBAS, Mathematisches Institut d. Univ., Beringstr. 4, 53115 Bonn, Germany

E. UTKIN, SISSA, Via Beirut 2-4, 34014 Trieste, Italia

G. VANELLA, Dipartimento di Matematica del Politecnico, 70100 Bari, Italia

E. VITALI, Dipartimento di Matematica, Via Abbiategrasso 215. 27100 Pavia, Italia

H. VON DER MOSEL, Mathematisches Institut d. Univ., Wegelerstr. 10, 53115 Bonn, Italia

D. WIENHOLTZ, Alfred-Bucherer-Str. 47, 53115 Bonn, Germany

1993 - 117. Integrable Systems and Quantum Groups   (LNM 1620)   Springer-Verlag
        118. Algebraic Cycles and Hodge Theory   (LNM 1594)
        119. Phase Transitions and Hysteresis   (LNM 1584)   "

1994 - 120. Recent Mathematical Methods in   (LNM 1640)   "
            Nonlinear Wave Propagation
        121. Dynamical Systems   (LNM 1609)   "
        122. Transcendental Methods in Algebraic   (LNM 1646)   "
            Geometry

1995 - 123. Probabilistic Models for Nonlinear PDE's   (LNM 1627)   "
        124. Viscosity Solutions and Applications   (LNM 1660)   "
        125. Vector Bundles on Curves. New Directions   (LNM 1649)   "

1996 - 126. Integral Geometry, Radon Transforms   (LNM 1684)   "
            and Complex Analysis
        127. Calculus of Variations and Geometric   LNM 1713
            Evolution Problems
        128. Financial Mathematics   LNM 1656   "

1997 - 129. Mathematics Inspired by Biology   LNM 1714
        130. Advanced Numerical Approximation of   LNM 1697
            Nonlinear Hyperbolic Equations
        131. Arithmetic Theory of Elliptic Curves   LNM 1716r   "
        132. Quantum Cohomology   to appear   "

1998 - 133. Optimal Shape Design   to appear
        134. Dynamical Systems and Small Divisors   to appear
        135. Mathematical Problems in Semiconductor   to appear
            Physics
        136. Stochastic PDE's and Kolmogorov Equations   LNM 1715
            in Infinite Dimension
        137. Filtration in Porous Media and Industrial   to appear
            Applications

1999 - 138. Computional Mathematics driven by Industrual
            Applicationa   to appear
        139. Iwahori-Hecke Algebras and Representation
            Theory   to appear
        140. Theory and Applications of Hamiltonian
            Dynamics   to appear

141. Global Theory of Minimal Surfaces in Flat
     Spaces   to appear
142. Direct and Inverse Methods in Solving
     Nonlinear Evolution Equations   to appear

FONDAZIONE C.I.M.E.
CENTRO INTERNAZIONALE MATEMATICO ESTIVO
INTERNATIONAL MATHEMATICAL SUMMER
CENTER

"Computational Mathematics driven by Industrial Applications"

is the subject of the first 1999 C.I.M.E. Session.

The session, sponsored by the Consiglio Nazionale delle Ricerche (C.N.R.), the Ministero dell'Università e della Ricerca Scientifica e Tecnologica (M.U.R.S.T.) and the European Community, will take place, under the scientific direction of Professors Vincenzo CAPASSO (Università di Milano), Heinz W. ENGL (Johannes Kepler Universitaet, Linz) and Doct. Jacques PERIAUX (Dassault Aviation) at the Ducal Palace of Martina Franca (Taranto), from 21 to 27 June, 1999.

Courses

a) Paths, trees and flows: graph optimisation problems with industrial applications (5 lectures in English) Prof. Rainer BURKARD (Technische Universität Graz)

Abstract

Graph optimisation problems play a crucial role in telecommunication, production, transportation, and many other industrial areas. This series of lectures shall give an overview about exact and heuristic solution approaches and their inherent difficulties. In particular the essential algorithmic paradigms such as greedy algorithms, shortest path computation, network flow algorithms, branch and bound as well as branch and cut, and dynamic programming will be outlined by means of examples stemming from applications.

References

1) R. K. Ahuja, T. L. Magnanti & J. B. Orlin, *Network Flows: Theory, Algorithms and Applications*, Prentice Hall, 1993

2) R. K. Ahuja, T. L. Magnanti, J.B.Orlin & M. R. Reddy, *Applications of Network Optimization*. Chapter 1 in: Network Models (Handbooks of Operations Research and Management Science, Vol. 7), ed. by M. O. Ball et al., North Holland 1995, pp. 1-83

3) R. E. Burkard & E. Cela, *Linear Assignment Problems and Extensions*, Report 127, June 1998 (to appear in Handbook of Combinatorial Optimization, Kluwer, 1999).

Can be downloaded by anonymous ftp from
ftp.tu-graz.ac.at, directory/pub/papers/math

4) R. E. Burkard, E. Cela, P. M. Pardalos & L. S. Pitsoulis, *The Quadratic Assignment Problem*, Report 126 May 1998 (to appear in Handbook of Combinatorial Optimization, Kluwer, 1999). Can be downloaded by anonymous ftp from ftp.tu-graz.ac.at, directory /pub/papers/math.

5) E. L. Lawler, J. K. Lenstra, A. H. G.Rinnooy Kan & D. B. Shmoys (Eds.), *The Travelling Salesman Problem*, Wiley, Chichester, 1985.

b) New Computational Concepts, Adaptive Differential Equations Solvers and Virtual Labs (5 lectures in English) Prof. Peter DEUFLHARD (Konrad Zuse Zentrum, Berlin).

Abstract

The series of lectures will address computational mathematical projects that have been tackled by the speaker and his group. In all the topics to be presented novel mathematical modelling, advanced algorithm developments. and efficient visualisation play a joint role to solve problems of practical relevance. Among the applications to be exemplified are:

1) Adaptive multilevel FEM in clinical cancer therapy planning;

2) Adaptive multilevel FEM in optical chip design;

3) Adaptive discrete Galerkin methods for countable ODEs in polymer chemistry;

4) Essential molecular dynamics in RNA drug design.

References

1) P. Deuflhard & A Hohmann, *Numerical Analysis. A first Course in Scientific Computation*, Verlag de Gruyter, Berlin, 1995

2) P. Deuflhard et al *A nonlinear multigrid eigenproblem solver for the complex Helmoltz equation*, Konrad Zuse Zentrum Berlin SC 97-55 (1997)

3) P. Deuflhard et al. *Recent developments in chemical computing*,
Computers in Chemical Engineering, **14**, (1990),pp.1249-1258.

4) P. Deuflhard et al. (eds) *Computational molecular dynamics: challenges, methods, ideas*, Lecture Notes in Computational Sciences and Engineering, vol.4 Springer Verlag, Heidelberg, 1998.

5) P.Deuflhard & M. Weiser, *Global inexact Newton multilevel FEM for nonlinear elliptic problems*, Konrad Zuse Zentrum SC 96-33, 1996.

c) Computational Methods for Aerodynamic Analysis and Design. (5 lectures in English) Prof. Antony JAMESON (Stanford University, Stanford).

Abstract

The topics to be discussed will include: - Analysis of shock capturing schemes, and fast solution algorithms for compressible flow; - Formulation of aerodynamic shape optimisation based on control theory; - Derivation of the adjoint equations for compressible flow modelled by the potential Euler and Navies-Stokes equations; - Analysis of alternative numerical search procedures; - Discussion of geometry control and mesh perturbation methods; - Discussion of numerical implementation and practical applications to aerodynamic design.

d) Mathematical Problems in Industry (5 lectures in English) Prof. Jacques-Louis LIONS (Collège de France and Dassault Aviation, France).

Abstract

1. Interfaces and scales. The industrial systems are such that for questions of reliability, safety, cost no subsystem can be underestimated. Hence the need to address problems of scales, both in space variables and in time and the crucial importance of modelling and numerical methods.

2. Examples in Aerospace Examples in Aeronautics and in Spatial Industries. Optimum design.

3. Comparison of problems in Aerospace and in Meteorology. Analogies and differences

4 Real time control. Many methods can be thought of. Universal decomposition methods will be presented.

References

1) J. L. Lions, *Parallel stabilization hyperbolic and Petrowsky systems*, WCCM4 Conference, CDROM Proceedings, Buenos Aires, June 29- July 2, 1998.

2) W. Annacchiarico & M. Cerolaza, *Structural shape optimization of 2-D finite elements models using Beta-splines and genetic algorithms*, WCCM4 Conference, CDROM Proceedings, Buenos Aires, June 29- July 2, 1998.

3) J. Periaux, M. Sefrioui & B. Mantel, *Multi-objective strategies for complex optimization problems in aerodynamics using genetic algorithms*, ICAS '98 Conference, Melbourne, September '98, ICAS paper 98-2.9.1

e) Wavelet transforms and Cosine Transform in Signal and Image Processing (5 lectures in English) Prof. Gilbert STRANG (MIT, Boston).

Abstract

In a series of lectures we will describe how a linear transform is applied to the sampled data in signal processing, and the transformed data is compressed (and quantized to a string of bits). The quantized signal is transmitted and then the inverse transform reconstructs a very good approximation to the original signal. Our analysis concentrates on the construction of the transform. There are several important constructions and we emphasise two: 1) the discrete cosine transform (DCT); 2) discrete wavelet transform (DWT). The DCT is an orthogonal transform (for which we will give a new proof). The DWT may be orthogonal, as for the Daubechies family of wavelets. In other cases it may be biorthogonal - so the reconstructing transform is the inverse but not the transpose of the analysing transform. The reason for this possibility is that orthogonal wavelets cannot also be symmetric, and symmetry is essential property in image processing (because our visual system objects to lack of symmetry). The wavelet construction is based on a "bank" of filters - often a low pass and high pass filter. By iterating the low pass filter we decompose the input space into "scales" to produce a multiresolution. An infinite iteration yields in the limit the scaling function and a wavelet: the crucial equation for the theory is the refinement equation or dilatation equation that yields the scaling function. We discuss the mathematics of the refinement equation: the existence and the smoothness of the solution, and the construction by the cascade algorithm. Throughout these lectures we will be developing the mathematical ideas, but always for a purpose. The insights of wavelets have led to new bases for function spaces and there is no doubt that other ideas are waiting to be developed. This is applied mathematics.

References

1) I. Daubechies, *Ten lectures on wavelets*, SIAM, 1992.

2) G. Strang & T. Nguyen, *Wavelets and filter banks*, Wellesley-Cambridge, 1996.

3) Y. Meyer, *Wavelets: Algorithms and Applications*, SIAM, 1993.

Seminars

Two hour seminars will be held by the Scientific Directors and Professor R. Mattheij.

1) **Mathematics of the crystallisation process of polymers.** Prof. Vincenzo CAPASSO (Un. di Milano).

2) **Inverse Problems: Regularization methods, Application in Industry.** Prof. H. W. ENGL (Johannes Kepler Un., Linz).

3) **Mathematics of Glass.** Prof. R. MATTHEIJ (TU Eindhoven).

4) **Combining game theory and genetic algorithms for solving multi-objective shape optimization problems in Aerodynamics Engineering.** Doct. J. PERIAUX (Dassault Aviation).

Applications

Those who want to attend the Session should fill in an application to C.I.M.E Foundation at the address below, **not later than April 30**, 1999. An important consideration in the acceptance of applications is the scientific relevance of the Session to the field of interest of the applicant. Applicants are requested, therefore, to submit, along with their application, a scientific curriculum and a letter of recommendation. Participation will only be allowed to persons who have applied in due time and have had their application accepted. CIME will be able to partially support some of the youngest participants. Those who plan to apply for support have to mention it explicitly in the application form.

Attendance

No registration fee is requested. Lectures will be held at Martina Franca on June 21, 22, 23, 24, 25, 26, 27. Participants are requested to register on June 20, 1999.

Site and lodging

Martina Franca is a delightful baroque town of white houses of Apulian spontaneous architecture. Martina Franca is the major and most aristocratic centre of the "Murgia dei Trulli" standing on an hill which dominates the well known Itria valley spotted with "Trulli" conical dry stone houses which go back to the 15th century. A masterpiece of baroque architecture is the Ducal palace where the workshop will be hosted. Martina Franca is part of the province of Taranto, one of the major centres of Magna Grecia, particularly devoted to mathematics. Taranto houses an outstanding museum of Magna Grecia with fabulous collections of gold manufactures.

Lecture Notes

Lecture notes will be published as soon as possible after the Session.

Arrigo CELLINA
CIME Director

Vincenzo VESPRI
CIME Secretary

Fondazione C.I.M.E. c/o Dipartimento di Matematica ?U. Dini? Viale Morgagni, 67/A - 50134 FIRENZE (ITALY) Tel. +39-55-434975 / +39-55-4237123 FAX +39-55-434975 / +39-55-4222695 E-mail CIME@UDINI.MATH.UNIFI.IT

Information on CIME can be obtained on the system World-Wide-Web on the file HTTP: //WWW.MATH.UNIFI.IT/CIME/WELCOME.TO.CIME

FONDAZIONE C.I.M.E.
CENTRO INTERNAZIONALE MATEMATICO ESTIVO
INTERNATIONAL MATHEMATICAL SUMMER
CENTER

"Iwahori-Hecke Algebras and Representation Theory"

is the subject of the second 1999 C.I.M.E. Session.

The session, sponsored by the Consiglio Nazionale delle Ricerche (C.N.R.), the Ministero dell'Università e della Ricerca Scientifica e Tecnologica (M.U.R.S.T.) and the European Community, will take place, under the scientific direction of Professors Velleda BALDONI (Università di Roma "Tor Vergata") and Dan BARBASCH (Cornell University) at the Ducal Palace of Martina Franca (Taranto), from June 28 to July 6, 1999.

Courses

a) Double HECKE algebras and applications (6 lectures in English)
Prof. Ivan CHEREDNIK (Un. of North Carolina at Chapel Hill, USA)
Abstract:

The starting point of many theories in the range from arithmetic and harmonic analysis to path integrals and matrix models is the formula:

$$\Gamma(k+1/2) = 2\int_0^\infty e^{-x^2}x^{2k}dx.$$

Recently a q-generalization was found based on the Hecke algebra technique, which completes the 15 year old Macdonald program.

The course will be about applications of the double affine Hecke algebras (mainly one-dimensional) to the Macdonald polynomials, Verlinde algebras, Gauss integrals and sums. It will be understandable for those who are not familiar with Hecke algebras and (hopefully) interesting to the specialists.

1) *q-Gauss integrals.* We will introduce a q-analogue of the classical integral formula for the gamma-function and use it to generalize the Gaussian sums at roots of unity.

2) *Ultraspherical polynomials.* A connection of the q-ultraspherical polynomials (the Rogers polynomials) with the one-dimensional double affine Hecke algebra will be established.

3) *Duality.* The duality for these polynomials (which has no classical counterpart) will be proved via the double Hecke algebras in full details.

4) *Verlinde algebras.* We will study the polynomial representation of the 1-dim. DHA at roots of unity, which leads to a generalization and a simplification of the Verlinde algebras.

5) *$PSL_2(\mathbf{Z})$-action.* The projective action of the $PSL_2(\mathbf{Z})$ on DHA and the generalized Verlinde algebras will be considered for A_1 and arbitrary root systems.

6) *Fourier transform of the q-Gaussian.* The invariance of the q-Gaussian with respect to the q-Fourier transform and some applications will be discussed.

References:

1) *From double Hecke algebra to analysis*, Proceedings of ICM98, Documenta Mathematica (1998).

2) *Difference Macdonald-Mehta conjecture*, IMRN:10, 449–467 (1997).

3) *Lectures on Knizhnik-Zamolodchikov equations and Hecke algebras*, MSJ Memoirs (1997).

b) Representation theory of affine Hecke algebras

Prof. Gert HECKMAN (Catholic Un., Nijmegen, Netherlands)

Abstract.

1. The Gauss hypergeometric equation.
2. Algebraic aspects of the hypergeometric system for root systems.
3. The hypergeometric function for root systems.
4. The Plancherel formula in the hypergeometric context.
5. The Lauricella hypergeometric function.
6. A root system analogue of 5.

I will assume that the audience is familiar with the classical theory of ordinary differential equations in the complex plane, in particular the concept of regular singular points and monodromy (although in my first lecture I will give a brief review of the Gauss hypergeometric function). This material can be found in many text books, for example E.L. Ince, Ordinary differential equations, Dover Publ, 1956. E.T. Whittaker and G.N. Watson, A course of modern analysis, Cambridge University Press, 1927.

I will also assume that the audience is familiar with the theory of root systems and reflection groups, as can be found in N. Bourbaki, Groupes et algèbres de Lie, Ch. 4,5 et 6, Masson, 1981. J. E. Humphreys, Reflection groups and Coxeter groups, Cambridge University Press, 1990. or in one of the text books on semisimple groups.

For the material covered in my lectures references are W.J. Couwenberg, Complex reflection groups and hypergeometric functions, Thesis Nijmegen, 1994. G.J. Heckman, Dunkl operators, Sem Bourbaki no 828, 1997. E.M. Opdam, Lectures on Dunkl operators, preprint 1998.

c) Representations of affine Hecke algebras.

Prof. George LUSZTIG (MIT, Cambridge, USA)

Abstract

Affine Hecke algebras appear naturally in the representation theory of p-adic groups. In these lectures we will discuss the representation theory of affine Hecke algebras and their graded version using geometric methods such as equivariant K-theory or perverse sheaves.

References.

1. V. Ginzburg, *Lagrangian construction of representations of Hecke algebras*, Adv. in Math. 63 (1987), 100-112.

2. D. Kazhdan and G. Lusztig, *Proof of the Deligne-Langlands conjecture for Hecke algebras.*, Inv. Math. 87 (1987), 153-215.

3. G. Lusztig, *Cuspidal local systems and graded Hecke algebras, I*, IHES Publ. Math. 67 (1988),145-202; II, in "Representation of groups" (ed. B. Allison and G. Cliff), Conf. Proc. Canad. Math. Soc.. 16, Amer. Math. Soc. 1995, 217-275.

4. G. Lusztig, *Bases in equivariant K-theory, Represent. Th.*, 2 (1998).

d) Affine-like Hecke Algebras and p-adic representation theory

Prof. Roger HOWE (Yale Un., New Haven, USA)

Abstract

Affine Hecke algebras first appeared in the study of a special class of representations (the spherical principal series) of reductive groups with coefficients in p-adic fields. Because of their connections with this and other topics, the structure and representation theory of affine Hecke algebras has been intensively studied by a variety of authors. In the meantime, it has gradually emerged that affine Hecke algebras, or slight generalizations of them, allow one to understand far more of the representations of p-adic groups than just the spherical principal series. Indeed, it seems possible that such algebras will allow one to understand all representations of p-adic groups. These lectures will survey progress in this approach to p-adic representation theory.

Topics:

1) Generalities on spherical function algebras on p-adic groups.

2) Iwahori Hecke algebras and generalizations.

3) - 4) Affine Hecke algebras and harmonic analysis

5) - 8) Affine-like Hecke algebras and representations of higher level.

References:

J. Adler, *Refined minimal K-types and supercuspidal representations*, Ph.D. Thesis, University of Chicago.

D. Barbasch, *The spherical dual for p-adic groups*, in Geometry and Representation Theory of Real and p-adic Groups, J. Tirao, D. Vogan, and J. Wolf, eds, Prog. In Math. 158, Birkhauser Verlag, Boston, 1998, 1 - 20.

D. Barbasch and A. Moy, *A unitarity criterion for p-adic groups*, Inv. Math. 98 (1989), 19 - 38.

D. Barbasch and A. Moy, *Reduction to real infinitesimal character in affine Hecke algebras*, J. A. M. S.6 (1993), 611- 635.

D. Barbasch, *Unitary spherical spectrum for p-adic classical groups*, Acta. Appl. Math. 44 (1996), 1 - 37.

C. Bushnell and P. Kutzko, *The admissible dual of GL(N) via open subgroups*, Ann. of Math. Stud. 129, Princeton University Press, Princeton, NJ, 1993.

C. Bushnell and P. Kutzko, *Smooth representations of reductive p-adic groups*: Structure theory via types, D. Goldstein, *Hecke algebra isomorphisms for tamely ramified characters*, R. Howe and A. Moy, *Harish-Chandra Homomorphisms for p-adic Groups*, CBMS Reg. Conf. Ser. 59, American Mathematical Society, Providence, RI, 1985.

R. Howe and A. Moy, *Hecke algebra isomorphisms for GL(N) over a p-adic field*, J. Alg. 131 (1990), 388 - 424.

J-L. Kim, *Hecke algebras of classical groups over p-adic fields and supercuspidal representations,I, II, III*, preprints, 1998.

G. Lusztig, *Classification of unipotent representations of simple p-adic groups*, IMRN 11 (1995), 517 - 589.

G. Lusztig, *Affine Hecke algebras and their graded version*, J. A. M. S. 2 (1989), 599 - 635.

L. Morris, *Tamely ramified supercuspidal representations of classical groups, I, II*, Ann. Ec. Norm. Sup 24, (1991) 705 - 738; 25 (1992), 639 - 667.

L. Morris, *Tamely ramified intertwining algebras*, Inv. Math. 114 (1994), 1 - 54.

A. Roche, *Types and Hecke algebras for principal series representations of split reductive p-adic groups*, preprint, (1996).

J-L. Waldspurger, *Algebres de Hecke et induites de representations cuspidales pour GLn*, J. reine u. angew. Math. 370 (1986), 27 - 191.

J-K. Yu, *Tame construction of supercuspidal representations*, preprint, 1998.

Applications

Those who want to attend the Session should fill in an application to the Director of C.I.M.E at the address below, **not later than** April 30, 1999.

An important consideration in the acceptance of applications is the scientific relevance of the Session to the field of interest of the applicant.

Applicants are requested, therefore, to submit, along with their application, a scientific curriculum and a letter of recommendation.

Participation will only be allowed to persons who have applied in due time and have had their application accepted.

CIME will be able to partially support some of the youngest participants. Those who plan to apply for support have to mention it explicitely in the application form.

Attendance

No registration fee is requested. Lectures will be held at Martina Franca on June 28, 29, 30, July 1, 2, 3, 4, 5, 6. Participants are requested to register on June 27, 1999.

Site and lodging

Martina Franca is a delightful baroque town of white houses of Apulian spontaneous architecture. Martina Franca is the major and most aristocratic centre of the Murgia dei Trulli standing on an hill which dominates the well known Itria valley spotted with Trulli conical dry stone houses which go back to the 15th century. A masterpiece of baroque architecture is the Ducal palace where the workshop will be hosted. Martina Franca is part of the province of Taranto, one of the major centres of Magna Grecia, particularly devoted to mathematics. Taranto houses an outstanding museum of Magna Grecia with fabulous collections of gold manufactures.

Lecture Notes

Lecture notes will be published as soon as possible after the Session.

Arrigo CELLINA
CIME Director

Vincenzo VESPRI
CIME Secretary

Fondazione C.I.M.E. c/o Dipartimento di Matematica U. Dini Viale Morgagni, 67/A - 50134 FIRENZE (ITALY) Tel. +39-55-434975 / +39-55-4237123 FAX +39-55-434975 / +39-55-4222695 E-mail CIME@UDINI.MATH.UNIFI.IT

Information on CIME can be obtained on the system World-Wide-Web on the file HTTP: //WWW.MATH.UNIFI.IT/CIME/WELCOME.TO.CIME.

FONDAZIONE C.I.M.E.
CENTRO INTERNAZIONALE MATEMATICO ESTIVO
INTERNATIONAL MATHEMATICAL SUMMER
CENTER

"Theory and Applications of Hamiltonian Dynamics"

is the subject of the third 1999 C.I.M.E. Session.

The session, sponsored by the Consiglio Nazionale delle Ricerche (C.N.R.), the Ministero dell'Università e della Ricerca Scientifica e Tecnologica (M.U.R.S.T.) and the European Community, will take place, under the scientific direction of Professor Antonio GIORGILLI (Un. di Milano), at Grand Hotel San Michele,Cetraro (Cosenza), from July 1 to July 10, 1999.

Courses

a) Physical applications of Nekhoroshev theorem and exponential estimates (6 lectures in English)

Prof. Giancarlo BENETTIN (Un. di Padova, Italy)

Abstract

The purpose of the lectures is to introduce exponential estimates (i.e., construction of normal forms up to an exponentially small remainder) and Nekhoroshev theorem (exponential estimates plus geometry of the action space) as the key to understand the behavior of several physical systems, from the Celestial mechanics to microphysics.

Among the applications of the exponential estimates, we shall consider problems of adiabatic invariance for systems with one or two frequencies coming from molecular dynamics. We shall compare the traditional rigorous approach via canonical transformations, the heuristic approach of Jeans and of Landau–Teller, and its possible rigorous implementation via Lindstet series. An old conjecture of Boltzmann and Jeans, concerning the possible presence of very long equilibrium times in classical gases (the classical analog of "quantum freezing") will be reconsidered. Rigorous and heuristic results will be compared with numerical results, to test their level of optimality.

Among the applications of Nekhoroshev theorem, we shall study the fast rotations of the rigid body, which is a rather complete problem, including degeneracy and singularities. Other applications include the stability of elliptic equilibria, with special emphasis on the stability of triangular Lagrangian points in the spatial restricted three body problem.

References:

For a general introduction to the subject, one can look at chapter 5 of V.I. Arnold, VV. Kozlov and A.I. Neoshtadt, in *Dynamical Systems III*, V.I. Arnold Editor (Springer, Berlin 1988). An introduction to physical applications of Nekhorshev theorem and exponential estimates is in the proceeding of the Noto School "Non-Linear Evolution and Chaotic Phenomena", G. Gallavotti and P.W. Zweifel Editors (Plenum Press, New York, 1988), see the contributions by G. Benettin, L. Galgani and A. Giorgilli.

General references on Nekhoroshev theorem and exponential estimates: N.N. Nekhoroshev, Usp. Mat. Nauk. **32**:6, 5-66 (1977) [Russ. Math. Surv. **32**:6, 1-65

(1977)]; G. Benettin, L. Galgani, A. Giorgilli, Cel. Mech. **37**, 1 (1985); A. Giorgilli and L. Galgani, Cel. Mech. **37**, 95 (1985); G. Benettin and G. Gallavotti, Journ. Stat. Phys. **44**, 293-338 (1986); P. Lochak, Russ. Math. Surv. **47**, 57-133 (1992); J. Pöschel, Math. Z. **213**, 187-216 (1993).

Applications to statistical mechanics: G. Benettin, in: *Boltzmann's legacy 150 years afrer his birth*, Atti Accad. Nazionale dei Lincei **131**, 89-105 (1997); G. Benettin, A. Carati and P. Sempio, Journ. Stat. Phys. **73**, 175-192 (1993); G. Benettin, A. Carati and G. Gallavotti, Nonlinearity **10**, 479-505 (1997); G. Benettin, A. Carati e F. Fassò, Physica D **104**, 253-268 (1997); G. Benettin, P. Hjorth and P. Sempio, *Exponentially long equilibrium times in a one dimensional collisional model of a classical gas*, in print in Journ. Stat. Phys.

Applications to the rigid body: G. Benettin and F. Fassò, Nonlinearity **9**, 137-186 (1996); G. Benettin, F. Fassò e M. Guzzo, Nonlinearity **10**, 1695-1717 (1997).

Applications to elliptic equilibria (recent nonisochronous approach): F. Fassò, M. Guzzo e G. Benettin, Comm. Math. Phys. **197**, 347-360 (1998); L. Niederman, *Nonlinear stability around an elliptic equilibrium point in an Hamiltonian system*, preprint (1997). M. Guzzo, F. Fasso' e G. Benettin, Math. Phys. Electronic Journal, Vol. **4**, paper 1 (1998); G. Benettin, F. Fassò e M. Guzzo, *Nekhoroshev-stability of L4 and L5 in the spatial restricted three-body problem*, in print in Regular and Chaotic Dynamics.

b) KAM-theory (6 lectures in English)

Prof. Hakan ELIASSON (Royal Institute of Technology, Stockholm, Sweden)

Abstract

Quasi-periodic motions (or invariant tori) occur naturally when systems with periodic motions are coupled. The perturbation problem for these motions involves small divisors and the most natural way to handle this difficulty is by the quadratic convergence given by Newton's method. A basic problem is how to implement this method in a particular perturbative situation. We shall describe this difficulty, its relation to linear quasi-periodic systems and the way given by KAM-theory to overcome it in the most generic case. Additional difficulties occur for systems with elliptic lower dimensional tori and even more for systems with weak non-degeneracy.

We shall also discuss the difference between initial value and boundary value problems and their relation to the Lindstedt and the Poincaré-Lindstedt series.

The classical books Lectures in Celestial Mechanics by Siegel and Moser (Springer 1971) and Stable and Random Motions in Dynamical Systems by Moser (Princeton University Press 1973) are perhaps still the best introductions to KAM-theory. The development up to middle 80's is described by Bost in a Bourbaki Seminar (no. 6 1986). After middle 80's a lot of work have been devoted to elliptic lower dimensional tori, and to the study of systems with weak non-degeneracy starting with the work of Cheng and Sun (for example *"Existence of KAM-tori in Degenerate Hamiltonian systems"*, J. Diff. Eq. 114, 1994). Also on linear quasi-periodic systems there has been some progress which is described in my article *"Reducibility and point spectrum for quasi-periodic skew-products"*, Proceedings of the ICM, Berlin volume II 1998.

c) The Adiabatic Invariant in Classical Dynamics: Theory and applications (6 lectures in English).

Prof. Jacques HENRARD (Facultés Universitaires Notre Dame de la Paix, Namur, Belgique).

Abstract

The adiabatic invariant theory applies essentially to oscillating non-autonomous Hamiltonian systems when the time dependance is considerably slower than the oscillation periods. It describes "easy to compute" and "dynamicaly meaningful" quasi-invariants by which on can predict the approximate evolution of the system on very large time scales. The theory makes use and may serve as an illustration of several classical results of Hamiltonian theory.

1) Classical Adiabatic Invariant Theory (Including an introduction to angle-action variables)

2) Classical Adiabatic Invariant Theory (continued) and some applications (including an introduction to the "magnetic bottle")

3) Adiabatic Invariant and Separatrix Crossing (Neo-adiabatic theory)

4) Applications of Neo-Adiabatic Theory: Resonance Sweeping in the Solar System

5) The chaotic layer of the "Slowly Modulated Standard Map"

References:

J.R. Cary, D.F. Escande, J.L. Tennison: Phys.Rev. A, 34, 1986, 3256-4275

J. Henrard, in "*Dynamics reported*" (n=B02- newseries), Springer Verlag 1993; pp 117-235)

J. Henrard: in "*Les méthodes moderne de la mécanique céleste*" (Benest et Hroeschle eds), Edition Frontieres, 1990, 213-247

J. Henrard and A. Morbidelli: Physica D, 68, 1993, 187-200.

d) Some aspects of qualitative theory of Hamiltonian PDEs (6 lectures in English).

Prof. Sergei B. KUKSIN (Heriot-Watt University, Edinburgh, and Steklov Institute, Moscow)

Abstract.

I) Basic properties of Hamiltonian PDEs. Symplectic structures in scales of Hilbert spaces, the notion of a Hamiltonian PDE, properties of flow-maps, etc.

II) Around Gromov's non-squeezing property. Discussions of the finite-dimensional Gromov's theorem, its version for PDEs and its relevance for mathematical physics, infinite-dimensional symplectic capacities.

III) Damped Hamiltonian PDEs and the turbulence-limit. Here we establish some qualitative properties of PDEs of the form <non-linear Hamiltonian PDE>+<small linear damping> and discuss their relations with theory of decaying turbulence

Parts I)-II) will occupy the first three lectures, Part III - the last two.

References

[1] S.K., *Nearly Integrable Infinite-dimensional Hamiltonian Systems*. LNM 1556, Springer 1993.

[2] S.K., *Infinite-dimensional symplectic capacities and a squeezing theorem for Hamiltonian PDE's*. Comm. Math. Phys. 167 (1995), 531-552.

[3] Hofer H., Zehnder E., *Symplectic invariants and Hamiltonian dynamics*. Birkhauser, 1994.

[4] S.K. *Oscillations in space-periodic nonlinear Schroedinger equations*. Geometric and Functional Analysis 7 (1997), 338-363.

For I) see [1] (Part 1); for II) see [2,3]; for III) see [4]."

e) An overview on some problems in Celestial Mechanics (6 lectures in English)

Prof. Carles SIMO' (Universidad de Barcelona, Spagna)

Abstract

1. Introduction. The N-body problem. Relative equilibria. Collisions.

2. The 3D restricted three-body problem. Libration points and local stability analysis.

3. Periodic orbits and invariant tori. Numerical and symbolical computation.

4. Stability and practical stability. Central manifolds and the related stable/unstable manifolds. Practical confiners.

5. The motion of spacecrafts in the vicinity of the Earth-Moon system. Results for improved models. Results for full JPL models.

References:

C. Simò, *An overview of some problems in Celestial Mechanics*, available at http://www-ma1.upc.es/escorial .

Click of "curso completo" of Prof. Carles Simó

Applications

Deadline for application: **May 15, 1999.**

Applicants are requested to submit, along with their application, a scientific curriculum and a letter of recommendation.

CIME will be able to partially support some of the youngest participants. Those who plan to apply for support have to mention it explicitly in the application form.

Attendance

No registration fee is requested. Lectures will be held at Cetraro on July 1, 2, 3, 4, 5, 6, 7, 8, 9, 10. Participants are requested to register on June 30, 1999.

Site and lodging

The session will be held at Grand Hotel S. Michele at Cetraro (Cosenza), Italy. Prices for full board (bed and meals) are roughly 150.000 italian liras p.p. day in a single room, 130.000 italian liras in a double room. Cheaper arrangements for multiple lodging in a residence are avalaible. More detailed information may be obtained from the Direction of the hotel (tel. +39-098291012, Fax +39-098291430, email: sanmichele@antares.it.

Further information on the hotel at the web page www.sanmichele.it

Arrigo CELLINA
CIME Director

Vincenzo VESPRI
CIME Secretary

Fondazione C.I.M.E. c/o Dipartimento di Matematica U. Dini Viale Morgagni, 67/A
- 50134 FIRENZE (ITALY) Tel. +39-55-434975 / +39-55-4237123 FAX
+39-55-434975 / +39-55-4222695 E-mail CIME@UDINI.MATH.UNIFI.IT

Information on CIME can be obtained on the system World-Wide-Web on the file
HTTP: //WWW.MATH.UNIFI.IT/CIME/WELCOME.TO.CIME.

FONDAZIONE C.I.M.E.
CENTRO INTERNAZIONALE MATEMATICO ESTIVO
INTERNATIONAL MATHEMATICAL SUMMER
CENTER
"Global Theory of Minimal Surfaces in Flat Spaces"

is the subject of the fourth 1999 C.I.M.E. Session.

The session, sponsored by the Consiglio Nazionale delle Ricerche (C.N.R.), the Ministero dell'Università e della Ricerca Scientifica e Tecnologica (M.U.R.S.T.) and the European Community, will take place, under the scientific direction of Professor Gian Pietro PIROLA (Un. di Pavia), at Ducal Palace of Martina Franca (Taranto), from July 7 to July 15, 1999.

Courses

a) Asymptotic geometry of properly embedded minimal surfaces (6 lecture in English)

Prof. William H. MEEKS, III (Un. of Massachusetts, Amherst, USA).

Abstract:

In recent years great progress has been made in understanding the asymptotic geometry of properly embedded minimal surfaces. The first major result of this type was the solution of the generalized Nitsch conjecture by P. Collin, based on earlier work by Meeks and Rosenberg. It follows from the resolution of this conjecture that whenever M is a properly embedded minimal surface with more than one end and $E \subset M$ is an annular end representative, then E has finite total curvature and is asymptotic to an end of a plan or catenoid. Having finite total curvature in the case of an annular end is equivalent to proving the end has quadratic area growth with respect to the radial function r. Recently Collin, Kusner, Meeks and Rosenberg have been able to prove that any middle end of M, even one with infinite genus, has quadratic area growth. It follows from this result that middle ends are never limit ends and hence M can only have one or two limit ends which must be top or bottom ends. With more work it is shown that the middle ends of M stay a bounded distance from a plane or an end of a catenoid.

The goal of my lectures will be to introduce the audience to the concepts in the theory o f properly embedded minimal surfaces needed to understand the above results and to understand some recent classification theorems on proper minimal surfaces of genus 0 in flat three-manifolds.

References

1) H. Rosenberg, *Some recent developments in the theory of properly embedded minimal surfaces in E*, Asterisque **206**, (19929, pp. 463-535;

2) W. Meeks & H. Rosenberg, *The geometry and conformal type of properly embedded minimal surfaces in E*, Invent.Math. **114**, (1993), pp. 625-639;

3) W. Meeks, J. Perez & A. Ros, *Uniqueness of the Riemann minimal examples*, Invent. Math. **131**, (1998), pp. 107-132;

4) W. Meeks & H. Rosenberg, *The geometry of periodic minimal surfaces*, Comm. Math. Helv. **68**, (1993), pp. 255-270;

5) P. Collin, *Topologie et courbure des surfaces minimales proprement plongees dans E*. Annals of Math. **145**, (1997), pp. 1-31;

6) H. Rosenberg, *Minimal surfaces of finite type*, Bull. Soc. Math. France **123**, (1995), pp. 351-359;

7) Rodriquez & H. Rosenberg, *Minimal surfaces in E with one end and bounded curvature*, Manusc. Math. **96**, (1998), pp. 3-9.

b) Properly embedded minimal surfaces with finite total curvature (6 lectures in English)

Prof. Antonio ROS (Universidad de Granada, Spain)

Abstact:

Among properly embedded minimal surfaces in Euclidean 3-space, those that have finite total curvature form a natural and important subclass. These surfaces have finitely many ends which are all parallel and asymptotic to planes or catenoids. Although the structure of the space M of surfaces of this type which have a fixed topology is not well understood, we have a certain number of partial results and some of them will be explained in the lectures we will give.

The first nontrivial examples, other than the plane and the catenoid, were constructed only ten years ago by Costa, Hoffman and Meeks. Schoen showed that if the surface has two ends, then it must be a catenoid and López and Ros proved that the only surfaces of genus zero are the plane and the catenoid. These results give partial answers to an interesting open problem: decide which topologies are supported by this kind of surfaces. Ros obtained certain compactness properties of M. In general this space is known to be noncompact but he showed that M is compact for some fixed topologies. Pérez and Ros studied the local structure of M around a nondegenerate surface and they proved that around these points the moduli space can be naturally viewed as a Lagrangian submanifold of the complex Euclidean space.

In spite of that analytic and algebraic methods compete to solve the main problems in this theory, at this moment we do not have a satisfactory idea of the behaviour of the moduli space M. Thus the above is a good research field for young geometers interested in minimal surfaces.

References

1) C. Costa, *Example of a compete minimal immersion in $\mathbb{R}^3$ of genus one and three embedded ends*, Bull. SOc. Bras. Math. **15**, (1984), pp. 47-54;

2) D. Hoffman & H. Karcher, *Complete embedded minimal surfaces of finite total curvature*, R. Osserman ed., Encyclopedia of Math., vol. of Minimal Surfaces, **5-90**, Springer 1997;

3) D. Hoffman & W. H. Meeks III, *Embedded minimal surfaces of finite topology*, Ann. Math. **131**, (1990), pp. 1-34;

4) F. J. Lòpez & A. Ros, *On embedded minimal surfaces of genus zero*, J. Differential Geometry **33**, (1991), pp. 293-300;

5) J. P. Perez & A. Ros, *Some uniqueness and nonexistence theorems for embedded minimal surfaces*, Math. Ann. **295** (3), (1993), pp. 513-525;

6) J. P. Perez & A. Ros, *The space of properly embedded minimal surfaces with finite total curvature*, Indiana Univ. Math. J. **45** 1, (1996), pp.177-204.

c) Minimal surfaces of finite topology properly embedded in E (Euclidean 3-space).(6 lectures in English)

Prof. Harold ROSENBERG (Univ. Paris VII, Paris, France)

Abstract:

We will prove that a properly embedded minimal surface in E of finite topology and at least two ends has finite total curvature. To establish this we first prove that each annular end of such a surface M can be made transverse to the horizontal planes

(after a possible rotation in space), [Meeks-Rosenberg]. Then we will prove that such an end has finite total curvature [Pascal Collin]. We next study properly embedded minimal surfaces in E with finite topology and one end. The basic unsolved problem is to determine if such a surface is a plane or helicoid when simply connected. We will describe partial results. We will prove that a properly immersed minimal surface of finite topology that meets some plane in a finite number of connected components, with at most a finite number of singularities, is of finite conformal type. If in addition the curvature is bounded, then the surface is of finite type. This means M can be parametrized by meromorphic data on a compact Riemann surface. In particular, under the above hypothesis, M is a plane or helicoid when M is also simply connected and embedded. This is work of Rodriquez- Rosenberg, and Xavier. If time permits we will discuss the geometry and topology of constant mean curvature surfaces properly embedded in E.

References

1) H. Rosenberg, *Some recent developments in the theory of properly embedded minimal surfaces in E*, Asterique **206**, (1992), pp. 463-535;

2) W.Meeks & H. Rosenberg, *The geometry and conformal type of properly embedded minimal surfaces in E*, Invent. **114**, (1993), pp.625-639;

3) P. Collin, *Topologie et courbure des surfaces minimales proprement plongées dans E*, Annals of Math. **145**, (1997), pp. 1-31

4) H. Rosenberg, *Minimal surfaces of finite type*, Bull. Soc. Math. France **123**, (1995), pp. 351-359;

5) Rodriquez & H. Rosenberg, *Minimal surfaces in E with one end and bounded curvature*, Manusc. Math. **96**, (1998), pp. 3-9.

Applications

Those who want to attend the Session should fill in an application to the C.I.M.E Foundation at the address below, **not later than** May 15, 1999.

An important consideration in the acceptance of applications is the scientific relevance of the Session to the field of interest of the applicant.

Applicants are requested, therefore, to submit, along with their application, a scientific curriculum and a letter of recommendation.

Participation will only be allowed to persons who have applied in due time and have had their application accepted.

CIME will be able to partially support some of the youngest participants. Those who plan to apply for support have to mention it explicitly in the application form

Attendance

No registration fee is requested. Lectures will be held at Martina Franca on July 7, 8, 9, 10, 11, 12, 13, 14, 15. Participants are requested to register on July 6, 1999.

Site and lodging

Martina Franca is a delightful baroque town of white houses of Apulian spontaneous architecture. Martina Franca is the major and most aristocratic centre of the Murgia dei Trulli standing on an hill which dominates the well known Itria valley spotted with Trulli conical dry stone houses which go back to the 15th century. A masterpiece of baroque architecture is the Ducal palace where the workshop will be hosted. Martina Franca is part of the province of Taranto, one of the major centres of Magna Grecia, particularly devoted to mathematics. Taranto houses an outstanding museum of Magna Grecia with fabulous collections of gold manufactures.

Lecture Notes

Lecture notes will be published as soon as possible after the Session.

Arrigo CELLINA
CIME Director

Vincenzo VESPRI
CIME Secretary

Fondazione C.I.M.E. c/o Dipartimento di Matematica U. Dini Viale Morgagni, 67/A - 50134 FIRENZE (ITALY) Tel. +39-55-434975 / +39-55-4237123 FAX +39-55-434975 / +39-55-4222695 E-mail CIME@UDINI.MATH.UNIFI.IT

Information on CIME can be obtained on the system World-Wide-Web on the file HTTP: //WWW.MATH.UNIFI.IT/CIME/WELCOME.TO.CIME.

FONDAZIONE C.I.M.E.
CENTRO INTERNAZIONALE MATEMATICO ESTIVO
INTERNATIONAL MATHEMATICAL SUMMER CENTER

"Direct and Inverse Methods in Solving Nonlinear Evolution Equations"

is the subject of the fifth 1999 C.I.M.E. Session.

The session, sponsored by the Consiglio Nazionale delle Ricerche (C.N.R.), the Ministero dell'Università e della Ricerca Scientifica e Tecnologica (M.U.R.S.T.) and the European Community, will take place, under the scientific direction of Professor Antonio M. Greco (Università di Palermo), at Grand Hotel San Michele,Cetraro (Cosenza), from September 8 to September 15, 1999.

a) Exact solutions of nonlinear PDEs by singularity analysis (6 lectures in English)

Prof. Robert CONTE (Service de physique de l'état condensé, CEA Saclay, Gif-sur-Yvette Cedex, France)

Abstract

1) Criteria of integrability : Lax pair, Darboux and Bäcklund transformations. Partial integrability, examples. Importance of involutions.

2) The Painlevé test for PDEs in its invariant version.

3) The "truncation method" as a Darboux transformation, ODE and PDE situations.

4) The one-family truncation method (WTC), integrable (Korteweg-de Vries, Boussinesq, Hirota-Satsuma, Sawada-Kotera) and partially integrable (Kuramoto-Sivashinsky) cases.

5) The two-family truncation method, integrable (sine-Gordon, mKdV, Broer-Kaup) and partially integrable (complex Ginzburg-Landau and degeneracies) cases.

6) The one-family truncation method based on the scattering problems of Gambier: BT of Kaup-Kupershmidt and Tzitzéica equations.

References

References are divided into three subsets: prerequisite (assumed known by the attendant to the school), general (not assumed known, pedagogical texts which would greatly benefit the attendant if they were read before the school), research (research papers whose content will be exposed from a synthetic point of view during the course).

Prerequisite bibliography.

The following subjects will be assumed to be known : the Painlevé property for nonlinear ordinary differential equations, and the associated Painlevé test.

Prerequisite recommended texts treating these subjects are

[P.1] E. Hille, *Ordinary differential equations in the complex domain* (J. Wiley and sons, New York, 1976).

[P.2] R. Conte, *The Painlevé approach to nonlinear ordinary differential equations, The Painlevé property, one century later*, 112 pages, ed. R. Conte, CRM series in mathematical physics (Springer, Berlin, 1999). Solv-int/9710020.

The interested reader can find many applications in the following review, which should not be read before [P.2] :

[P.3] A. Ramani, B. Grammaticos, and T. Bountis, *The Painlevé property and singularity analysis of integrable and nonintegrable systems*, Physics Reports 180 (1989) 159–245.

A text to be avoided by the beginner is Ince's book, the ideas are much clearer in Hille's book.

There exist very few pedagogical texts on the subject of this school.

A general reference, covering all the above program, is the course delivered at a Cargèse school in 1996 :

[G.1] M. Musette, *Painlevé analysis for nonlinear partial differential equations*, The Painlevé property, one century later, 65 pages, ed. R. Conte, CRM series in mathematical physics (Springer, Berlin, 1999). Solv-int/9804003.

A short subset of [G.1], with emphasis on the ideas, is the conference report

[G.2] R. Conte, *Various truncations in Painlevé analysis of partial differential equations*, 16 pages, Nonlinear dynamics : integrability and chaos, ed. M. Daniel, to appear (Springer? World Scientific?). Solv-int/9812008. Preprint S98/047.

Research papers.

[R.2] J. Weiss, M. Tabor and G. Carnevale, *The Painlevé property for partial differential equations*, J: Math. Phys. 24 (1983) 522–526.

[R.3] Numerous articles of Weiss, from 1983 to 1989, all in J. Math. Phys. [singular manifold method].

[R.4] M. Musette and R. Conte, *Algorithmic method for deriving Lax pairs from the invariant Painlevé analysis of nonlinear partial differential equations*, J. Math. Phys. 32 (1991) 1450–1457 [invariant singular manifold method].

[R.5] R. Conte and M. Musette, *Linearity inside nonlinearity: exact solutions to the complex Ginz-burg-Landau equation*, Physica D 69 (1993) 1–17 [Ginzburg-Landau].

[R.6] M. Musette and R. Conte, *The two–singular manifold method, I. Modified KdV and sine-Gordon equations*, J. Phys. A 27 (1994) 3895–3913 [Two–singular manifold method].

[R.7] R. Conte, M. Musette and A. Pickering, *The two–singular manifold method, II. Classical Boussinesq system*, J. Phys. A 28 (1995) 179–185 [Two–singular manifold method].

[R.8] A. Pickering, *The singular manifold method revisited*, J. Math. Phys. 37 (1996) 1894–1927 [Two–singular manifold method].

[R.9] M. Musette and R. Conte, *Bäcklund transformation of partial differential equations from the Painlevé-Gambier classification, I. Kaup-Kupershmidt equation*, J. Math. Phys. 39 (1998) 5617–5630. [Lecture 6].

[R.10] R. Conte, M. Musette and A. M. Grundland, *Bäcklund transformation of partial differential equations from the Painlevé-Gambier classification, II. Tzitzéica equation*, J. Math. Phys. 40 (1999) to appear. [Lecture 6].

b) Integrable Systems and Bi-Hamiltonian Manifolds (6 lectures in English)

Prof. Franco MAGRI (Università di Milano, Milano, Italy)

Abstract

1) Integrable systems and bi-hamiltonian manifolds according to Gelfand and Zakharevich.

2) Examples: KdV, KP and Sato's equations.

3) The rational solutions of KP equation.

4) Bi-hamiltonian reductions and completely algebraically integrable systems.

5) Connections with the separabilty theory.

6) The τ function and the Hirota's identities from a bi-hamiltonian point of view.

References

1) R. Abraham, J.E. Marsden, *Foundations of Mechanics*,Benjamin/Cummings, 1978

2) P. Libermann, C. M. Marle, *Symplectic Geometry and Analytical Mechanics*, Reidel Dordrecht, 1987

3) L. A. Dickey, *Soliton Equations and Hamiltonian Systems*, World Scientific, Singapore, 1991, Adv. Series in Math. Phys Vol. 12

4) I. Vaisman, *Lectures on the Geometry of Poisson Manifolds*, Progress in Math., Birkhäuser, 1994

5) P. Casati, G. Falqui, F. Magri, M. Pedroni (1996), *The KP theory revisited. I,II,III,IV.* Technical Reports, SISSA/2,3,4,5/96/FM, SISSA/ISAS, Trieste, 1995

c) Hirota Methods for non Linear Differential and Difference Equations (6 lectures in English)

Prof. Junkichi SATSUMA (University of Tokyo, Tokyo, Japan)

Abstract

1) Introduction;

2) Nonlinear differential systems;

3) Nonlinear differential-difference systems;

4) Nonlinear difference systems;

5) Sato theory;

6) Ultra-discrete systems.

References.

1) M.J.Ablowitz and H.Segur, *Solitons and the Inverse Scattering Transform*, (SIAM, Philadelphia, 1981).

2) Y.Ohta, J.Satsuma, D.Takahashi and T.Tokihiro, " Prog. Theor. Phys. Suppl. No.94, p.210-241 (1988)

3) J.Satsuma, *Bilinear Formalism in Soliton Theory*, Lecture Notes in Physics No.495, Integrability of Nonlinear Systems, ed. by Y.Kosmann-Schwarzbach, B.Grammaticos and K.M.Tamizhmani p.297-313 (Springer, Berlin, 1997).

d) Lie Groups and Exact Solutions of non Linear Differential and Difference Equations (6 lectures in English)

Prof. Pavel WINTERNITZ (Universitè de Montreal, Montreal, Canada) 3J7

Abstract

1) Algorithms for calculating the symmetry group of a system of ordinary or partial differential equations. Examples of equations with finite and infinite Lie point symmetry groups;

2) Applications of symmetries. The method of symmetry reduction for partial differential equations. Group classification of differential equations;

3) Classification and identification of Lie algebras given by their structure constants. Classification of subalgebras of Lie algebras. Examples and applications;

4) Solutions of ordinary differential equations. Lowering the order of the equation. First integrals. Painlevè analysis and the singularity structure of solutions;

5) Conditional symmetries. Partially invariant solutions.

6) Lie symmetries of difference equations.

References.

1) P. J. Olver, *Applications of Lie Groups to Differential Equations*, Springer,1993,

2) P. Winternitz, *Group Theory and Exact Solutions of Partially Integrable Differential Systems*, in Partially Integrable Evolution Equations in Physics, Kluwer, Dordrecht, 1990, (Editors R.Conte and N.Boccara).

3) P. Winternitz, in *"Integrable Systems, Quantum Groups and Quantum Field Theories"*, Kluwer, 1993 (Editors L .A. Ibort and M. A. Rodriguez).

Applications

Those who want to attend the Session should fill in an application to the C.I.M.E Foundation at the address below, **not later than May 30**, 1999.

An important consideration in the acceptance of applications is the scientific relevance of the Session to the field of interest of the applicant.

Applicants are requested, therefore, to submit, along with their application, a scientific curriculum and a letter of recommendation.

Participation will only be allowed to persons who have applied in due time and have had their application accepted.

CIME will be able to partially support some of the youngest participants. Those who plan to apply for support have to mention it explicitly in the application form.

Attendance

No registration fee is requested. Lectures will be held at Cetraro on September 8, 9, 10, 11, 12, 13, 14, 15. Participants are requested to register on September 7, 1999.

Site and lodging

The session will be held at Grand Hotel S. Michele at Cetraro (Cosenza), Italy. Prices for full board (bed and meals) are roughly 150.000 italian liras p.p. day in a single room, 130.000 italian liras in a double room. Cheaper arrangements for multiple lodging in a residence are avalaible. More detailed informations may be obtained from the Direction of the hotel (tel. +39-098291012, Fax +39-098291430, email: sanmichele@antares.it.

Further information on the hotel at the web page www.sanmichele.it

Lecture Notes

Lecture notes will be published as soon as possible after the Session.

Arrigo CELLINA
CIME Director

Vincenzo VESPRI
CIME Secretary

Fondazione C.I.M.E. c/o Dipartimento di Matematica U. Dini Viale Morgagni, 67/A - 50134 FIRENZE (ITALY) Tel. +39-55-434975 / +39-55-4237123 FAX +39-55-434975 / +39-55-4222695 E-mail CIME@UDINI.MATH.UNIFI.IT

Information on CIME can be obtained on the system World-Wide-Web on the file HTTP: //WWW.MATH.UNIFI.IT/CIME/WELCOME.TO.CIME.

Vol. 1670: J. W. Neuberger, Sobolev Gradients and Differential Equations. VIII, 150 pages. 1997.

Vol. 1671: S. Bouc, Green Functors and G-sets. VII, 342 pages. 1997.

Vol. 1672: S. Mandal, Projective Modules and Complete Intersections. VIII, 114 pages. 1997.

Vol. 1673: F. D. Grosshans, Algebraic Homogeneous Spaces and Invariant Theory. VI, 148 pages. 1997.

Vol. 1674: G. Klaas, C. R. Leedham-Green, W. Plesken, Linear Pro-p-Groups of Finite Width. VIII, 115 pages. 1997.

Vol. 1675: J. E. Yukich, Probability Theory of Classical Euclidean Optimization Problems. X, 152 pages. 1998.

Vol. 1676: P. Cembranos, J. Mendoza, Banach Spaces of Vector-Valued Functions. VIII, 118 pages. 1997.

Vol. 1677: N. Proskurin, Cubic Metaplectic Forms and Theta Functions. VIII, 196 pages. 1998.

Vol. 1678: O. Krupková, The Geometry of Ordinary Variational Equations. X, 251 pages. 1997.

Vol. 1679: K.-G. Grosse-Erdmann, The Blocking Technique. Weighted Mean Operators and Hardy's Inequality. IX, 114 pages. 1998.

Vol. 1680: K.-Z. Li, F. Oort, Moduli of Supersingular Abelian Varieties. V, 116 pages. 1998.

Vol. 1681: G. J. Wirsching, The Dynamical System Generated by the 3n+1 Function. VII, 158 pages. 1998.

Vol. 1682: H.-D. Alber, Materials with Memory. X, 166 pages. 1998.

Vol. 1683: A. Pomp, The Boundary-Domain Integral Method for Elliptic Systems. XVI, 163 pages. 1998.

Vol. 1684: C. A. Berenstein, P. F. Ebenfelt, S. G. Gindikin, S. Helgason, A. E. Tumanov, Integral Geometry, Radon Transforms and Complex Analysis. Firenze, 1996. Editors: E. Casadio Tarabusi, M. A. Picardello, G. Zampieri. VII, 160 pages. 1998.

Vol. 1685: S. König, A. Zimmermann, Derived Equivalences for Group Rings. X, 146 pages. 1998.

Vol. 1686: J. Azéma, M. Émery, M. Ledoux, M. Yor (Eds.), Séminaire de Probabilités XXXII. VI, 440 pages. 1998.

Vol. 1687: F. Bornemann, Homogenization in Time of Singularly Perturbed Mechanical Systems. XII, 156 pages. 1998.

Vol. 1688: S. Assing, W. Schmidt, Continuous Strong Markov Processes in Dimension One. XII, 137 page. 1998.

Vol. 1689: W. Fulton, P. Pragacz, Schubert Varieties and Degeneracy Loci. XI, 148 pages. 1998.

Vol. 1690: M. T. Barlow, D. Nualart, Lectures on Probability Theory and Statistics. Editor: P. Bernard. VIII, 237 pages. 1998.

Vol. 1691: R. Bezrukavnikov, M. Finkelberg, V. Schechtman, Factorizable Sheaves and Quantum Groups. X, 282 pages. 1998.

Vol. 1692: T. M. W. Eyre, Quantum Stochastic Calculus and Representations of Lie Superalgebras. IX, 138 pages. 1998.

Vol. 1694: A. Braides, Approximation of Free-Discontinuity Problems. XI, 149 pages. 1998.

Vol. 1695: D. J. Hartfiel, Markov Set-Chains. VIII, 131 pages. 1998.

Vol. 1696: E. Bouscaren (Ed.): Model Theory and Algebraic Geometry. XV, 211 pages. 1998.

Vol. 1697: B. Cockburn, C. Johnson, C.-W. Shu, E. Tadmor, Advanced Numerical Approximation of Nonlinear Hyperbolic Equations. Cetraro, Italy, 1997. Editor: A. Quarteroni. VII, 390 pages. 1998.

Vol. 1698: M. Bhattacharjee, D. Macpherson, R. G. Möller, P. Neumann, Notes on Infinite Permutation Groups. XI, 202 pages. 1998.

Vol. 1699: A. Inoue,Tomita-Takesaki Theory in Algebras of Unbounded Operators. VIII, 241 pages. 1998.

Vol. 1700: W. A. Woyczyński, Burgers-KPZ Turbulence,XI, 318 pages. 1998.

Vol. 1701: Ti-Jun Xiao, J. Liang, The Cauchy Problem of Higher Order Abstract Differential Equations, XII, 302 pages. 1998.

Vol. 1702: J. Ma, J. Yong, Forward-Backward Stochastic Differential Equations and Their Applications. XIII, 270 pages. 1999.

Vol. 1703: R. M. Dudley, R. Norvaiša, Differentiability of Six Operators on Nonsmooth Functions and p-Variation. VIII, 272 pages. 1999.

Vol. 1704: H. Tamanoi, Elliptic Genera and Vertex Operator Super-Algebras. VI, 390 pages. 1999.

Vol. 1705: I. Nikolaev, E. Zhuzhoma, Flows in 2-dimensional Manifolds. XIX, 294 pages. 1999.

Vol. 1706: S. Yu. Pilyugin, Shadowing in Dynamical Systems. XVII, 271 pages. 1999.

Vol. 1707: R. Pytlak, Numerical Methods for Optical Control Problems with State Constraints. XV, 215 pages. 1999.

Vol. 1708: K. Zuo, Representations of Fundamental Groups of Algebraic Varieties. VII, 139 pages. 1999.

Vol. 1709: J. Azéma, M. Émery, M. Ledoux, M. Yor (Eds), Séminaire de Probabilités XXXIII. VIII, 418 pages. 1999.

Vol. 1710: M. Koecher, The Minnesota Notes on Jordan Algebras and Their Applications. IX, 173 pages. 1999.

Vol. 1711: W. Ricker, Operator Algebras Generated by Commuting Projections: A Vector Measure Approach. XVII, 159 pages. 1999.

Vol. 1712: N. Schwartz, J. J. Madden, Semi-algebraic Function Rings and Reflectors of Partially Ordered Rings. XI, 279 pages. 1999.

Vol. 1713: F. Bethuel, G. Huiksen, S. Müller, K. Steffen, Calculus of Variations and Geometric Evolution Problems. Cetraro, 1996. Editors: S. Hildebrandt, M. Struwe. VII, 293 pages. 1999.

Vol. 1714: O. Diekmann, R. Durrett, K. P. Hadeler, P. Maini, H. L. Smith, Mathematics Inspired by Biology. Martina Franca, 1997. Editors: V. Capasso, O. Diekmann. VII, 268 pages. 1999.

Vol. 1715: N. V. Krylov, M. Röckner, J. Zabczyk, Stochastic PDE's and Kolmogorov Equations in Infinite Dimensions. Cetraro, 1998. Editor: G. Da Prato. VIII, 239 pages. 1999.

Vol. 1716: J. Coates, R. Greenberg, K. A. Ribet, K. Rubin, Arithmetic Theory of Elliptic Curves. Cetraro, 1997. Editor: C. Viola. VIII, 260 pages. 1999.

4. Lecture Notes are printed by photo-offset from the master-copy delivered in camera-ready form by the authors. Springer-Verlag provides technical instructions for the preparation of manuscripts. Macro packages in T_EX, L^AT_EX2e, $L^AT_EX2.09$ are available from Springer's web-pages at

http://www.springer.de/math/authors/b-tex.html.

Careful preparation of the manuscripts will help keep production time short and ensure satisfactory appearance of the finished book.

The actual production of a Lecture Notes volume takes approximately 12 weeks.

5. Authors receive a total of 50 free copies of their volume, but no royalties. They are entitled to a discount of 33.3% on the price of Springer books purchase for their personal use, if ordering directly from Springer-Verlag.

Commitment to publish is made by letter of intent rather than by signing a formal contract. Springer-Verlag secures the copyright for each volume. Authors are free to reuse material contained in their LNM volumes in later publications: A brief written (or e-mail) request for formal permission is sufficient.

Addresses:

Professor F. Takens, Mathematisch Instituut,
Rijksuniversiteit Groningen, Postbus 800,
9700 AV Groningen, The Netherlands
E-mail: F.Takens@math.rug.nl

Professor B. Teissier, DMI, École Normale Supérieure
45, rue d'Ulm,
F-7500 Paris, France
E-mail: Teissier@ens.fr

Springer-Verlag, Mathematics Editorial, Tiergartenstr. 17,
D-69121 Heidelberg, Germany,
Tel.: *49 (6221) 487-701
Fax: *49 (6221) 487-355
E-mail: lnm@Springer.de